Sea Floor Exploration

Springer Oceanography

For further volumes:
http://www.springer.com/series/10175

Roger Hekinian

Sea Floor Exploration

Scientific Adventures Diving into the Abyss

Springer

Roger Hekinian
Saint-Renan
France

ISBN 978-3-319-03202-3 ISBN 978-3-319-03203-0 (eBook)
DOI 10.1007/978-3-319-03203-0
Springer Cham Heidelberg New York Dordrecht London

Library of Congress Control Number: 2013955911

Printed on acid-free paper

Springer is part of Springer Science+Business Media (www.springer.com)

*I dedicate this book to my children Aram,
Diran and Anna, and to the future,
which will be discovered
by my grandchildren*

Preface

I decided to write about my experience at sea on the 8th of October 2002, while sailing on board the German research vessel F.S. *SONNE.* Looking out at the ocean, I suddenly noticed that time was passing too quickly and I seemed to have more difficulty remembering previous cruises and discoveries. At that time, I thought this would have been my last cruise and I should summarize what I have learned during all these years.

Now that I have finished writing this text, I must admit that 2002 was certainly not the year of my last cruise. Indeed, since then I continued to go to sea every year until 2005. The Research Vessel F.S. *SONNE* is particularly important to me because I have navigated more than 14 months on this ship from 1986 to 2005. That is more time than I've spent on any other research vessel.

The history of our planet Earth is written in the rocks. The record left on the emerged continents indicates that water has been a major architect shaping our planet. Since the magma ocean forming our early planet more than 4 billion years ago cooled and gave rise to blue seawater and the atmosphere, Earth has become a haven for living organisms. Many species have evolved and disappeared, while other new ones have appeared and flourished ever since the first organisms, in the form of bacteria, were born in the earliest periods of Earth's existence.

As humans, our impact during our short time on Earth is insignificant. Humans cannot alter the course of Earth's evolution but rather they have to adapt. Our living planet is in constant motion; its outer shell is created through the transfer of molten mantle at spreading ridge axes and will eventually disappear inside the deep Earth to regenerate itself again.

This book concludes with my last diving experience looking at an area near the Tonga-Kermadec subduction zone, where old crust is being returned to the mantle to be re-melted once again. While old lithosphere is disappearing, new crust is being created to form a volcanic arc.

When looking at a map of the Earth it might appear that the ocean and the continents are immutable, and one might think that they have always been there, that nothing has changed since their creation. This is just an illusion due to the brief time we humans have spent on this planet, as opposed to the Earth's rocks, which have a geological time scale and are related to the beginning of the Universe, which occurred 15 billion years ago.

I have written this book to give the reader the fruit of my experience and some of the information we have gleaned during my 45 years as a marine geologist and volcanologist. If you close this book a bit wiser about the mysteries of our planet, I will be pleased to think that I have fulfilled my work as an author.

Acknowledgments

I would like to thank my wife for her help and encouragement in writing this book, and for her patience and encouragement during my years of activity as a scientist. Having lived though so many exciting experiences while learning about our planet, my retirement has been like a prison where I've chafed at the confinements of living a "normal life" each day. My wife says I am like a dormant volcano, just waiting for another chance to explode, to explore, and to meet the new challenges that life has to offer. Maybe she is right.

This book would not have been possible without the collaboration and help of my colleagues, many of whom were involved in the sea-going expeditions I have participated in over the years. I am particularly indebted to the captains and officers that were responsible for the research vessels and the submersibles. The ships' crews, and the groups involved in the maintenance and the operational aspect of the submersible dives are also warmly acknowledged. I would also like to thank Prof. Rodey Batiza, Peter Stoffers, and Steven Scott who have taken time to comment parts of this work.

Contents

Chapter 1
An Overview of Oceanographic Exploration

Abstract Earth's scientists have been interested in the ocean as well as the continents for at least the last 150 years. However, "modern oceanography" started after World War II in 1946, when the world's politicians and scientists realized that in order to have supremacy over other countries and acquire knowledge of our planet it would be necessary to increase the capability of sea floor exploration. Private and government funds were provided to enhance underwater technology. Ocean-going ships began to carry out regular and more extensive expeditions on all the major oceans and seas. Also, we now know that the modern oceans and seas are relatively young (150–170 million years old) when compared to the continental landmasses. Geological records found on the continents indicate that the ocean must have existed since our planet's creation 4.6 billion years ago, and it seems probable that the Earth has undergone several cycles of ocean formation and retrieval during the past 4 billion years.

Introduction

Most of the time, marine geologists work on the surface of the sea, at several kilometers distance from the sea floor, which allows their imagination to flourish with what might be happening beneath the sea's surface. It is only recently, during the late 1960s and 1970s, that specially organized deep-diving scientific expeditions with manned and unmanned vehicles were undertaken to directly observe and sample the sea floor.

Seawater covers an unknown world, which is difficult for man to access. Because of its physical state, seawater forms a screen and the darkness of the abysses creates a serious handicap for investigations by human eyes. Sunlight does not penetrate deeper than 100–200 m after which it diminishes in intensity and breaks down into different wavelengths so that the red rays (infra-rouge) are absorbed while the blue and the green rays penetrate more deeply. Nevertheless, most components of the sun's rays do not penetrate deeper than 150 m except for blue rays that could reach

R. Hekinian, *Sea Floor Exploration*, Springer Oceanography,
DOI: 10.1007/978-3-319-03203-0_1,

depths of 250 m. Illumination is difficult in the ocean environment since even the 2000-watt lights that are installed on deep-sea vehicles are only able to light up a short distance of 10–12 m when exploring the ocean floor.

Historical Background

Oceanography was born in Europe and gave rise to the great expeditions carried out by countries such as Denmark, Sweden, Germany, Holland, France and Great Britain during the 1880s and 1890s. Today, the leadership of modern oceanography still remains with European countries and the USA. In 1964, Georg Wüst published a paper on the major expeditions carried out between 1873 and 1960.

After World War II, the USA attracted many young and brilliant oceanographers. This was mainly the result of the United States' technological development. In fact, the earliest precision echo sounders, marine magnetometers for measuring the Earth's magnetic field, and the air-gun development for conducting seismological exploration were first developed in the USA. The capability of deep sea drilling started in 1965 with the JOIDES (Joint Oceanographic Deep Sea Drilling) program. Also, the most sophisticated navigation and bathymetric recording system (SEA-BEAM) was conceived in the USA. Most of these tools were developed for military purposes and were then integrated into basic scientific programs. Thus, oceanic research was motivated by curiosity but also by our desire to acquire basic knowledge of the deep-sea environment in order to maintain "supremacy of the seas".

The earliest organized worldwide oceanographic expedition started in 1872 on board the British sailing ship *HMS Glomar Challenger*, and was called the Murray expedition. This expedition spent 713 days at sea between 1872 and 1876. The first manganese nodules were discovered during this expedition. It was also during this expedition that several islands such as the St Paul's Rocks in the equatorial Atlantic, St. Paul and Amsterdam islands in the Indian ocean and Easter island in the south Pacific were visited and studied. At that time, the scientific team and sailors had to bring their vessel to a full stop in order to measure the ocean depths using ropes hanging overboard. A total of 372 measurements were taken over 3 years time. They also discovered that a submerged mountain range might exist in the Mid Atlantic Ocean. Now, 140 years later, measuring methods have become more sophisticated so a sea floor surface of about 70,000 km^2 can be mapped in less than 10 days (see Chap. 3).

Ocean Explorers and Pioneers in Oceanography

What is it that makes human beings want to challenge the unknown? Maybe this is due to their curiosity and their desire to explore new frontiers or perhaps it's just a hope for adventure or a drive to surpass themselves. Whatever the reason, modern explorers spend considerable time trying to understand and explain the environment in which we live and this type of "discovery" is an ongoing process that will never end as long as human beings exist.

One of the most prestigious naturalists as well as oceanographers of the last century was ***Alexander von Humboldt,*** founder of modern geography and meteorology. He was the person who drew the first isothermal map of the ocean. His findings later helped to discover the existence of different density currents in the oceans due to changes in water salinity and temperature. The present day *Humboldt Foundation* was initiated with funding by the post World War II German government to promote scientific exchange between the western nations' scientific communities; it was this same Humboldt Foundation that enabled me to continue my cooperative programs with German institutions after I was obliged to retire at the age of 65 from IFREMER (*Institut Français de Recherche pour l'Exploitation de la Mer*).

In 1910, a German meteorologist named ***Alfred Wegener*** and a US geologist called ***Franck Taylor*** suggested that "continental drift" might be a possible explanation for the shape and apparent fit of the European—African continents with the North and South American continents. After World War II, long-range exploration of the deep ocean floor was undertaken by the major US oceanographic centers.

It was in the 1960s that ***Harry Hess*** and ***Robert Dietz*** suggested that the sea floor might be spreading from the area of the Mid-Oceanic Ridge axis. ***Henry William Menard*** is another of these milestone scientists who has contributed to the study of the Pacific islands and seamounts since the 1950s. At that time, we knew very little about the growth and formation of submarine volcanoes and their implications in forming oceanic islands.

Harry Hess hypothesized that continental drift is closely related to the presence of convection currents in the mantle, which he claimed were the force for moving the continents. The proof of continental drift came from the measurements of the earth's magnetic field (Vine and Matthews 1963) (see Chap. 2). The magnetic field is what controls the compass needle, directing it to the north. However, it has been established that the magnetism of our planet has jumped from north to south and vice versa several times in the Earth's past history. When volcanic rocks were erupted on the Mid-Oceanic Ridge axis, they cooled rapidly and their iron-rich minerals (magnetite) solidified in the direction of the magnetic field (the direction of the compass needle) at the time of eruption. An observation of the changing pattern of magnetic attraction from North to South and vice versa provides us with a means for seeing the evolution of these erupted rocks over time and space.

In 1963, a Canadian geologist named ***Tuzo Wilson*** published an article in the Canadian Journal of Physics where he suggested that the alignment of the Hawaiian Island chain was the consequence of the lithospheric plate's displacement over a fixed thermal source that he called a "plume". It was also Wilson (1963, 1965), after looking at the Mid-Atlantic Ridge's east-west trending fracture zones, who suggested that the junction of the spreading ridge and the fracture zones is the site of intense tectonic activity and called these structures "transform faults". Indeed, these faults are the site of most of the world's major earthquakes observed along the Mid-Atlantic Ridge (see Chap. 8). Wilson inspired many scientists to continue developing the theory of continental plate motions in relation to the hotspot volcanic areas located within oceanic basins.

Maurice Ewing was the first director of the Lamont-Doherty Geological Observatory (LDO). He was a tall man with white hair, dark-eyebrows, and round glasses, who looked more like a farmer than an intellectual, even if his nickname was "Doc". He was a geophysicist and he developed the seismic method of exploration at sea, which become the principal tool used by oceanographers to explore the deep structures under the oceans. ***Joe Lamar Worzel*** was also a geophysicist who worked closely with Doc Ewing. Joe was a short, good looking, fast-talking man, a hard worker at sea, and he was always on the run, as if he were trying to catch up with lost time.

Lamont-Doherty Geological Observatory (LDO) of Columbia University (founded in 1949) was one of the leading Oceanographic institutions in the 1960s. Even if the techniques were not as advanced as they are now, nevertheless the Lamont ships, *R.V. Vema* in the early 1950s (1953) and the *R.V. Robert D. Conrad* in the 1960s, were notorious for their reputation of spending most of their time at sea. However, it was in the 1940s on the Woods Hole sailing ship *Atlantis* that "Doc" Ewing first studied the structure of the sea floor when he explored and mapped the Mid-Atlantic Ridge with its high mountain range.

Bruce Heezen and ***Marie Tharp***, could be found either at sea or in the Oceanography Building of the Lamont-Doherty Observatory (LDO) between 1958 to the early 1960s. They were the first ones to draft an ocean floor structural map showing the major features of the World's sea floor as it might look if all the water were drained out. The compilation of data was made from single-channel depth echosounding. In 1959, they and Maurice Ewing published "*The Floor of the Ocean*" which accompanied a series of physiographic maps of the ocean floor.

Earthquakes and other instabilities on sedimented slopes trigger turbidity currents on the sea floor. Bruce Heezen pioneered the study of turbidity currents and other research concerning the origin of submarine canyons in the deep sea. He also determined that a 1925 earthquake on the East Coast of the USA was the cause of the breaks along trans-Atlantic submarine telephone cables. In addition, he was interested in bottom photography so Heezen and ***Charlie Hollister*** were among the first scientists to publish a comprehensive atlas illustrating the sea floor landscape, in 1971.

Bruce Heezen believed in on-site submarine exploration and he was one of the first people to use manned submersibles for scientific purposes. His work in 1974–1975 consisted in making scientific observations on the submarine flanks of the Hawaiian Island in the underwater continuation of the active Kilauea volcano. He demonstrated the existence of magma injection through fissural eruption extending in a submarine environment. Using the *Deep Sea Vessel (DSV) Turtle*, Heezen also pioneered the observation of the subduction zone of the Middle America Trench underneath the coastal range off Mexico. Bruce died during one of his diving expeditions in 1977, inside the NR-1 (a nuclear powered U.S. Navy submersible). He was observing the sea floor along the Rekjanes Ridge south of the Azores on his last day of life.

I have a particular debt of gratitude to Bruce Heezen for drawing me a sketch map of the Indian Ocean that I published in my first scientific paper. He was also the person who encouraged me to publish one of my earliest papers in marine geology entitled: *Rocks from the Mid-Oceanic Ridge in the Indian Ocean.*

Charles Fray also called *Chuck,* is a marine geologist whom I particularly admired and I would like to remember him because of his humanity and his capacity for handling difficult situations at sea. He was a very good chief scientist and was extremely appreciated by all the crew and scientific groups on board. Chuck was my direct adviser while I was working in the core laboratory at LDO, with ***Roy Capo***. Roy was a geologist and the meticulous curator of the "core Lab"; his responsibility was cataloging the drill cores and rock samples which had been collected since the early 1950s but which were scattered in the different buildings of Lamont.

I had the privilege of sailing with Chuck Fray while he was chief scientist during one of my first legs on board the *R.V. R.D. Conrad.* This was my first cruise on an oceanographic vessel. At that time, Chuck was one of the few chief scientists from Lamont who used to take time for sharing and discussing what the cruise objectives were with the crew as well as with the scientists. One day Chuck told me: "Roger, if you want to have a successful cruise you need the crew and the technicians on your side. They are doing the most difficult work." This piece of advice still rings a bell up to now, and every time I've gone to sea, my first task has always been to give everyone the feeling that we are all on the same team, doing team work, and that it is in everybody's interest that everything goes as well as possible.

I regarded Chuck as one of the best chief scientists that I have ever worked with and from whom I have learned a great deal. Another person who was very close to me soon after he became responsible for the core laboratory (in 1965) was Roy Capo. Later on, he was my best man when I married Ginny. I want to remember Roy for his kindness and helpful advice during my early years as a marine geologist.

In 1962 when I arrived at work at Lamont-Doherty, I was a European man who was tossed into the American way of life. Having immigrated to the USA after obtaining a "Doctorate" (masters degree) in Geology from the University of Pisa, Italy, I applied for a position at Lamont-Doherty, but was only given a job as a student assistant on the RV *Conrad.* Being on board and having a chance to associate with men such as Chuck, Roy and Joe Worzel has helped shaped my attitude towards deep-sea research from the very beginning.

The atmosphere on board the Lamont research vessels was fast-paced and intense. Our goal was to gather data as quickly as possible. At that time, all the information gathered on board was immediately forwarded to Doc Ewing by telex, and during each port of call, a package of data was mailed or carried back to Lamont for study. Sometimes Doc would join the ship to make sure everything was going the way he expected, and then he himself would carry the new data or samples back to the Observatory.

Colleagues and Collaborators

I would like to write a few words and have a special thought for some of my friends and colleagues who have had a major impact on my scientific life but who are now no longer living.

Jean Francheteau was a marine geophysicist who worked with me at CNEXO and IFREMER from when we both arrived in 1970–1982, and we continued our scientific collaboration and friendship even when he went on to work for other institutes in Paris, or at the University of Brest. He had a thick head of hair and a drooping mustache, so he looked a bit like a Mexican freedom fighter, as well as a French nobleman. Born in Pornichet, France, Jean obtained his PhD in oceanography in 1970 from the Scripps Institution of Oceanography, in La Jolla, California. He was the author of more than 90 publications and his outstanding contribution to Earth Sciences inspired many researchers in the field of marine geology. However the most important aspect of his influence on me was due to his qualities as a human being. He was enthusiastic, honest and a strong believer that conducting good science should be done without any interference due to political compromises or leadership. He believed that Earth study starts in the field with the exploration of the sea floor. It was very enlightening for me to spend time at sea with him. We shared more than 11 sea-going expeditions, and our families are best of friends. I will always remember Jean as an explorer and adventurer, ready to seek new horizons. Unfortunately he left us when he was still too young, at the age of 67, in July 2010.

Guy Paquet was nicknamed "Nounours" (Teddy Bear) perhaps because he could sometimes appear to be as fierce as a grizzly, but he was really as soft as a child's toy. He terminated his long career in the French merchant marine as captain of the N.O. *JEAN CHARCOT*. On board he was sometimes moody, often pleasant and at the same time harsh and opinionated. Nevertheless, he was truly kind and very concerned about the fate of his fellow men. Built like a rock, with the neck of a rugby player, he was born in Brive la Gaillarde, which is also a region famous for its rugby team. Guy was very much appreciated by all his crew members, and I will always remember his ability to navigate our oceanographic vessels under extremely difficult conditions, including a hurricane which chased our ship across the Atlantic during the September 1976 Post-Famous scientific cruise.

Daniel Bideau was my student and also a collaborator on many scientific endeavors. He was a brilliant scientist who left us far too early when he was at the peak of his career, at the age of 55. We used to share the same lab, our desks faced each other in our shared office, and over the years we exchanged many thoughts and ideas about our Earth and our lives.

My work has brought me into contact with several other colleagues and friends to whom I am very appreciative for their support and unconditional friendship during my scientific career.

David Needham is a marine geologist who graduated from Columbia University and whom I met while I was at Lamont-Geological observatory in 1964.

I spent part of my first sea-going experience at sea with David on board the R.V. *R.D. Conrad*. Later on, he came to work in France, so our lives have been intertwined for nearly half a century.

Captain Hartmut Andresen was one of the captains of the German Research vessel *F.S. Sonne* prior to his retirement. He was a typical explorer and eager for new adventures and new sensations. He was slender, well over 6 feet tall, and often wore a headband rather than a sea captain's cap on his head. He was always the first to take the zodiac out to sea to go to test the feasibility of landing in unknown area. We went together to visit Pitcairn Island, which will remain one of the high points of my memories of going at sea.

Vincent Renard is a geophysicist who graduated from Rice University before completing his studies at Lamont-Doherty in New York. He joined the group of CNEXO organized by Le Pichon in 1970 along with David Needham and myself. Born in Belgium, Vincent was chief scientist on several ocean-going expeditions. He has been among the best chief scientists that I have shared cruises with, perhaps because Vincent is a discrete and friendly person who is very sensitive to people and who is always ready to share his time and expertise, in order to help others. As well as being good in communicating with the various groups on board the ships, Vincent is also very knowledgeable about the different techniques and instruments used at sea. Competence and communication are two keys to being a successful chief scientist.

Xavier Le Pichon is a French geophysicist who still wears round glasses on his gentle-looking round face. We met at Lamont-Doherty, and later he invited me to join his group after he founded the French marine geology group of CNEXO (Centre National d'Exploitation des Oceans) in 1968. Xavier is a pragmatic person who knows how to hide his emotions. At the same time, he is altruistic and ready to help people in need. I am very grateful to Xavier because he introduced me to underwater exploration in 1971 during the "FAMOUS Project". I also have a great admiration for Xavier because, he, along with Jason Morgan, is one of the pioneers who advanced the theory of plate tectonics based on observations on the sea floor.

Jason Morgan is a person who has been a part of my life since he and his family came for a sabbatical in France, in 1972. We discovered we were born the same week of the same year, and our wives and children always appreciated seeing each other as if we were all part of the same family. Although Jason and I have never published a paper together, the conversations we have shared over the years are certainly an important part of my life and work as an undersea geologist.

Peter Stoffers, who was originally trained as a sediment geochemist, was a professor of marine geology and Department Head at the University of Kiel in Germany. Peter is a fervent believer in scientific research and has the personality of a stately professor who feels that rules need to be obeyed, but he is also a gentle and understanding father figure for his students. Peter has been very keen in helping young scientists succeed in the field of marine geology. His honesty, humanity and capacity for uniting knowledgeable people and for obtaining funding for his many projects are extremely important qualities. I believe Peter was the scientist who was granted the largest number of cruises in Germany during his career and I am very grateful to him for his help in creating the international "hotspot project" during the early 1980s.

Félix Avedik, Bertrand Sichler, Yaoling Niu, Steve Scott, Nicolas Binard, Rodey Batiza, *and* ***Jill Karsten***, are just a few of the names of other colleagues who have impacted my life on a professional and personal level. In fact, in looking over the list of my publications on my Curriculum Vitae, I realize just how important the co-authors of my papers were during a few months or during my entire career and my lifetime.

I think Chuck Fray got it absolutely right when he told me how important it is to work with others, not for yourself, and to aim for the good of the science you are trying to accomplish. My publications wouldn't have been possible without the collaboration and friendship I have shared with so many of the other people listed in my bibliography.

Lamont-Doherty Earth Observatory (LDEO)

Among the most prestigious oceanographic institutions in the world is the Lamont-Doherty Oceanographic Observatory of Columbia University in New York, often called "Lamont" for short. A beacon of oceanography and scientific research, located on the on Hudson River about 30 km north of Manhattan on the Palisades cliffs, Lamont is essentially oriented towards the fields of geophysics, magnetism, sea floor structure and morphology, sedimentology and geochemistry.

I was born in 1935 in Marseilles, France. Nothing in my background indicated that I would become involved in such an exciting field as undersea geology. My parents were refugees who had been welcomed to France after the Armenian genocide in 1915, and my father was a prisoner of war during World War II, so he was absent from my life for 5 years. I was raised in an Armenian speaking home where my parents, grandparents, and my Uncle and Aunt all lived together. When I was only 13, my family decided to send me away to an Armenian Catholic secondary boarding school in Venice Italy, and after high-school I continued my studies, in land geology, at the University of Pisa, where I obtained a Masters Degree in 1962.

During my busy, early years adapting to various cultures and languages, I always hoped to live a life of adventure like Marco Polo or Charles Lindberg, so it was only natural that I would be attracted to the new frontier which oceanography represented in the early 1960s.

When I arrived in the USA in the summer of 1962, I was hoping to find a job in the field of geology. Through the help of my cousins Birj and Diran Deckmejian, who shared their living space with me, and thanks to my cousin Albert Barsamian, who contacted someone he knew, I was granted an interview at Lamont with Charles Fray. Chuck was looking for a geologist to spend time at sea; even though my English wasn't very good, he hired me for a trial period of 1 year. My mission was to spend several months at a time at sea with the research team from Lamont

Geological Observatory, called "*Lamont*". In 1963, I signed a contract involving a 6 month-long cruise on a sea-going expedition. However before going to sea I first had to spend some time in the core-laboratory where sediment samples were being described and stored. This part of the job was right up my alley and I was comfortable with the work, which consisted in determining some physical parameters for the samples, such as their porosity, density, color and type of sediment.

In the summer of 1964, I took a flight from New York to Alaska, to one of the most distant U.S. islands called Adak, located in the Aleutian arc, where I was to join the research vessel (R.V.) *Robert D. Conrad*. Two days later, we left the island on a grey misty morning, first heading northward to the Bearing Sea. About 2 weeks later we turned south then south-southeast towards western Canada in the northern Pacific Ocean where we sailed along the coast of Vancouver and then Oregon.

I started to be seasick and uncomfortable as soon as we left Adak's dock. The chief scientist (Joe Worzel) kept rushing into my cabin to ask me to go out on deck and keep on working because, according to him, that was the only way to feel better. He was right and he surely met well, but at the time I did not want to see him nor did I wish to leave my bunk to go outside on deck. I only felt good if I were lying in a horizontal position without moving. But at the same time I felt guilty, and I knew I that had to get up because there was a job that had to be done. In fact, I was very much aware that even if I felt like I was dying from seasickness, I had to take my watch and replace my colleagues on deck.

Our job consisted in creating sound waves in the water column by throwing lighted explosives overboard. This was acoustic measuring at its most primitive. A half-pound TNT stick was held by one hand against our stomachs while the fuse was inserted into the other end of the dynamite stick with our other hand, and then we were supposed to light the wick. Immediately, the TNT stick with its lighted fuse had to be thrown overboard without delay, or we could lose a hand or worse. This action of dropping lighted sticks of TNT overboard had to be repeated every 30 s, so that as the ship advanced, the explosions could be heard at its stern. The acrid, chemical smell of a lighted fuse was definitely extremely unpleasant for some one suffering from seasickness the way I was.

Chronic Seasickness

The most difficult time for anyone at sea is when he or she suffers from being seasick. Your body is in a state of total exhaustion and extreme fatigue due to constant vomiting. You would like to die but it's impossible. You have no will power to go to work, no appetite, and the idea of eating or drinking anything, even a glass of water or a saltine cracker, is beyond imagination. You feel terrible since you can't control your own weakness and you are in an intense state of suffering. This is called "Chronic Seasickness" and this is how I felt. It was humiliating and it was awful. After about one full month at sea with this constant feeling of seasickness, I felt exhausted, since I had totally lost my appetite and I was also loosing my strength. This is when I decided that I would never again go to sea, and as soon as I put my feet on land I would abandon ship and get a different job.

Then 1 day, about a month out of port, somewhere south of Alaska in the North Pacific, we encountered a terrible storm. We met up with a cyclone with a sea-state of 12 on the Beaufort scale and with winds up to 71 kn/h and waves reaching up to 19 m (55 feet) in height. I was astonished to notice that I suddenly felt comfortable and was starting to feel good about being on board. It was a magic sensation of being reborn at sea. Now, because of the storm, the rest of the crewmen were green and nauseous, but I felt normal. I felt like working and eating, and my energy returned. We passed into the eye of the cyclone where the sea was as calm as a lake for as far as my eyes could see, up to the horizon, about 15 km away. I could even look up into the eye of the cyclone and admire the blue sky through a hole in the clouds above my head. After nearing the other side of the cyclone's eye, the sea again became wild and stormy as we crossed out of the calm waters. But I was feeling great and even enjoyed the food being cooked in the galley.

We transited through the Panama Canal and we did more work, coring and dredging in the Caribbean Sea, and then we headed north-northeast into the Atlantic Ocean where we sailed and took cores and measurements until we arrived at Bermuda Island in the winter of 1964.

Back to Lamont

After more than 6 months at sea spent on the R.V. *R.D.* CONRAD, I returned to Lamont and was told to go and see Doctor Ewing, the director. I was informed that Doc wanted to talk to me but I did not know what it was all about, and I thought that I might even be fired. When I went to see him, his secretary, Marjorie, who was a lovely and pleasant person, introduced me to her boss. I was very stressed about meeting with Doctor Ewing who was a tall, strong looking man with white hair, heavy eyebrows and a thin smile on his face.

As soon as I saw him, Doc's first question was: "Did you enjoy your cruise?"

I told him that I had been sick in my bunk for at least one entire month and that other than that discomfort, I had managed to survive. I realized later that his chief scientist, Joe Worzel, had probably informed him about all this. I also mentioned that for me, ocean exploration was a new frontier and I liked the idea of exploring the unknown. Then, there was a pause in our conversation before he asked me "What do you want to do now?"

I replied that I would like to continue studying in any field related to the geology of the ocean floor. To my surprise, he picked up the phone on his desk and called the Dean of the Admissions Office at Columbia University in Manhattan. "I have a new student for you, could you see him?" That is how I got enrolled in studying marine geology in 1964.

I obtained a scholarship from Columbia University and I went back to my work at Lamont in November 1964. I was working in the core lab with Roy Capo and under the supervision of my advisor, Charles Fray.

My first scientific publication was written with Neil Opdyke and Hekinian (1967). It was among the first studies of its kind dealing with the measurement of the intensity of magnetization of rocks collected from the ocean floor. Neil was well known in the field of Earth magnetism and 1 day he approached me while I was looking at thin sections of rocks under the microscope and asked, "What do you think about doing some magnetic measurements on the rocks that you are looking at?" During our discussion it turned out that nothing had been done about measuring the properties of magnetization in oceanic rocks. Neil said he would take care of the geophysical aspect if I were willing to do the petrographic work. I was thrilled with the idea and mostly with the fact of being able to collaborate and put my knowledge to use.

Since in the 1960s, many sea-going expeditions had been routinely measuring the magnetic field of the Earth along large transects and Neil and I thought that maybe our approach could help to find out what the magnetic intensity of different rock types might be. This study had an important application for determining what kind of rocks are the most intensely magnetized and has also served to help in prospecting to find ore bodies.

Cnexo-Ifremer

One of the major oceanographic centers in Europe is the French *Centre National d'Exploitation des Oceans* (CNEXO) that was founded in 1968. The center changed its name and became "IFREMER" (I*nstitut Français de Recherche pour l'Exploitation de la Mer*) after 1984 when the national fishery institute became part of the center. IFREMER is located on the cliffs of the town of Plouzané, looking out on the *Rade de Brest* (Brest roadstead) near the city of Brest. When we lifted our eyes from our desks and stood to stretch our backs, we could see the many commercial and military vessels entering or leaving the port of this western city in northern Brittany.

Personal Feelings and Experience

I believe that if someone spends more than 30 years in the same institute for more than 8 h a day, this place should eventually become a second home and even a shelter. My office at IFREMER was a place where I was comfortable and also a place where I exercised a certain amount of power. I felt powerful, or empowered, because I was hoping to orient future research in oceanographic exploration and the idea of making new discoveries totally fulfilled my life.

Along with most of my scientific colleagues, it was a pleasure to work discovering the mysteries of the ocean floor while taking part in missions at sea and doing research in our laboratories back at IFREMER. The unknown was a challenge and we were ready to make compromises regarding our family life since part of our research meant being far away from home in order to collect samples and data, which also satisfied our need for adventure.

My experience with IFREMER started in January 1971. The first Director (*Président Directeur General,* PDG, which is equivalent to the term CEO in English) of CNEXO was *Monsieur* **Yves La Prairie.** A former French Navy officer, La Prairie had fought during World War II in the exiled French Army of Charles De Gaulle. La Prairie's family origins were from Brittany and he was passionate about the ocean and its mystery. This one time Navy Officer was also a poet, a writer and a philosopher interested in ocean science and eager to facilitate new discoveries. A trustworthy person with a capacity for listening to others and for having confidence in the team that he had chosen to build this new institution, I must say that Yves La Prairie is the only Director of our institute that I had the pleasure to meet and appreciate before he left us. It was fortunate that he was the first PDG of CNEXO, since he is certainly responsible for the fact that IFREMER acquired a good international reputation. I kept in contact with La Prairie even after he left the center, but all the other PDGs of our institute merely passed through, after spending an average of 4 years in place. I never even saw most of them or I only met them during official meetings to hear their speeches when they first took their positions.

My major concern and criticism with IFREMER was that there was often a serious lack of communication and personal contact between the various "chiefs" and the lower level employees, like myself, who could be considered as being mere "Indians". This is the result of the strong vertical hierarchy that exists in France, and this *modus operandi* seems to be continuing as an accepted system for conducting every-day work at the center, even now.

For example, at my level of being a senior scientist, in order to pass a message to the top echelon I would have had to go through five different "chiefs" or program directors: the group leader, then the head of the department, then the director for research (in Brest), then the science director (in Paris) and the general director of the institute of Brest, who might, if he had time, eventually inform the PDG in Paris. This was also true for our research projects and proposals that had to be accepted and approved at several different levels. The projects, or many of their ideas, were often extremely diluted before any eventual approval, so a person might easily become discouraged after a few attempts at initiating a scientific proposal.

However I must say this was not the case for me, because I always stuck to my primary ideas and made a point of not capitulating. In other words, because I am a stubborn guy who thinks things through before making a proposal, most of my suggestions for missions or collaboration were eventually approved in a form I could accept as well. However, in some cases, I was obliged to work with people outside my IFREMER home.

My choice of this career was due to the fact that I always wanted to attain new horizons and explore the unknown. Oceanography was a change from the routine of land-based geology or laboratory work, which often involved mapping or re-mapping or analyzing and re-analyzing the same area and material. Even so, there were still the serious barriers of finding funding for our work! Scientific proposals for projects that required a great deal of financial investment were difficult to write and successfully obtain, and sometimes there were complicated issues to be resolved.

"Fundamental research at sea was and still is one of the key aspects for discovery and it increases scientific knowledge by means of understanding the mechanisms that make our planet Earth change and evolve." This message had to be constantly reiterated by the scientists of our department in order to enable their projects to be accepted by the various funding agencies at all the different levels of the IFREMER administration. Any proposal had to go through several persons who were directly or indirectly responsible for its approval.

In the end, in my case, it was thanks to outside institutions such as the *Institut Physique du Globe* (IPG) in Paris and the University of Kiel in Germany that most of the cruises that I participated on during the last 15 years of my career were funded. I still find this fact difficult to accept. I was actually very disappointed that my own research institution IFREMER seemed to refuse or appeared to make very little effort to support and encourage my research after Xavier Le Pichon and Yves La Prairie left the center.

From its beginning in 1968 until about 1974, the department of marine sciences of CNEXO-IFREMER was able to meet its goal of carrying out high quality fundamental research in the growing field of oceanography. The personnel hired at the time all had previous experience at sea and had worked in various international institutions prior to coming to the institute. This was thanks to the determination of Xavier Le Pichon who had the trust of La Prairie and who was therefore able to successfully direct competitive programs in marine geology. Unfortunately, after 1974, things slowly changed, but luckily I had established some contacts with other groups so my research could continue according to my defined goals.

Scientific Projects

To conduct a scientific project one has to be aware that science does not have any frontiers and in order to understand the nature and evolution of the Earth we are sometimes obliged to explore remote and uncharted areas. It is often difficult to convince funding agencies that understanding scientific processes cannot be achieved by exploring our own back yard. Major discoveries have often been made when data were gathered and evaluated after many expeditions from different environments.

There is always a logical, scientific approach when proposing and conducting ocean-going investigations. For example, one of the first aims, after fine-tuning the theories of crust-lithosphere spreading and plate tectonics, was to acquire some ground-true geological evidence. For this reason, the Department of Marine Sciences at IFREMER, along with other French and international institutions, were constantly trying to set up and to conduct cooperative projects to investigate the spreading ridge systems where tectonic plates move away from each other. Indeed, this cooperation with outside groups began in the early 1970s and has proved to be vital for conducting our cruises and also for complementing the fields that were not covered by the scientific program of IFREMER. In oceanography, the field of exploration is large and specific targets were fairly easy to choose from among the extensive, still unstudied areas of the ocean floor.

In the early days (1970s) when CNEXO started to take shape and the laboratory for earth science research started to become operational, exploration of the northern Mid-Atlantic ridge provinces also started to take on importance. Since crust-lithosphere has been observed to spread at different rates, it seemed logical to discover the mechanisms of spreading and their implications for remodeling the crust. The other aspect that interested the scientists at the time was the importance of the rigid plate motions, which are often punctuated by active volcanic activity. This was based on the theory of hotspots in oceanic basins and along spreading ridges giving rise to volcanism. We needed to verify that the presence of these hotspots is the result of deep-seated hot convection currents in the Earth's Mantle, which cause magma to reach the surface and form large volcanoes and volcanic chains, such as the Hawaiian volcanoes.

When a scientific research project is being developed, knowing details about the use of instrumentation and various ocean-going vehicles is another key for success. One of the first requirements is to have good bathymetric coverage of the area chosen. If the need for detailed geological investigation is necessary, then a deep-diving operation could be proposed. Depending on the goals, this diving proposal should follow a strategy that uses the tools best adapted for the operation. Usually when a site is new and unexplored, it is better to use a combination of tools in a logical sequence.

The goal of fundamental science evolves in function of a long-term vision on how a particular project could impact the world's scientific community and attract more investigation in the future. Scientists from various disciplines will often study data as well as the rocks, sediment or animals recovered from the ocean depths, over a period of several years. It may well happen that different specialists contribute towards a new discovery for a project by revisiting an area, which was previously explored. A typical example is the *FAMOUS* project, which provided the first comprehensive study of a spreading ridge segment in the North Atlantic. Other projects such as the *Cyamex, Geocyatherm* and *Geocyarise* cruises on spreading ridge segments with faster spreading rates, such as the East Pacific Rise, permitted us to determine the importance of rapidly evolving volcanism and other geological phenomena. The project on intra-plate hotspot volcanism (VIP = *Volcanisme Intra-Plaque* in French) within oceanic basins was also one of the most productive scientific programs involving several international institutions and providing a great amount of input into the study of volcanic chains and islands (See Chap. 9).

Preparing an Expedition

The preparation of a sea-going expedition is often a one-person show, which is undertaken by the person responsible for the idea or the topic of a proposal. Being able to imagine your own expedition is a creative endeavor that starts from a kind of "dream" and then slowly takes shape to become a reality on paper. However the most difficult task is to organize the expedition once the proposal has been approved and funded: Where can we go? What equipment do we need to use? Who should be chosen to participate on the scientific team?

Even today, the Earth's ocean floor remains practically unexplored and it still is very difficult to sell a scientific program to most funding agencies. The Space Program is more attractive to government agencies and to the general public than ocean floor exploration. Maybe this is the result of not being able to excite public interest and/or our politicians about the veritable need to understand everything about the ocean's depths. Also, since the ocean is right beside us, people might think that that there will always be time to deal with it later on, or they may even believe that we already have determined everything we need to know, which is far from the fact.

To obtain appropriate funding for an oceanic program is an extremely difficult task and requires considerable effort, energy and patience. These days in Europe, a complicated administrative system has been put together to give grants to various projects. These proposals need to involve several institutes from different countries, which is always scientifically rewarding. But the system which has been put together is quite bureaucratic and heavy so it is difficult for a scientist to handle his own research and at the same time to write a proposal, do the administrative work, and lead the project to its conclusion if it has somehow managed to be accepted.

The need of a lawyer or specialized person familiar in dealing with the European bureaucratic system may become necessary. In any case, it is a long and tedious job to present a project to such a bureaucratic organization and scientists often become discouraged. Most of the time, projects are halted at the institutional level because of their cost. Indeed, oceanic expeditions are very expensive and they offer no (or very little) immediate economic return.

Only a highly developed country can afford the rich-man's game of ocean floor exploration. I am inclined to feel that if a country cannot afford oceanography, then it is better to not even try. It is not wise to cut corners in order to reduce the budget of exploration because in the end, the return of knowledge is priceless. I believe that if a country cannot pay the price for oceanographic research, then this nation should not pretend to do it for economic reasons. In my opinion, it is very important that government agencies and those people who are responsible for government-run oceanographic institutions encourage undersea research only if they want to increase our basic knowledge of our water covered planet, which is unique in the Universe, or at least, unique in our Solar System.

> ***What will be won in knowledge and understanding of our home in the universe will be worth far more than the cost (Ericson and Wollin 1967).***

The necessities of time, funding and qualified personnel are often difficult to bring together for sea going expeditions, especially since the expeditions are obliged to last for a period of several weeks. The choice of the scientific theme or topic for a project must be of worldwide interest and include a large number of scientific disciplines in order to be able to collect as much information as possible during a given cruise.

The choice of the team members needs to include people who are genuinely interested in the project and their specializations should also be adapted to the needs of the cruise. Several criteria could be used to choose a sea-going team of scientists. Candidates must:

(a) be competent in the field to which the cruise will be oriented.
(b) have had previous experience at sea if possible, but this is not a prerequisite.
(c) be sociable and be able to work on a team, be good at sharing competence and data on board. Indeed it is important for the group on board to communicate any findings and work they are carrying out at the same time as they are at sea.
(d) be open to a multi-disciplinary approach and willing to help their fellow scientists.
(e) be in good physical condition in order to be comfortable during the journey, and to avoid any risks of reducing the time of a cruise at sea due to illness.

The choice of the equipment such as ships, submersibles, unmanned vehicles or other apparatus to be used in order to achieve the scientific goals of the mission is of great importance as well. The appropriate gear should be selected for each scientific cruise on the basis of the goals to be achieved, and this requires that the chief scientist be sure of his choices and objectives. Selecting appropriate equipment will make a considerable difference in the outcome of the cruise and the various choices should be clearly explained when the project of any sea-going expedition is first proposed.

The project has to be evaluated by a scientific committee. Once the evaluation is made and the project has been accepted, then the real work will start. This is the beginning of a long marathon. The project must be taken into consideration by a "program committee", which deals with all the cruise requests that have been accepted by the scientific committee. Their role is to coordinate a country's different expeditions within a time frame of about a year. This also helps to decide on a research ship's overall route and the areas where the various oceanographic vessels will be operating in the next 12 months. Once the cruise is accepted and the timing fixed, there will be a meeting with the ship's captain and the engineers who are to be involved, in order to decide on the final operational plan for the cruise.

Many projects and cruises are dropped because of discouragement after several unsuccessful attempts at organizing all this. This even happened to me when I proposed a diving cruise in the Garrett Transform Fault (south Pacific Ocean). It was, however, a good thing that I did not give up, even after the project was rejected four times, otherwise, the world's scientific community would never have been able to work with the information that we acquired during our cruise (See Chap. 8).

Other problems with sea-going expeditions are that the scientists are at the mercy of their instruments' performance as well as unpredictable environmental conditions. Sometimes the instruments do not perform as expected or there are bad-weather conditions that will prevent the team from doing the job they were assigned. Many things could go wrong at sea and one should always be prepared to have alternative plans in order to minimize any loss of time. That is why I have always believed that a sea-going program should be 200 % prepared in order to have a 70–80 % rate of success.

Life on Board a Research Vessel

A person who wants to do oceanography should have the desire to accomplish things, a love of adventure and the unknown, and be ready to be flexible and make compromises. Once on board, a member of the scientific group should be aware that a mission's success relies on teamwork and each person has a responsibility so if he is not able to do something, then someone else will have to do it for him. Also, one should be aware of the time schedule and be able to make quick decisions. With improved techniques of data acquisition on board, it is important to properly record all information in order for it to be used for further study on shore. I believe that oceanography cannot only be taught at school, but it is also learned at sea. Of course it is important to earn degrees in oceanography at our universities, however the ultimate achievement is to learn through having experience at sea.

The environment on board is that of a small community and becoming acquainted quickly with the other fellow scientists and crew is helpful for a team's moral and might even have positive effects on the results. Mealtimes are the moment when people are often gathered together and can spend time talking about everything. The fact that we are cut-off from the rest of the world may mean that small events and details on board can assume overblown proportions.

It is important to be careful when walking on board a vessel. When the sea is rough, accidents can happen very easily. At night, usually it is not safe to walk on the deck alone. Even in the daytime, if a wave should wash you into the sea, it is very hard for a ship to come back on its course to recover a crewmember lost overboard. It will take a minimum of 20–45 min to turn the ship around and return for a rescue, even when visual contact with the person at sea is maintained. This is an optimistic estimation under good weather conditions, with no high waves and during daylight.

Going to the bridge where most of the command operation takes place might not be a good idea if a delicate maneuver is underway unless you are expected to be there on-duty. The best thing is to ask the captain if you could visit the bridge. It often happens that someone is not welcome and has arrived at the bridge at the wrong moment, but you will probably be aware of your error because of the mood of the on-duty group working there. However, the officer-on-watch is sometimes happy to talk with someone, if nothing serious is going on, so it is best ask permission before visiting the bridge.

It is commonly believed that the most important man on board is the captain, but I do not think it is true, unless there is an emergency or a delicate maneuver to be accomplished. In my opinion, on a daily basis, the most important people on board are the cook and his crew. Indeed, the people working in the galley have the art and the possibility of keeping the rest of the crew and their guests happy during a journey. The galley becomes the best place to go to meet with others and discuss life and everyday problems. The boson (chief of the crew members) is another person that you really want to keep on your good side. He seems to be the fellow who can make things work properly and smoothly as far as the maintenance of all the material and apparatus, which the inexperienced scientists coming on board seem to have difficulty handling in a proper way.

There is always noise on board a ship and one has no choice but to get used to it. It seems that the crew is constantly painting and repainting the ship, and scratching-off old paint is a tedious, noisy job that has to be performed if we don't want to see a rusty, floating wreck at sea, like some of the other commercial ships we have encountered on our missions.

Captain's table When embarking on French vessels, it was an honor to be invited to the captain's table. Usually the meal lasted for more than an hour. This was a custom during the early 1970s until 1980. Four-course meals and drinks were served as if the Captain and his guests were sitting in a three-star restaurant, with the chief steward wearing a white jacket serving the delicious dishes. After the meal, a cigar used to be a commonplace, enjoyable way to finish the experience. Of course, this might upset some people's queasy stomachs if they were not used to being on a ship and if the weather conditions were not conducive to remaining inside for a long period of time. I used to be apprehensive about an invitation to the captain's table because I knew I would be uncomfortable and yet feel I needed to be too polite to request permission to excuse myself. Afterwards, when you feel better, you always regret having had stomach troubles that prevented you from enjoying meals with the ship's captain.

The science headquarters (*called the "PC—Poste de Controle"* in French) is the area where all the scientific information and real-time data acquisition are stored and where meetings take place. This is where we go to conduct our scientific work and data recovery and for any on-board decision-making.

Team Work

To be an ocean-going scientist one has to learn how to work on a team. The chief scientist responsible for the scientific project has the duty of bringing together the different types of personnel: the crew, the officers, the diving group and the scientists. During ocean-going expeditions several groups are usually involved such as the seamen, the engineers and the technical and scientific team members. It is very important that a good relationship between the various groups be created at the beginning of each cruise. Even if this seems to be obvious for many, it is often difficult to set-up a cooperative environment and then it becomes easy for working conditions to deteriorate quickly, in turn causing everyone's life on board to be miserable.

Creating a united team of the crew, scientists, ship's officers, etc. is a difficult task and few people are capable of handling this properly and doing a good job. However, if the team is not united and motivated, there is a big risk of jeopardizing the outcome of the cruise. The most important duty of a chief scientist is to handle any difficult situations, such as the failure of an instrument and/or finding time for equipment repair or any other delays that are often unpredictable at sea.

The role of the chief scientist is a difficult task because he/she has to deal with several groups of people often coming from different institutions, all of whom are eager to obtain the maximum amount of information from a particular cruise. It is important for the Chief Scientist to inform the officers and crewmembers about the cruise objectives so they don't see their ship as being invaded by the transiting scientists, who they might sometimes consider as being unruly guests rather then working partners. The scientists are called "passengers" to distinguish them from the more permanent crew and officers who spend longer periods of time at sea. An experienced chief scientist often has to follow the logic of unwritten pre-established rules during the preparation of the cruise. However, it is sometimes necessary that he follow his own intuition when confronted with an unexpected situation. For example, in order to decide on the choice of a diving site in an unexplored region, it would be a wise idea to take into consideration all the possible details that might jeopardize the expedition. In order to have the maximum return for a mission's results, it is absolutely necessary to create good cooperation between all the groups.

Communicating Scientific Results

The recognition of authorship in oceanography is always biased and sometimes does not give credit to all the people involved in data acquisition and data processing. Scientists are sometimes politicians or businessmen without scruples, or they could be insensitive people who ignore the needs of being appreciated on the part of another human being. These types of authors believe the world is only about themselves, and others are nonexistent, therefore these unscrupulous scientists often make every move to be the first in publishing a result, no matter who must pay the price.

Some scientists have built their reputation by pretending that they were the "first to recognize" a process or make a discovery just by ignoring what others have accomplished in the past. Some American scientists are notorious for ignoring their international partners. I believe one reason for doing so is due to the fact that many U.S. colleagues depend on their research papers for getting more money to continue their science, which is how they are supposed to be making a living, and it is a fact that they badly need funding in order to survive. Even if all this is true, it is still no excuse for ignoring the hard work of their other team members.

Also, the publication of a scientific communication related to oceanic studies differs from scientific research done at shore in experimental labs or during geological field studies on land. While field geology on land can be accomplished with modest funding and involves a relatively small amount of personnel, ocean-going expeditions involve huge amounts of personnel and incredible amounts of material and logistics. In fact, a scientific project at sea is not at all a one-man show. Oceanographic research is the result of the work accomplished by a large number of people and requires a great deal of money, and above all, this kind of science involves a huge team effort.

This sense of teamwork should not terminate at the end of the cruise, but rather it needs to continue on shore when the laboratory studies and interpretation of data are shared between the various partners. Also, it is a common practice that all the participants on a mission will be "co-authors" of the first paper written after an expedition. That is why many papers publishing data from sea going expeditions might involve up to 10–15 authors. Criticism for such practices has been made by other colleagues and even by a scientific journal (EOS), which published an editorial claiming that the long list of names cited as authors for a scientific paper is misleading and often does not reflect the scientific input. I totally disagree with this statement because I strongly believe that scientific results obtained at sea definitely rely on "teamwork". For me, a mission's final success will depend on the expertise of each person chosen for the team. A minimum form of acknowledgment for his group and their work on the part of a chief scientist is to have his entire team listed as participants in the first publication after a cruise.

Also, we should be aware that data acquisition and maximum data processing at sea is the best way to proceed for maximum efficiency. From my personal experience I have found it to be difficult and time consuming to go back to reprocess data on shore once a cruise has ended. This is often unproductive because we may have forgotten much of the information obtained, which we have shared with other colleagues while at sea. Because of the large amount of information obtained in a relatively short amount of time (1 or 2 months), it is difficult and annoying to be obliged to re-view and/or reprocess the data when we have left the ship and are again on shore.

I strongly believe that post cruise data-processing is less complete and less precise when compared to that which is done immediately on board while our minds and our energy are focused on the mission at hand. This is especially true for the interpretation of images from the sea floor and for mapping done by manned and unmanned engines, such as submersibles and robots.

References

Ericson DV, Wollin G (1967) The ever changing sea. A. Knopf Inc., Random House of Canada Limited, New York, p 354

Hekinian R (1968) Rocks from the Mid-Oceanic Ridge in the Indian Ocean. Deep Sea Res 15:195–213

Heezen BC, Ewing M, Tharp M (1959) The floor of the oceans. Geol Soc Spec Paper 65:122p

Heezen BC, Hollister CD (1971) The face of the deep. Oxford University Press, Oxford, 159, P

Opdyke ND, Hekinian R (1967) Magnetic properties of some igneous rocks from the Mid-Atlantic Ridge. J Geophys Res 78(8):2257–2260

Vine FJ, Matthews DH (1963) Magnetic anomalies over oceanic ridges. Nature 199:947

Wilson JT (1963) A possible origin of the Hawaiian Islands. Can J Phys 41:863–870

Wilson JT (1965) A new class of faults and their bearing on continental drift. Nature 207:343–347

Wüst G (1964) The major deep-sea expeditions and research vessels 1873–1960. Progress in Oceanography vol. 2, Pergamon Press, Oxford pp. 1–52

Chapter 2
Our Haven, Planet Earth

Abstract Planet Earth was formed from the agglomeration of solid bodies. Earth's iron-nickel enriched core segregated from a silicate mantle as early 30 million years after its formation 4.1–4.7 billion years ago. Earth is the third planet in our solar system circling around the Sun in an elliptical orbit at a distance of 147–152 million kilometers. Such a distance from the Sun enables our planet's temperature to be hospitable to animal and plant life. Among the nine planets of our solar system, Earth is the only one whose surface environment is adapted for water to exist in its three states: solid, liquid and gas. The elements necessary for life consisting essentially of oxygen, carbon, hydrogen, nitrogen and sulfur, are also found in comets, stars and probably on other planets in the Universe. 71 % of the Earth's surface is covered by water, which means that when seen from space, our home could be called the "Blue Planet".

The universe has expanded and cooled from an extremely hot and dense state after the "Big Bang" occurred about 15 billion years ago (Hazen 2012). The term "Big Bang" refers to the beginning of the Universe, when a release of energy triggered a rapid expansion of primordial elementary particles. The regions with particles having a high density also had a gravitational potential to attract more matter from their surroundings and to grow even denser. Thus, the beginning of the galaxies started as matter began to accumulate in various confined regions of the universe. Today, our universe is also littered with blocks of dead stars. The larger and more massive stars have exploded in supernovas and now form relics of denser stars. Some of these massive stars have even become black holes. Other blocks and dust have agglutinated together and contracted under gravitational forces to form hot fireballs like the Sun. Our Sun and our solar system are located at the edge of the Milky Way Galaxy.

R. Hekinian, *Sea Floor Exploration*, Springer Oceanography,
DOI: 10.1007/978-3-319-03203-0_2,

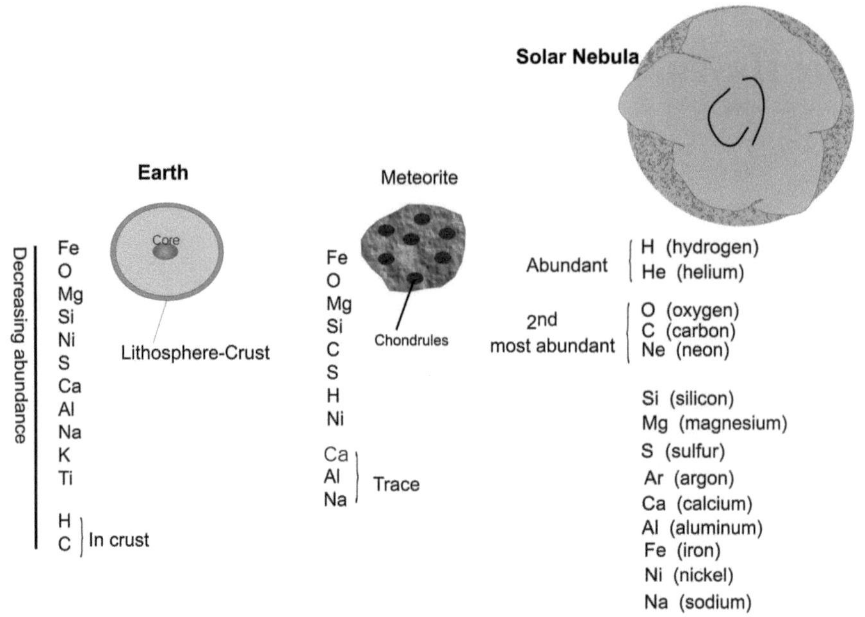

Fig. 2.1 The schematic representation infers the creation of planet Earth from our Solar Nebula. The major-element compositional distribution with their relative abundance is shown (after White 2013)

The Birth of Planet Earth

About 4.7 billion years ago, dust and rocky blocks gravitated around the Sun and formed spherical bodies of variable size (Fig. 2.1). Our home planet, Planet Earth, must have begun to be formed with the accretion of dust and large blocks that collided and agglomerated together. This agglomerated sphere attracted more material from the stellar system and its gravitational forces generated heat so the material began to melt. As melting occurred, particle differentiation began, therefore the heavier components such iron (Fe), magnesium (Mg) and nickel (Ni) sank into a magma pool towards the center of the planet while a silicate liquid made up essentially of silicon (Si), aluminum (Al) and sodium (Na) rose towards the surface.

There are many similarities between our planet and others in distant galaxies, called exoplanets, because they are planets located beyond our solar system that are orbiting around other stars. In terms of similarities, we could consider a planet's density. In our own solar system, for instance, Mercury (5.42 g/cm^3), Venus (5.25 g/cm^3) and Earth (5.52 g/cm^3) have similar densities when compared to that of Mars (3.94 g/cm^3). The average density of a planet depends on its mass. The bulk density of a planet is the ratio of the mass of a planet compared to its volume. The variation in density between planets of similar size is related to a change in the composition of the material forming each planet. The relationship between density and mineralogical and/or chemical composition is related to the way the atomic structure of the matter is arranged (Broecker 1985).

Since the density of a planet depends on its mass, the "denser" or "heavier" a planet is, the stronger its gravitational pull will be. Thus, since Venus and Earth are larger than Mercury, gravity's attraction on their surface will also be greater. The pressure generated by a planet's gravity will cause the atoms forming the planet's interior to contract in size. If a planet's mass is increased, then its gravity will increase and the contraction of the atoms forming the planet's material will also increase.

The information we have concerning the composition of the stars, planets, their satellites (or moons) and comets has been obtained from astronomical observations conducted from Earth and, more recently, from observations obtained by spacecraft launched from Earth to explore our galaxy. The astronomical observations on the elements composing other celestial bodies are based on studies of light refracted by a prism, in order to separate the light's different colors into a spectrum. Each color of the prismatic light characterizes an element. Laboratory experiments to define the various elements consist in calibrating the passage of light from an electrical arc through a known gas mixture.

The most abundant elements in a star's interior are hydrogen (H) and helium (He), and the second most abundant elements are oxygen (O), carbon (C) and neon (Ne) (Fig. 2.1). Volatiles in comets consist essentially of water and minor amounts (about 1 %) of other constituents such as CO (carbon monoxide), CO_2 (carbon dioxide), methanol, formaldehyde, ammonia and hydrogen sulfide (HS_2).

Compared to stars or comets, the most abundant elements forming the largest planets are oxygen, magnesium, silicon and iron with nickel, sulfur, calcium, aluminum, sodium, potassium, and titanium, in decreasing abundance. The abundance of other elements drops rapidly, and becomes so insignificant that all the other elements represent no more than one percent of a planet's entire composition (White 2013). Due to these compositional variations, it is not surprising to observe that when we are on Earth, we are walking around on a solid crust which is mainly made up of elements such as O, Fe, Mg, Si and Al forming crystalline compounds, which we call minerals and rocks.

Another direct source of knowledge about the composition of our planet comes from the meteorites, which have fallen on Earth. These "space rocks" are essentially made up of nickel, iron, silica, magnesium and oxygen, formed from molten liquids which also contain small amounts of long-life radio-isotopes, which can be used like a clock. By measuring the concentration of the radioactive elements and the time necessary for one radioactive isotope to decay and change into another element, it is then possible to define the date of the meteorites' formation. It was determined that the meteorites from our galaxy were formed about 4.6 billion years ago (Myers and Crowley 2000). No rocks as old as this have as yet been discovered on Earth, but analyzing a zircon compound in a metamorphic formation in Canada, some scientists have determined an age of 4.1 billion years (Myers and Crowley 2000). If, as according to astronomers, the age of the

Universe is around 15 billion years, then the Earth was probably formed 4.1 billion years ago from the same matter as that found on other planets (Dalrymple 2001). The continued exploration of our planet can provide us with an opportunity to better understand the extraterrestrial bodies in the rest of our universe.

Recently, more information on the composition of our galaxy has come from spacecraft sent into space to collect data and make in situ measurements of meteorites and comets. For example, the Comet Wild 2 images and dust samples collected from the NASA Stardust and Deep Impact missions gave us some information on the nature of the solid material in comets (From Internet: news.com.au). The presence of minerals called fosterite (Mg_2SiO_4) and enstatite ($MgSiO_3$) along with other phases have been detected in meteorites (Brownlee 2008), and these minerals are similar to the mineral composition found in our own Earth's mantle.

Our nearest neighbor, planet Mars with its lower density, is also one of the smallest planets in our solar system. It has been explored by robots carrying out geological and biological experiments. So far, little has been achieved to determine the existence of water (H_2O) such as we know it on Earth. Mars' atmosphere has a low density and it is very cold (−93 °C to 13 °C), so it is unlikely that water could exist in its liquid form.

In 2004, based on satellite observations on Mars, scientists detected the existence of an ice cap located on the South Pole of this planet using a spectrometer capable of detecting images in the range of infrared sensors. This ice cap is not made of water, but is formed of frozen CO_2 (Pappalardo et al. 1998). If liquid water (H_2O) could have been present on Mars in the past, this might suggest the existence of life on the planet. However, the thin CO_2-rich atmosphere does not prevent incoming UV (ultra violet) radiation from the Sun, which is fatal for living organisms.

On the other hand, on Europa, which is a moon circling around Jupiter, we have observed a liquid ocean covering its surface. Is this ocean similar to that of our Earth? The Galileo space probe permitted us to observe patchy structures on the surface of Europa that were interpreted by scientists as representing floating icebergs (Pappalardo et al. 1998).

It was in 2004 that the Cassini-Huygen spacecraft entered Saturn's orbit. Landing on Titan, one of the largest moons of Saturn, scientists interpreted evidence of frozen methane as being flat and rounded slabs (Pappalardo et al. 1998). NASA scientists have announced that they were able to photograph features resembling lakes of liquid hydrocarbons more than 200 km long and 70 km wide.

The Solid Earth's Interior

Probably the Earth's interior today does not reflect the composition and the structure that it must have had at the beginning. During the creation of the solar system, crystalline solids condensed from the solar nebula. These solids were swept up by gravitational attraction after which intense heating and melting took

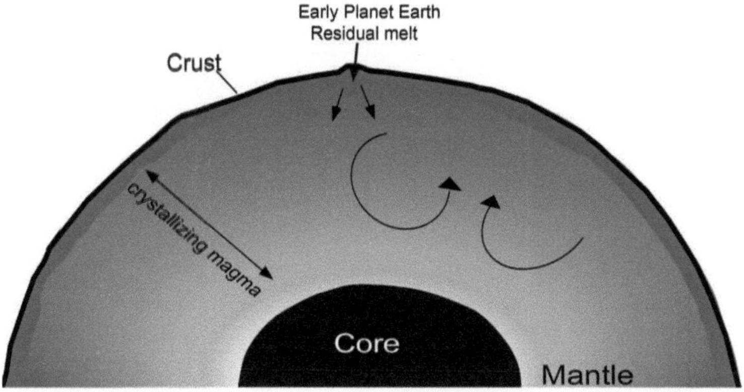

Fig. 2.2 Early-solidifying Earth after its formation shows thermal convection mixing the composition of lower and upper mantle from the base of the mantle upward. Various elements (mainly incompatible) will be diffused and transferred throughout the solidifying layers composed of different crystallizing phases. A reservoir of enriched heavy elements distilled from the early "magma ocean" resides within today's Earth, probably in the area of the core-mantle boundary at about 2900 km depth (after Boyet and Carlson 2005)

place on Earth. Based on isotopic element ratios measured on primitive meteorites and compared to present terrestrial samples, it was found that the Earth's mantle was a deep ocean magma during its first 30 million years (Boyet and Carlson 2005) (Figs. 2.2 and 2.3).

The early molten Earth most probably crystallized from the base of the mantle upward and Earth's surface was the residue of molten liquid sitting under a primordial crust. Similar to the other solid planets, the early dense iron contained in the molten magma settled at the center forming the metallic core overlaid by a peridotite (silica, aluminum, iron, magnesium and calcium) enriched mantle.

As the mantle continued to solidify, convection currents were generated and they were able to mix the upper and lower mantle material (Boyet and Carlson 2005). Reservoirs of enriched elements such as potassium, sodium, uranium, thorium and rare gases such as He (helium), Ne (neon), Ar (argon) and Xe (xenon) and their isotopes now mostly reside in a region found at the mantle-core boundary, at 2900 km depth. The heat generated by these radioactive elements and the heat within the molten core are responsible for creating the mantle's convection currents and for supplying energy and matter to hot magma plumes upwelling towards Earth's surface. Also, the early differentiation of the Earth's mantle interior must have been the result of at least two different sources to produce the most common volcanic rocks, which are mid-ocean ridge basalts (MORBs), and ocean island basalts (IOB) (Mukhopadhyay 2012).

Our present knowledge of the Earth's interior suggests that it is roughly like an onion made of successive concentrically arranged layers with increasing density towards its interior (Fig. 2.3). The deep Earth's interior is beyond our visual observation; the only evidence of its composition comes from meteorites falling on

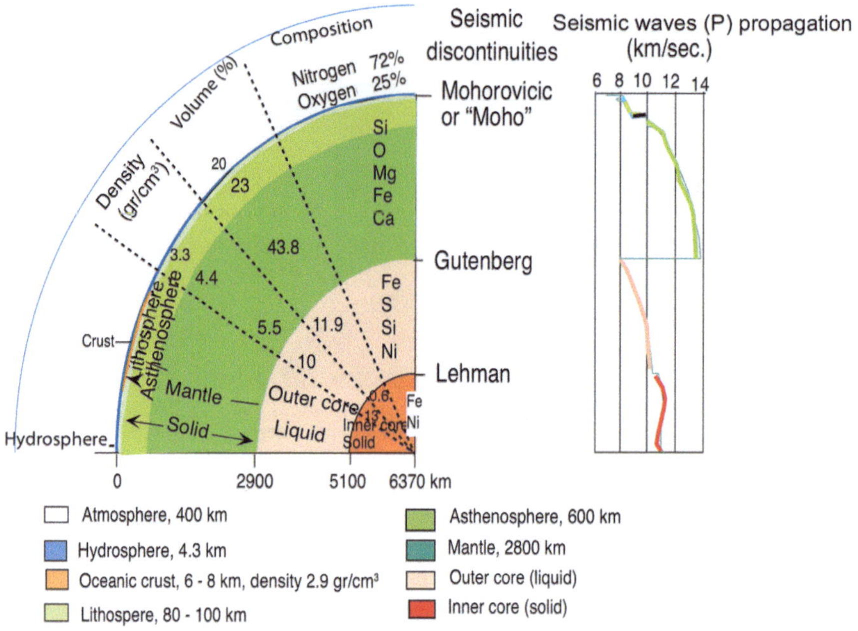

Fig. 2.3 The internal structure of the Earth shows different compositional layers. Seismic study reveals several discontinuities encountered in the interior of the Earth. Each discontinuity relates to physical, chemical and mineralogical changes (after Hekinian and Binard 2008)

Earth's surface, or from interpreting seismic wave propagation within rocks. Also, we have learned about our Earth's interior due to laboratory experimental work on the stability of mineral phases at different pressure and temperatures. Thus, laboratory experiments exerting high pressure and high temperatures on minerals in order to change their stability have permitted us to determine the density distribution at the interior of the planet (Green and Ringwood 1970). Another way of determining the composition of the Earth's interior is by measuring the density differences of the material encountered and then comparing this information to that obtained from laboratory experiments. The density differences can be gathered during earthquakes and/or any extensive man-made explosive activity (bomb testing) where the release of energy at a known source produces waves, which radiate in all directions and are identified on seismographs (from the Greek words *seismic* meaning earthquake and *graphos* meaning writing). The seismograph measures time and the ground motion of the Earth's interior. The ground motion is transmitted into the Earth by "wave propagation", similar to what is observed on the sea's surface when dropping a stone. A seismograph will measure the propagation of certain seismic waves called P-waves (Primary waves), which are compression waves, as well as expansion waves, like sound and other S-waves, which vibrate at right angles to the direction of travel, as do light waves. The P waves travel through both solids and liquids while the S-waves only travel through solid material. A geologist reads and interprets the reading of a "seismogram", just as a doctor is able to interpret our heartbeat when reading a cardiogram.

When comparing the seismic sound wave velocity transiting through a rock formation in relation to the density differences encountered, a seismologist is able to infer the composition of the material. Three major discontinuities have been found inside our Earth: (1) Core, (2) Mantle and (3) more rigid Lithosphere-crust (Fig. 2.3).

The Core

A 2,900 km deep "discontinuity", also called the *Gutenberg discontinuity*, at the border of the inner mantle and the outer edge of the core, is made up essentially of a mixture of iron and magnesium silicates. This discontinuity marks the separation between the Earth's mantle and its core. Earth's core is composed mainly of iron and lesser amounts of nickel (3–6 %), and is revealed by a sharp drop in the P-wave velocity from about 12 km/s to about 8 km/s which then increases again up to 10–11 km/second at the inner core boundary (5,100 km depth), also called the *Lehman discontinuity* after the Danish seismologist, Inge Lehman. The inner core, at 5100–6130 km depth, is solid and essentially made up of nickel and iron. Its density is around 9–13 g/cm^3 (Fig. 2.3). This solid inner core was formed during the accumulation of heavy metal material that sank from the outer core, which is less dense and is still in a molten state. It was Descartes, in 1644, who first proposed that Earth must have a liquid core or center.

The outer core consists of liquid metallic material made up of siderophile (iron loving) components. The siderophile elements such as the platinum group of elements (rhenium, osmium, iridium, palladium and ruthenium) are likely to be associated with the nickel-iron core. They are also called High Strength Elements (HSE) because they prefer to be associated with metals rather than silicates. Hence it is likely that they are more prevalent in the liquid outer-core than in the Earth's mantle.

Convection currents may have allowed material from the outer-core to be mixed into the lower mantle, and will later enhance the distribution of the platinum group elements towards the upper mantle and then enable them to appear in erupted lava on the Earth's surface. However, this hypothesis assumes that the platinum group elements were differentiated during the early formation of the solid inner-core to the liquid outer-core. Furthermore, we do not know about the real composition of the liquid outer-core and the crystalline inner core. In addition, if elements have fractionated from the inner to the outer-core, this must have happened at the very beginning of Earth's formation due to what is observed by the long-lasting half-life of the platinum radioisotope (^{190}Pt). In fact, these chemical hypotheses are in conflict with geophysical theories based on Earth's heat flow history, which is used in order to explain the evolution of the inner-outer core system and which suggested that the core was formed later than Earth's beginning, and only occurred about 1 billion years ago (Labrosse et al. 2001; Melbom 2008).

The lower boundary between the core and the mantle is called the D" layer and is about 200–300 km thick. This is the area where the temperature of the Earth has the highest gradient and changes from 2200 °C at its base, closest to the solid, cooler core center, to 3400 °C at the top of the Core's D" layer. This is the region where iron-rich liquids interact with oxides and silicates in the mantle. It is an unstable and heterogeneous region responsible for generating the hot mantle plumes capable of rising towards the Earth's surface to form hotspot volcanoes. (See Chap. 9) (Fig. 2.3).

If the core were leaking, we should be able to retrace this fact through the presence of the High Strength Elements (HSE), which have accumulated inside the Core. The isotopic fingerprint of these HSE sometimes accompanies magmatic upwelling in some of the larger plumes generating hotspot volcanism, such as on the Hawaiian volcanoes. It is has been determined that mantle derived materials exposed through volcanic eruption are heterogeneous in their radio-isotopes and in their light-incompatible-element contents, which are more similar to the material found in lithosphere-mantle volcanism and which do not seem to require material or energy input from the deeper-lying outer-core.

The Mantle

The Earth's *upper and lower mantle* occurs above the major seismic *Gutenberg discontinuity* located at 2,900 km deep at the Earth's core boundary (Fig. 2.3). The mantle reveals another seismic discontinuity at 640–700 km depth separating the *upper mantle* from the *lower mantle*. The *upper mantle* is subdivided into the *lithosphere* (>100 km deep) and the *Asthenosphere,* whose lower limit is at about 670–700 km deep. The temperature at the base of the mantle is around 3000–3700 °C but is <1200 °C in the upper mantle region.

The composition of the interior of our layered Earth has been extrapolated from laboratory experiments on minerals that form rocks as well as from what we have observed in meteorites landing on Earth from outer space. Thus, it is inferred that the upper mantle must have a composition close to the mineral association of olivine–spinel (and/or garnet)–pyroxene forming a rock called a *lherzolite* as well as to the composition of the mineral known as peridotite, which is stable within the first 50–100 km depth inside the Earth.

At about 670 km depth in the lower mantle, there is a mineral phase change where synthetic perovskite (Mg, Fe, $CaTiO_3$) becomes stable until about 2,900 km depth. This perovskite could be the most abundant mineral phase found on Earth and is similar in composition to the enstatite mineral found in comets.

Most minerals of the Earth's upper mantle contain hydrogen, which is structurally bound as hydroxide (OH). The OH concentration in each mineral species is variable, so in some cases it may reflect the geological environment of a mineral's formation. Of the major mantle minerals, pyroxenes are the most hydrous, containing from 200 to 500 parts per million (ppm) H_2O by weight (Bell and Rossman 1992) while garnet and olivine contain only 1–70 ppm (written as H_2O = 1–70 ppm).

The Lithosphere-Crust

The interior of the Earth is further subdivided into the *lithosphere,* which is about 60–100 km thick. The *lithosphere* (from the Greek words *lithos* meaning rock and *sphere*, meaning ball) includes the outside layer, called the *crust,* which is a more rigid layer and the underlying *asthenosphere,* which is more ductile (See Chap. 4) (Fig. 2.3). In the outer part of the asthenosphere, the rocks are colder (<1000 °C) and more rigid therefore they deform elastically under loads and eventually break due to brittle failure. The continents consist essentially of granitic crust. Granite consists of quartz (a pure silica oxide), plus grey feldspar (K and Na aluminum silicate) and white colored plagioclase (Ca, Na, aluminum silicate) as well as darker iron-magnesium bearing minerals, which could include pyroxene and mica or amphibole. The granitic crust under the continents could be as much as 60 km thick. Underneath the oceans, the outside layer of crust is <6–10 km thick and consists of *extrusives* (volcanic rocks) or *intrusives* (dykes or sills).

The lower part of the lithosphere that intrudes into the upper mantle at about 100 km deep consists of peridotite (see Chap. 4). The lower boundary of the lithosphere (called the asthenosphere) is composed of a less rigid and more plastic (ductile) material than the upper-crust lithospheric layer.

In 1909, a Croatian seismologist named Andrija Mohorovicic discovered a seismic discontinuity located at about 35 km under the Balkan mountains, during his study of earthquakes. This has been called the *Mohorovicic discontinuity* and is commonly thought to be the lower level of the *crust.* The difference between the crust and the deeper lithosphere is observed by differences in seismic wave travel-times, which are related to the composition of the rocks.

Underneath the ocean, the difference between the lithosphere and the crust is arbitrarily defined. Generally the *crust* is defined as the portion of the most outer layer of the Earth, about 6–10 km thick, that is created during the upwelling of magma originating from the partial melting of the mantle. The magma transiting through conduits and magma reservoirs will undergo crystal-liquid fractionation and give rise to a solidified volcanic shell (see Chap. 5).

Crustal thickness varies between what is beneath the continents (estimated at about 60–80 km thick) to underneath the ocean where it could vary from 6 to 25 km depending on the constructional features. In the oceans, the thicker crust is located under islands, which have been built during repeated volcanic eruptions, enabling seamounts to rise to the ocean's surface. The lithosphere is really two different zones: a more ductile asthenosphere and the solid crust. In geology, the two zones are often combined together to represent the outermost shell of our planet. The crust's relative rigidity (brittle nature) corresponds to a relatively cooler region, and is involved in the mechanisms of tectonic plate motion. The lithosphere's crust has broken up into several rigid plates, which move with respect to each other according to a common pole of rotation. Each rigid plate is like a "raft" that is floating on the more ductile layer of the lithosphere, i.e. the asthenosphere.

The lithosphere-crust has changed continuously over geological time. This is due to the slow convective flows carrying mantle material and heat, which are responsible for the renewal of the Earth's outer layer (See Chap. 5). These changes have also been called the *Wilson Cycle* in honor of Tuzo Wilson (1972) who developed the concept that continental masses have been formed by the opening and closing of the Earth's oceans as a natural consequence of plate motions. How many times have the oceans and continents been recycled? How far back in Earth's history can the surface evidence of this convective cycle be extended? This has certainly occurred many times, but it is hard to be more specific on the basis of our present geological record.

Water on Earth

Water abounds in the Universe, however the origin of water on Earth is speculative. On Earth, 96.5 % of the planet's surface water is found in the oceans, 1.7 % in groundwater, 1.7 % in glaciers and the ice caps of Antarctica and Greenland, a small fraction in other large water bodies, and 0.001 % is found in the air as vapor or clouds, formed of solid and liquid water particles suspended in air (Fig. 2.4) as well as in precipitation. Only 2.5 % of the Earth's water is fresh water, and 98.8 % of that water is found in ice and in groundwater. <0.3 % of all fresh water is located in rivers, lakes, and the atmosphere, and an even smaller amount of the Earth's freshwater (0.003 %) is contained within biological bodies and manufactured products.

How this water formed the oceans is not easy to grasp. Convay's hypothesis (1943) claims that water condensed from the primeval atmosphere. Thus, the genesis of the air and ocean is due to planetary gas release during the capture, concentration and accumulation of particles forming a layered Earth composed of solid, liquid and gas. The blue color of the ocean is due to the fact that the red wavelengths of sunlight are absorbed quickly while the blue rays are transmitted through the water column.

Another hypothesis is that the hydrosphere was formed at the beginning of the Earth's formation from degassing during volcanic activity and then by gravitational differentiation, when volatiles concentrated on the surface. It is also evidenced that the release of volatiles from the Earth's mantle is another source of gas and liquid in the hydrosphere and atmosphere. Furthermore, it must be noted that the volatile release, which could condense and form the oceans, will react

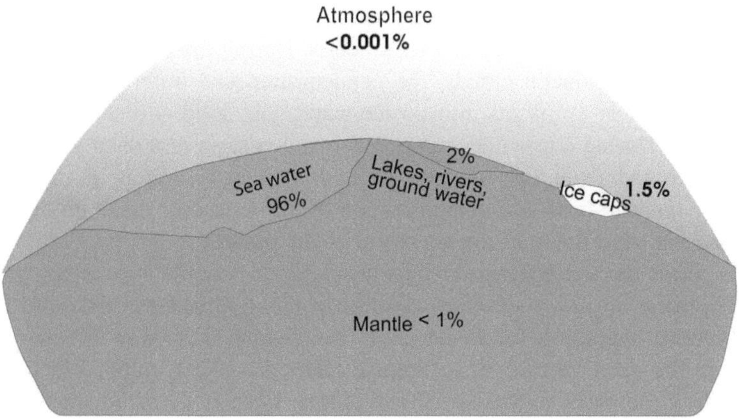

Fig. 2.4 Global distribution of water in the Earth shows that a small portion of water is stored in the atmosphere in the form of confined vapor. This vapor is essentially in the lower atmosphere. 96 % of the Earth's water is in the oceans, and <2 % in the solid mantle and lithosphere. About 2 % is found in lakes and rivers and 1.5 % makes up the ice caps. (The drawing was inspired after Broecker 1985, p. 203.)

with the alkali components (i.e. K^-, Na^-, ammonia, Cl^-) of the crust to form the atmosphere.

Volcanic eruption is one way for the Earth to release volatiles (gases) and various elements such as sulfur, methane, hydrogen, carbon dioxide, hydrogen chloride, and hydrogen sulfide that are present in magma (See Chap. 5). Whether volcanic activity has provided sufficient water to account for all the hydrosphere is a matter of speculation.

For any planet to be able to sustain life in any form, liquid water must be available in abundance. Most of the water on Earth is found in the Oceans and about 2 % in solid Earth (See Chap. 5), while the rest is located in rivers, lakes and ground water (3 %), and in ice caps (1.3 %) (Fig. 2.4). Only a very small content (<0.001 %) of water in the form of vapor is stored in the Asthenosphere. Seawater contains gas in solution such as carbon dioxide, hydrogen sulfide, oxygen, nitrogen, helium, argon, and neon, in order of their abundance. In addition, seawater has dissolved more than fifty elements and compounds of which the major constituents consist of Na^+, Sr, K^+, Mg^{2+}, Ca^{2+}, Br, $SO^{2-}{}_4$, HCO^{-3}, F^- and Cl^-, with a pH of about 7.8.

The origin of seawater and other volatiles such nitrogen, carbon dioxide and chloride ions may be two fold:

(1) These compounds are not abundant in the rocks forming our planet. Carbonaceous chondrites (meteorites) contain minerals that have water in their lattice; hence, by analogy, we believe that the rocks in the interior of our mantle must also have bonded-water in their mineral lattices. In order to understand their origin, we must turn to an extraterrestrial source. The sun and extraterrestrial material such as comets and meteorites could give us some clue on the origin of

these compounds. It is known that the tails and the cores of comets traveling in space are formed of frozen ice and rock debris. During the formation of our solar system from extragalactic clouds, the condensation and differentiation of interstellar matter made up of gas, mostly ice-water, and solid dust took place in the *solar nebula.* During a gravitational collapse and heating of a solar nebula, water could be formed from the reaction of hydrogen and oxygen. Eventually the increasing concentration and agglomeration of the solid dust made up of silicates and iron must have initiated the accretion of the planets.

Thus, from the extraterrestrial hypothesis, it is possible to explain the concentric layering of matter forming the atmosphere, hydrosphere and solid Earth.

(2) Another hypothesis for the origin of the seawater is the slow release of water trapped in the solid Earth over geological time. Water was trapped in the solid-Earth during the early formation of our planet. Even if the solid interior of the Earth's mantle has in average of <2 % water trapped in its different mineral phases, because of its volume (>40 % of the total Earth) the mantle remains the most important reservoir of water. The Earth's mantle contains more than 5 times the amount of water found in the oceans. The fragments of meteorites, mainly the carbonaceous chondrites found on earth, contain water in their minerals such as clay, hydrated silicate and carbon oxides forming organic compounds. These meteorites have a close composition to that of the Earth's mantle. If such material was heated during the early formation of our planet, water would have been released. During episodic geological events due to radioactive heating and or gravitational forces generating heat, water could be released from the lattice of the silicate minerals and from other compounds of iron bearing phases in addition to forming the raw material for our air, our oceans and life (Turikian 1968).

A study of zircons has found that liquid water must have existed as long ago as 4.4 billion years, very soon after the formation of the Earth. This requires the presence of an atmosphere.

Why Seawater is Salty

The main salt compounds found in seawater have originated from the slow alteration and decomposition of rocks. Rivers running into the ocean carry some of the constituents such as sodium, potassium and silica. Other elements such chlorine (hydrochloric acid) and magnesium are likely to be of volcanic origin. The fact that its salinity and seawater's composition has not changed considerably over time is due to the balance of material entering the ocean with that which is precipitating into sediment or is ingested by various organisms. Excess sodium will form zeolite and clay minerals in sediment. Chlorine forms complexes with silica and other nutrients (potassium, nitrogen, carbon) that are essential for life and can be ingested by organisms. Chlorine forms chlorite-bearing minerals such as apatite and sodalite, which are found in sedimentary rocks containing organic material as well as in volcanic rocks.

Seawater also conducts electricity because of its diluted-salt content. It is a good sound propagator, which helps us to use sound-propagation to map the ocean floor and to conduct underwater seismic experiments. Because of the dissolved salts and mineral contents, seawater freezes at about −4 °C. The composition of seawater with its major dissolved compounds is: Cl^- (55.05 %), SO^{2-} (7.68 %), HCO^- (0.41 %), Na^+ (30.61 %), Mg^{2+} (3.69 %) and Ca^{2+} (1.16 %).

How Deep Can Seawater Penetrate into the Earth's Lithosphere?

The penetration of seawater into the solid lithosphere is inferred from seismic data given during earthquakes generated in the ductile portion of the lithosphere. By analogy, the solid lithosphere is like a thick piece of wood that dries and breaks. Under some constraints it can generate fissures through which liquid penetrates. Generally such depths are <20 km deep in most submarine regions except in the subducting trench area where the depth of penetration could be several 100 km. The study of tectonic processes taking place in converging margins, such as underneath island-arcs and in the active margins bounding oceanic regions, can help us to understand continental accretion. It is under the modern Oceans that the internal structure of the Earth can best be studied without taking into consideration the constraints of the large tectonic events taking place during continental growth.

Hydrosphere-Atmosphere Interaction

The interaction between the ocean and the atmosphere has a direct implication on the climate changes observed on the Earth's surface. It is a dynamic process depending on many factors such as convection currents carrying deep water to shallow levels, the flow of rivers to the ocean, and the wind carrying various sized particles. Thus, the gas escaping from the crust and mantle is introduced into seawater and transported to the surface where it will eventually be discharged into the atmosphere. Helium, carbon dioxide, hydrogen and methane gas are found in the water column and then released into the atmosphere above their site of production. When measuring the content of seawater at different depths in the water column, chemists are able to find the exact source of the gas escaping on the ocean floor. These sources are multiple. They are due to the alteration of oceanic rocks, to the circulation of magmatic melts that give rise to volcanic eruptions as well as to hydrothermal fluid discharge into the seawater.

Between the hydrosphere and the atmosphere, it is necessary to maintain an equilibrium in order for life to exist. This equilibrium is due to the so-called carbon cycle, which involves carbon oxide (CO_2), the bicarbonates (HCO^{3-}) and the organic carbon forming animal tissues. Also sulfur dioxide (SO_2) is added to the atmosphere due to biological activity in seawater.

The oceans are the major reservoir for carbon storage (>4 × 10^{13} tons) while the atmosphere has much less (8 × 10^9 tons). The study of ocean water circulation is a difficult task because the time of mixing between different water masses is longer (thousands of years) than in the atmosphere. Thus the study of the water masses and the input of gaseous phases inside the ocean is a vital issue for humanity in order for us to understand and eventually prevent pollution and the green house effect.

Because of the numbers of their atoms and their lattice configuration, gases such as CO_2 and H_2O have a larger capacity for retaining the infrared rays from the sun than do oxygen and nitrogen. The equilibrium between them will be perturbed if the concentration of these gaseous phases is modified. It appears that for about a century this equilibrium has been being modified because of the massive introduction of CO_2 into the atmosphere due to human activities. However, the ocean has the capacity to absorb a large quantity of carbon from the atmosphere in the form of organic material, which can be stored away from the atmosphere in sediment for long geological periods. However, in subaerial regions the various forms of carbon compounds will stay in contact with the atmosphere and might influence climate change.

Formation of Continents

The Earth is in constant motion and has continued to change since its creation 4.7 billion years ago. Some of this evolution is witnessed during a human's lifetime, such as when volcanic eruptions, earthquakes or tsunamis disrupt daily activities or create major catastrophes.

It is difficult to draw the shape of early-formed continents. When the molten Earth solidified, solid crust started to form isolated pieces of landmass called "cratons". The cratons are thick portions of today's continent formed from older pieces of continents. The oldest rock formations or cratons exposed on the surface of the Earth are Archean in age (3.8–4.4 billion years and older) (Myers and Crowley 2000). Archean (2,500 million years ago) rocks have been seen in Greenland, on the Canadian Shield, in Western Australia, and in South Africa. These first continental-size land mass are also called "Ur" (Hazen 2012). At the beginning of Earth's creation, there were no solid landmasses until late in the Archean period, when small "proto-continents" were formed. These proto-continents probably formed as "patchy hotspots" during the cooling and differentiation of the "magma ocean" giving rise to mafic and felsic rocks.

The world's oceans have probably existed from the beginning of our planet's history. Evidence suggesting the presence of ancient oceans has been gathered from the discovery of metamorphosed sedimentary rocks from Canada and Australia (Wilde et al. 2001). Sediments containing fossils are deposited on the sea floor and solidified with time. The study of our modern sea floor will help scientists understand what happened in ancient oceans, traces of which can now be found on Earth's older continental with areas of exposed sedimentary rocks.

The "modern sea floor" is younger than 180 million years. Today's present continents and oceans have not always had the same appearance and the attachment and breaking-up of *Pangaea* and *Gondwanaland* might not have been a unique event. Indeed, it is not excluded that other continents previously existed and subsequently disappeared into the interior of the Earth. This may have occurred several times prior to formation of *Pangaea*. Extrapolating from our knowledge of the present arrangement of our planet's landmasses, we might be able to predict what will happen to Earth's continents in the future. For instance, if we imagine that the process of continental drift continues in the same pattern as now, it is not excluded that the re-closing of the oceans and seas will create another unique continent, or in other words, we may one day have a second *Pangaea*.

In geology, sometimes we refer to the continents as being the "granitic crust". The term granite is used to define the assemblage of silica-enriched rocks that are lighter in weight (or less dense) than basalts or fresh peridotites; granite is what constitutes the main component of the continental lithosphere-crust. The origin of a "granitic crust" is related to the melting and upwelling of mantle material which first gives rise to a basaltic liquid. The basaltic melt resides in large magma chambers where the differentiation between heavy and light elements forming the compounds will take place (See Chap. 5). The lighter compounds, made up essentially of silica, potassium and sodium, will move upward inside the magmatic reservoir and upon crystallization, they will form quartz (silica oxide) and feldspar (silicates of calcium, sodium, potassium and aluminum) which are the primary minerals found in a granite crust.

At the beginning of its formation, the Earth was a molten ball that slowly solidified. The early solidification of our planet began at the base of the mantle near the core, and continued upward. The various compounds found in the mantle were diffused and transferred throughout the solidifying layers as "blobs" of agglomerated solids floating on top of denser partially molten material (Fig. 2.2).

The age of the earliest continental material is hard to evaluate and it is often related to the presence of life on Earth. However, the fossil records on continental terrain have sometimes been obliterated during rock alteration due to weathering or metamorphism (rock transformation). Indications of the age range for a primitive continent have been found among preserved fossil records to vary from 0.1 to about 3.8 billion years in Canada (Schidlowski 2001). Metamorphosed sedimentary rocks from western Australia have given an age of 4.4 billion years based on age dating of a detrital zircon found in metamorphosed sediment from the Narryer Gneiss Terrain (Wilde et al. 2001). Using "age dating", which can be calculated by measuring the rate of decay of radioactive isotopes (i.e. $^{187}Re - ^{187}Os$ decay) and on carbon (CO_2) isotopes, and also based on stratigraphic observations, scientists have concluded that the major granite-forming activities in continent building have been cyclic and episodic.

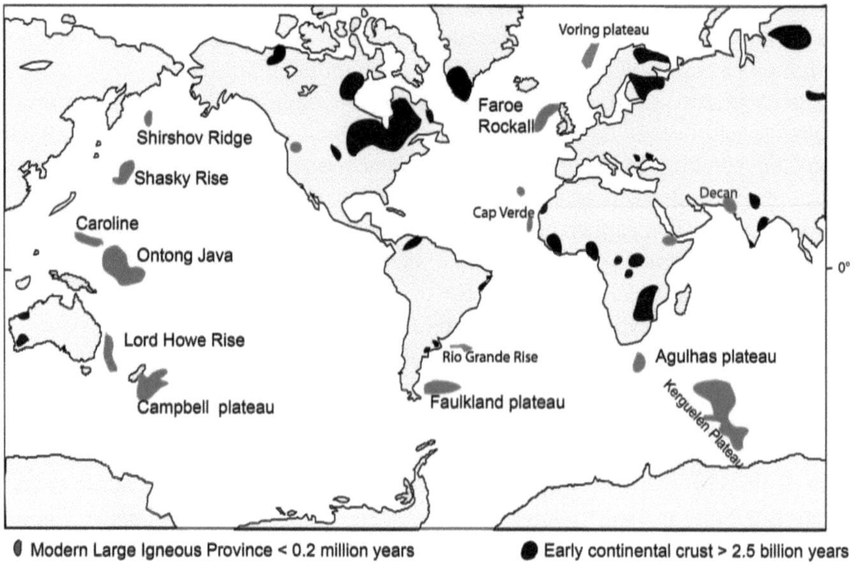

Fig. 2.5 The ancient (>0.2–2.5 billion years old) subaerial exposure of continental crust (*black*) is compared to younger (<0.2 billion years old) Large Igneous Provinces (*grey*)

The continental growth rate has been calculated to be about 3 km^3 a year during a period about 4–3 billion years ago but this rate of growth decreased to 0.8 km^3 a year after the early high growth-rate period (Dhuime et al. 2012). A decrease in the rate of continental accretion is probably due to the increase in subduction of newly formed lithospheric plates. The major granite-forming events have taken place sequentially with intermittent, quieter geological periods. When reading backward from the present, the major granite-forming events for building the crust can be dated in millions of years: 100, 400, 1,100, 1,500, 1,900, 2,600 and 3,400 million years (Gastil 1960; Engel and Engel 1964; Pearson et al. 2008). For example, the Alps in Europe, and the Andes, Rocky, Cascade and Appalachian mountains in the Americas are among the youngest examples of continental crust, since they are only about 100–200 million years old. The oldest granite crust which has been found in south Greenland, North America, Scandinavia, South Africa and in western Australia is older than 2.6–3.8 billion years (Hanika et al. 2012) (Fig. 2.5).

Wilde et al. (2001) used evidence from age dating on a detrital zircon ($ZrSiO_2$—zirconium silicate) found in the Narryer Gneiss Terrain, in Yilgarn Craton, in Western Australia, and have determined that this part of the Earth has an age of 4.4 billion years old. The Narryer Gneiss Terrain contains sedimentary rocks suggesting the presence of early water that interacted with ancient continental crust. The zircon's isotopic age goes back to the time when the zircon's magma separated from the mantle, at which point it crystallized. The oxygen isotope measurements on this zircon suggest that the crystal grew in a liquid environment. Thus, the presence of water implies that the surface temperature was cool enough to permit the evolution of life on Earth. In fact, life has been traced back to the Archean period, 3.5 billion years ago. Probably the early appearance of continental crustal nuclei as "blobs" of solid plates is comparable to what has been observed on ancient terrains exposed on all the larger continents, which are more than 3.5 billion years old (Fig. 2.5).

Large Igneous Provinces (LIP) as a Nucleus for Continent Accretion

It is likely that ancient rocky outcrops existed at the early stage of the Earth's formation when the "magma mantle" started to solidify and convective heat currents began to move matter from one part of the mantle to another (Fig. 2.5). These convective currents were the driving mechanism to bring lighter material to the surface and form crustal embryos of "continental islands". The early conditions for the formation of silica-rich continental rocks took place about 4 billion years ago (4000 Ma) and were probably comparable to the more recent (63–200 Ma) mechanisms giving rise to the Large Igneous Provinces (LIPs) observed on the Earth's surface.

LIPs are due to the upwelling of lava from large, hot mantle plumes. From geological records it is known that these LIP must have been the result of huge amounts of volcanism associated with the emission of deadly gases such as sulfur dioxide (SO_2) and carbon dioxide (CO_2) and coinciding with long periods of acid rain. Even if the toxic gases were eventually lost or consumed during rock weathering over the next few thousand years, nevertheless significant environmental damage must have occurred because of the large quantity of toxic elements emitted over a relatively short period of time during the closely spaced periods of volcanic eruptions.

These LIP are found today in the Kuroo volcanic region of South Africa (190 Ma), in the Deccan Trapp in India (63 Ma), in Uruanu in Japan (26 million years), in the Yellowstone area in northwestern USA (2.2 Ma), on the Kerguelen Plateau in the southern Indian Ocean (110 million years), and in the Parana region (128 Ma) on the South American continent as well as in the Edenteka trap in Namibia (Southwest Africa). All these continental Large Igneous Provinces are covered by flood basalts accompanied by large amounts of silica-rich volcanic rocks forming great volumes ($>10^4$ km^3) of eruptive material, which was later uplifted at the tectonic (lithospheric) plate margins to form continental crust.

Other, even older continental crust can be found in the Finno-Scadinavian provinces of Pechenga (2500 Ma), in South America (Uruguay) in the Rio de la Plata Craton (1790 Ma), and in Siberia (250 Ma). In today's ocean, Cretaceous (70–120 millions years old) and Jurassic (up to 168 Ma old) Large Oceanic Plateaus covered by extensive amounts of flood basalts, are found in today's submarine environments such as near Ontong, Java, and in the Maniki and Hikurangi plateaus, plus the Shatsky and the Chatham Rises in the southwest Pacific, and in the Caribbean igneous province (Ben-Avraham et al. 1981). In the eastern Caribbean, silica-enriched volcanism has been observed on the Aves Ridge off the northern coast of Venezuela.

These Large Oceanic Plateaus seem to be similar to the Large Igneous Provinces (LIP), thus they could be interpreted as being precursors of future continents. Seismic studies done on these Large Oceanic Plateaus indicate that they are made of a thicker crust (25–60 km in thickness), which is comparable to that found underneath continental regions. Other Large Oceanic Plateaus with "granitic" crust but which are much older, such as the Seychelles Islands in the Indian Ocean, which are Precambrian in age (>540 millions years), and the Aghulas Plateau south of Cape Town in South Africa, are considered as being submerged continental fragments.

In summary, silica-rich igneous rocks are an integral part of all continental Large Igneous Provinces (LIP), which erupted a great amount (>6500 km^3) of lava from the oldest Precambrian (>3000 millions years ago) period to the youngest Cenozoic times when the Large Oceanic Plateaus were created. Continental crust has been formed during several cycles of accretion around a more ancient or primitive «granitic nucleus» at different periods of Earth's history up to today. Eventually, the formation of continental crust and Large Igneous Provinces has contributed to depleting the Earth's mantle of its primordial incompatible (mostly lighter) elements.

Volcanic Arcs: Possible Sites of Crustal Accretion

Once mantle convection began and early "silica-rich" crust was created forming isolated nuclei (or patches of landmasses), the mechanism of continental crust formation continued to evolve as the rigid plates began to drift. Evidence of the existence of ancient intra-continental arcs is now lacking within the continents. It is however possible that the earliest continental crust was first constructed during the subduction of ancient lithosphere as early as the Precambrian period (more than 500 million years ago) because during plate tectonics and "continental drift", when the plates spread apart from the ridge axes, the crust along with its sediment, basaltic rocks and a portion of the upper mantle's intrusive rocks will be carried away until it encounters less dense material where it could be subducted (pushed underneath) the lighter material after a collision (see Chap. 10). If we look at the fore-arcs of today's subduction zones, we can see the accumulation of drifted sediment, which could later be uplifted during tectonic events. Modern volcanic formations (<150 million years old) have formed above subduction zones where volcanic plates are recycled back into the Earth's interior.

The total length of the arc margins where oceanic crust is being constructed is about the same as the length of the volcanic arcs themselves, and this length is estimated to be about 43,500 km (von Huenen and Sholl 1991).

Continental Drift

"Continental drift" is the movement of the Earth's continents in relationship to each other. When looking at the Earth's globe one is tempted to fit the South American continent with the West coast of Africa, and when looking closer towards the northern hemisphere, Europe and North America could also fit together with a small rotation with respect to each other.

The first hint of a theory for the drift of the continents was mentioned in 1620 by Francis Bacon, who suggested that the western hemisphere was once attached to Europe and Africa. However, the idea that the continents were once joined together was first put forward by the Flemish cartographer, Abraham Ortelius, in 1596. These ideas were more fully developed by the German astronomer, Alfred Wegener, in 1912. He proposed that all the continents were attached together 200 million years ago and he called this super-continent "Pangaea" (Fig. 2.6). Wegener also used paleontological records, which reported that fossils of some animal species found on both the American and African continents had similarities. It was also in the nineteenth century that Eduard Suess, a geologist from Austria, noted a close correspondence between certain geological formations on the southern hemisphere continents of Australia, Antarctica, southeast Asia and India. Based on his observations, Suess suggested that these areas were once part of an original, large landmass he named *Gondwana*.

Scientists did not accept Wegener's concept of continental drift until after his death, which occurred during an expedition to Greenland, in 1930. In fact, many scientists were not convinced about the drift of continents even up to the 1960's. However, the final irrefutable verification that the continents have been drifting apart came in 1963 when two British scientists, Vine and Matthews, reported the result of their investigations on the magnetic imprint left by the Earth's magnetic field at spreading ridge axes (See Chap. 7). It was at this point that the theory of Continental Drift was finally accepted.

Plate Tectonics

The "theory of plate tectonics", which involves the movement of the rigid segments (plates) of the Earth's lithosphere, was inspired by and evolved from Wegener's earlier hypothesis concerning continental drift. This theory has considerably changed our perception about how we view our living planet. The Plate Tectonic Theory was first presented in 1967 at the American Geophysical Union (AGU) meeting in Washington DC by Jason Morgan. According to his explanation, the surface of our planet is divided into several rigid plates that are moving with respect to each other. These rigid plates, which include the Crust and the Lithosphere, move above a less rigid substratum called the asthenosphere (Fig. 2.3). Jason Morgan and Xavier Le Pichon wrote and published the first

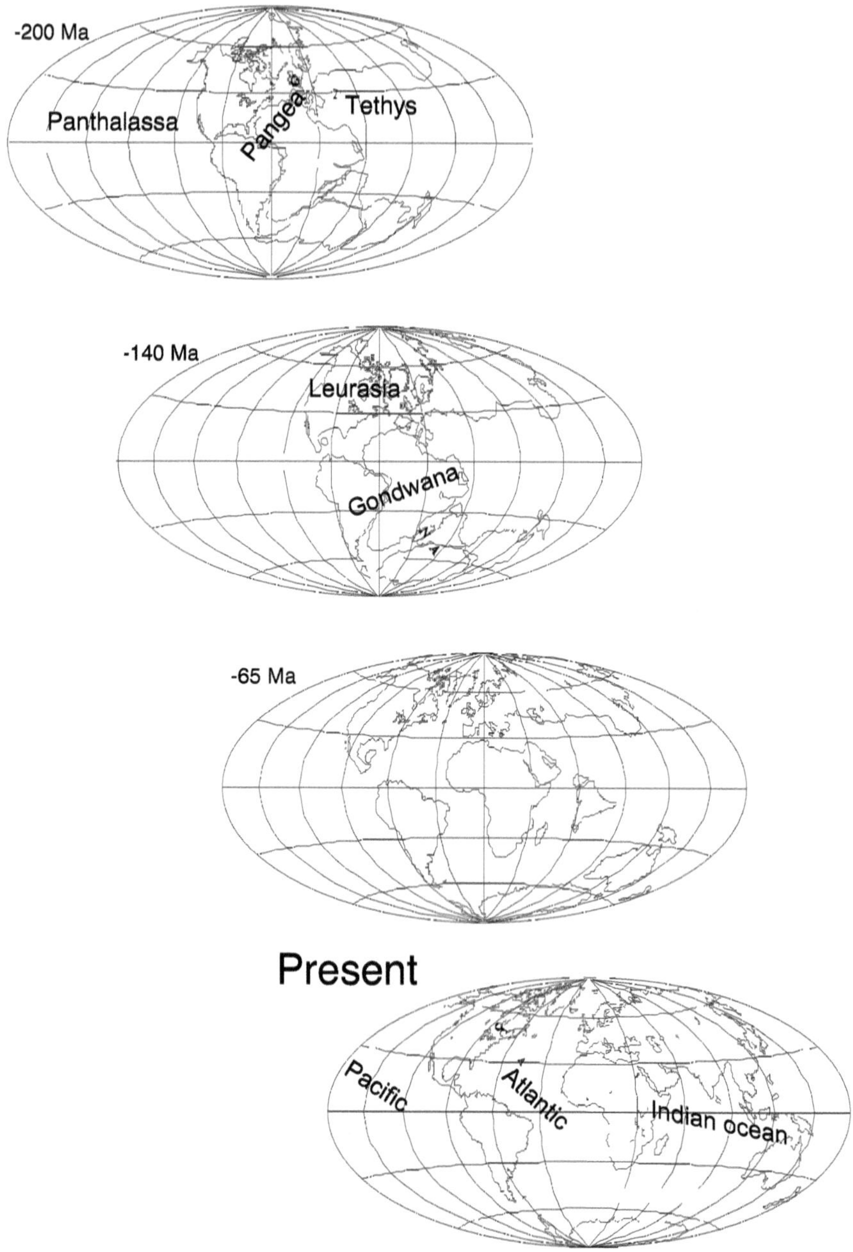

Fig. 2.6 The continents are formed from light material of granitic composition drifting on top of the asthenosphere. Early records show that the continents once formed a unique block that started to fragment under the effect of convection currents in the mantle. The continents become a giant puzzle that assembled themselves periodically, such as 200 million years ago. A unique ocean called *Pantalassa* surrounded the original continent called *Pangaea*

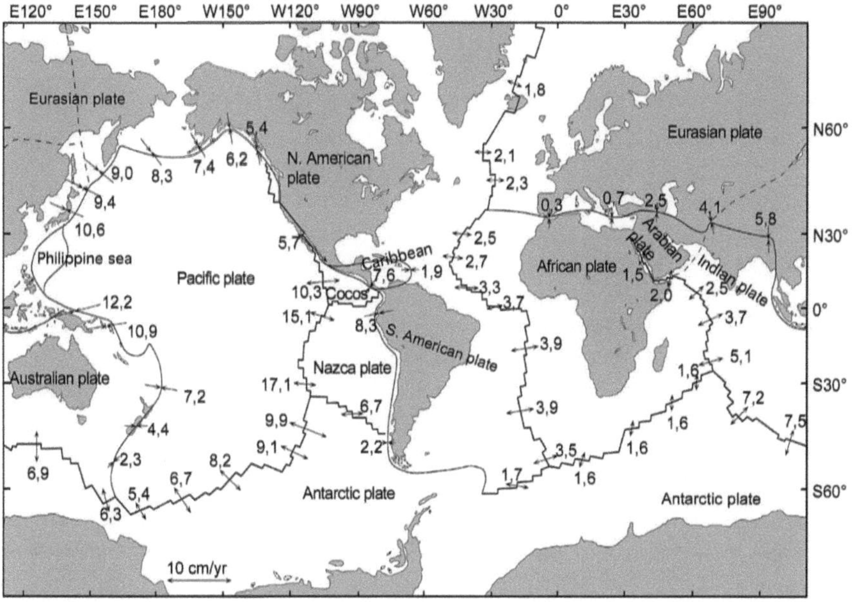

Fig. 2.7 Spreading ridge segments, major faults and subducting zones separate twelve major "rigid" plates. The *arrows* show the direction of tectonic plate motion and the spreading rates (after Hekinian and Binard 2008)

papers on plate tectonics in 1968. Spreading ridge systems, subduction zones and large fracture zones make up the boundaries of these plates (Fig. 2.7).

However the theory of plate tectonics does not explain, nor does it take into account, the interior of the oceanic basins (or "intraplate" regions) where volcanic activities have also formed large islands and underwater volcanoes called "seamounts". According to the "hotspot hypothesis", the motion of the lithospheric plates are the reason that many volcanic seamounts and islands are formed in a linear direction because the plates are sliding above and across huge sources of magma called "mantle plumes" (Morgan 1972) (See Chap. 9). Hotspot volcanic activity has triggered large eruptions throughout geological time and now scientists are certain that a good portion of the Earth's volcanic activity has been taking place in the ocean basins, also referred to as intraplate regions, which are found at a distance from the diverging plate boundaries (also referred to as "accreting plate boundaries" since volcanic material is accreting there, and as "spreading ridge axes" since the plates are "spreading apart from each" other in these areas).

Thus, volcanic activity is responsible for the processes involved in the movement of the lithospheric plates and this mainly takes place under the oceans, where volcanism is much greater than on the emerged continents. Every year more than 3 km^3 and as much as 18 km^3 (Hekinian and Binard 2008) of oceanic crust is created along the 70,000 km of spreading ridge axes around the world. So far we have counted <1,500 active volcanoes above sea level, and most of them are located in island arc provinces or on the continental margins of the circum-Pacific region. This number would have to be multiplied by 10,000 in order to have a close estimation for the amount of recent volcanism on the sea floor.

Oceanic crust is brittle and easily broken, especially in the areas of extensive fractures cutting 8–15 km deep within the lithosphere, as indicated by bathymetric studies and from monitoring seismic epicenters around the World. Earthquakes are a common phenomenon related to the readjustment of material during lithospheric stress. The material discharged in the water column during undersea volcanic eruptions such as helium, carbon dioxide and carbon monoxide, hydrogen, sulfur, methane etc. are carried by currents and/or absorbed by marine organisms. This volcanic discharge is of primary importance for determining the volcanic budget of our planet. Certain primary compounds such as carbon, hydrogen, and oxygen are the building blocks for a viable planet, which can support various life forms. Also, these discharge products could be of assistance in many aspects of human life. For example, if we consider the discovery that natural methane gas is seeping out onto the sea floor through bottom sediment, this could be of great interest since it is a large, new source of untapped energy. Because we have only limited knowledge concerning our ocean, it is expected that scientists will make many more new discoveries as sea floor exploration continues.

Mantle Convection Currents

Since its initial formation and subsequent cooling, our planet's interior has been transformed by means of convective currents of heat within the mantle. The heat flow (convective currents) is affected by heat lost across cold boundary layers inside the mantle and as well as through the lithosphere-crust sequences (outer shell of the Earth) (Figs. 2.3, 2.8). The convection currents form "rivers" of flowing matter and energy inside the mantle and will carry hotter and more viscous material towards the surface. The force driving this phenomenon of mantle convection is primarily due to the downward pull of gravity on the cold and dense lithosphere and the upward motion of hotter matter rising to the Earth's surface.

The increase of temperature with depth has been calculated to be about 20–25 °C/km below the plate boundary regions. The difference in temperature combined with the effect of density differences in the mantle is responsible for generating convection currents. The core-mantle boundary is about 3,500–3,700 °C, while the upper mantle-lithosphere is <1,000 °C. The convective flows inside our planet are comparable to what happens in the atmosphere where air currents are formed when an upward movement of hot air and a downward movement of cold air take place. Because of the constant changes inside the interior of the Earth, nothing has remained the same since the early formation of our planet. Convection currents are the driving force for generating various processes such as plate motion and mountain building.

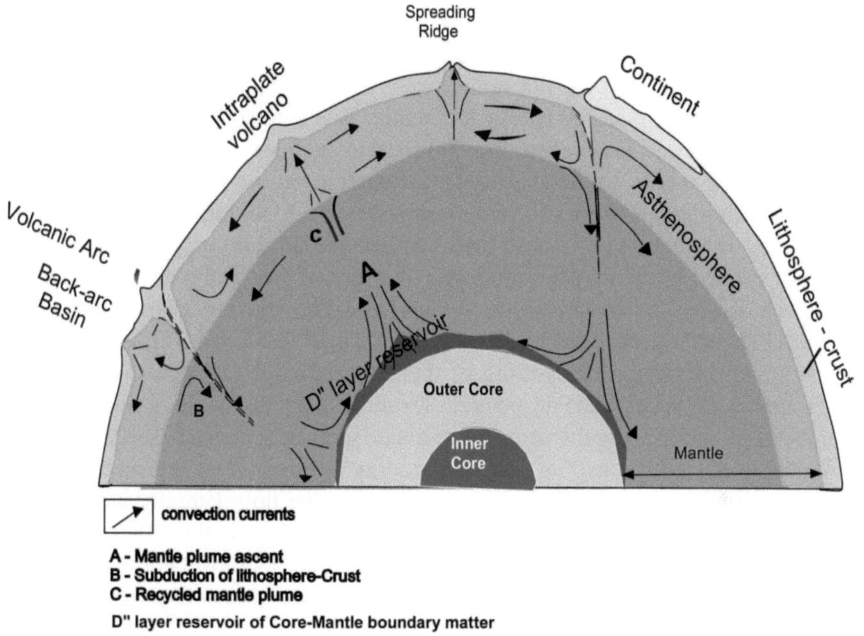

Fig. 2.8 Convection currents in the interior of the Earth show (*A*) the ascent of matter and energy forming hot mantle plumes, (*C*) recycled mantle plumes and magma creating spreading ridge segments as well as moving the lithospheric plates that will eventually plunge into the subduction zone (*B*) underneath continents and island arcs, which is where the lithosphere created at the spreading ridge axis descends into the mantle. A transition zone called the *D*" Layer marks the boundary between the Earth's mantle and the core, and is apparent due to the change in mineral crystalline structure

The temperature driven convection indicates that the interior of the earth is still extremely hot. This heat has been preserved since the early formation of the Earth and it is continuously being alimented by the energy release due to the radioactive decay of long-lived radioactive isotopes such as ^{236}U, ^{235}U, ^{238}U, ^{232}Th and ^{40}K. The average air temperature on the Earth's surface is about 13 °C. In the Earth's interior, the temperature differences between the core (inner and outer core) at 5,000–7,000 °C, the lower mantle at about 2,000 °C and the upper mantle at about 500–600 °C are responsible for generating the mechanism of heat convection and the subsequent transfer of matter throughout the Earth's interior (Wilson 2005; White 2013). The temperature increases with depth and there is also a relationship with the interior's increasing pressure. When critical depth has been reached, the rocks will approach their melting point (about 1000 °C) near the upper mantle (in the lithosphere), so they will become ductile and have lower seismic wave velocities. This coincides with the beginning of what is known as the *asthenosphere,* at 670–700 km depth, where convection currents are the dominant mode of heat repartition and heat flow. The thermal energy radiating from the core will be dissipated during its ascent in the mantle so it is not the same everywhere. In other words, thermal flux (heat flow) is irregularly distributed within the mantle before being dissipated through the lithosphere and the continental crust.

The ascent of deep-seated energy and matter is accompanied by a phenomenon of decompression. This decrease in pressure will increase the melting point temperature of the matter (mineral phases). Depending on the mineral structure and the surrounding temperature gradient, even just a small variation of only 10–30 °C is sufficient for enabling a rock to melt and become a liquid magma. The density decrease caused by heating will enhance the melted material's ability to rise to shallower depths and eventually to the surface. The melting takes place at the junction between mineral grains. The ability of a melt to ascend will depend on the porosity and the degree of fissuring of the surrounding area. Near the surface, a rising melt will accumulate in pockets forming large reservoirs (called magma chambers) at shallow depths of <500 m under spreading centers (see Chap. 5).

Probably not all the mantle's thermal energy is liberated at the surface of the Earth. Thermal energy and matter transfer could also flow laterally along major physical discontinuities marked by the boundary between the lithosphere and upper mantle. This generates weaknesses underneath the lithosphere, which could subsequently break and give rise to the observed major oceanic provinces such as fracture zones, spreading ridges and hotspots in oceanic basins.

Sea Floor Renewal

The major oceanic provinces are created along diverging plate boundaries at the spreading ridge axes, but also, to a lesser extent, in marginal basins formed behind island arcs, also called back-arc basins (see Chap. 10) (Figs. 2.8 and 2.9). The lateral dissipation of energy and matter driven by the mantle convective currents can move the lithosphere (tectonic plates) away from the heated region at spreading ridge axes towards a margin of colder and lighter density continents and/or older lithosphere at convergent margins where the heavier plates will sink back into the mantle to begin another new cycle of magma creation and subsequent volcanism.

The lithosphere (crust) moves away from its site of formation at a speed of a few centimeters per year (at a rate of <1 to more than 9 cm/year). The plates will cool progressively and will become thicker by incorporating a portion of the underlying asthenosphere. As the lithosphere moves and ages, it will become denser and eventually sink into the mantle at the subduction zones.

There is a geodynamic cycle of sea floor renewal. This cycle is relatively fast, over a period of about 346 million years. That is why the oceanic rocks are relatively young. The oldest oceanic sea floor located in the western Pacific is about 173 million years old, which is the geological Jurassic period. The renewal of the oceanic lithosphere occurs when a portion of the subsided lithosphere sinks into the Earth's mantle, regenerates itself and later reappears forming new material

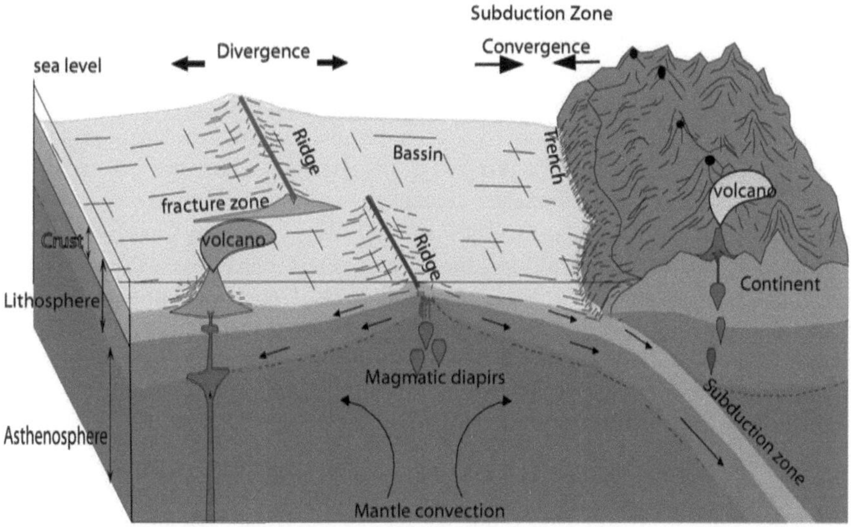

Fig. 2.9 The cycle of sea floor accretion and subduction shows that the sea floor created at the spreading ridge axis progressively ages as it is moved away from its point of origin (Hekinian and Binard 2008). At the same time the lithosphere cools, thickens and sinks into the mantle in the subduction zone. Convection cells (Fig. 2.8) drive the motions of matter and heat inside the mantle

on the surface of the Earth. This takes place during partial melting, which occurs during the extraction of upper mantle material found in the region of the asthenosphere. The magma that reaches the Earth's surface produces about 2–3 km^3 of volcanic rocks (mainly basalts) and about 15–20 km^3 of intrusive rocks (gabbros and various cumulates) per year along the planet's 70,000 km of spreading ridges. There is in equilibrium between the creation of new lithosphere at ridge axes and the subduction of the old lithosphere in oceanic trenches (subduction zones). For this reason, the Atlantic Ocean, which is bounded by accreting passive margins with no trenches or subduction zones, except for the region of the Caribbean with the Lesser Antilles arc, is considered as being an ocean "in expansion", while the Pacific Ocean, which is bordered by active margins (tectonically active subduction zones), is "decreasing its size" by about 0.5 km^3 per year. In other words, the Pacific is an ocean closing in on itself.

The underwater structures covered by the oceans comprise mainly the deep ocean basins (>4500 m depth) which represent about 30 % of the sea floor, the spreading Mid-Ocean Ridge segments (<4000 m depth) which cover about 60 % of the sea floor, and a smaller proportion (10 % of the total undersea provinces) which is the location of the back-arc basins associated with subduction zones and island arcs.

The exploration and description of these different oceanic provinces as presented in this book will help the reader to better understand the formation of our planet. The upper portion of the lithosphere with a solid crust that is <10 km thick (on the average) in oceanic regions is the most accessible to direct observations and thus is best known by marine geologists. During the last 120 years of investigations on land and in the deep sea, scientists have recognized several oceanic features based on their different types of environment and setting, and also based on their rock compositions. These major underwater structures include (1) spreading ridge systems, (2) fracture zones, (3) islands, underwater volcanoes and seamount chains, (4) deep trenches associated with subduction zones, (5) large volcanic plateaus which form the outer shell of our planet (see Chap. 4).

Thanks to deep-sea drilling operations, plus direct field observations of deep fracture zones and/or based on the upwelling of deep-seated molten material giving rise to volcanism on the surface, geologists have managed to probe the composition and structure of the lithosphere. Also, extrapolations have been made based on indirect studies involving seismic profiles and gravity anomaly measurements. However, we need to be careful when trying to define the stratigraphy of the crust. It is often found that the sequential variation of rock types is not uniformly distributed and sometimes the rocks recovered may vary considerably within the same structure. This is due to tectonic activities such as fracturing and fissuring as well as to the rheology (deformation) of the material forming the lithosphere and exposing different types of material. Diapiric upwelling of deep-seated material locally intruding the crust (see Chap. 4) gives rise to a large amount of tectonic instability and rock composition differences. In order to generalize the sequential compositional variation of a section of the lithosphere, it is necessary to have a good knowledge of the geological setting and the compositional variability of the rocks found within a given area.

Earth's Magnetism

Sailors are familiar with the effect of Earth's magnetic field and the preferential direction shown by the needle of a compass. The property of the magnetized needle in relationship to the Earth's magnetic field and the invention of the compass was probably first used by Chinese navigators and in fact, the first person recorded to have used the compass as a navigational tool was Zheng He (1371–1435), from the Yunnan province in China, who made seven ocean voyages between 1405 and 1433. (Mary Bellis, About.com Guide). William Gilbert wrote the earliest scientific treatises on magnetism in Latin, in 1600. However, the earliest evidence of Earth's field magnetization is likely to date back to the eleventh century (Jacobs 1967).

In modern science, the geomagnetic field has been used to explore the dynamics of the Earth's interior and for mapping mineral deposits. Furthermore, scientists have learned that the magnetic poles have jumped from north to south and vice versa several times during the Earth's history. Thus, the study of paleo-magnetism has given indications about the location of the poles during the emplacement of iron rich lava.

What is the Source of the Earth's Magnetic Field?

The natural magnetic field is generated by the Earth's electric-dynamo system. It is believed that 99 % of the Earth's magnetism and its energy is the result of the fluid motions in the Earth's core. The molten nickel and iron spread spinning in the outer core are capable of generating electrical currents where the mechanical energy is transformed into electromagnetic energy. This is the same as in a car battery where the generator transforms mechanical energy into electricity, and this is called the dynamo effect. A switch of polarity in a dynamo will flip the direction of the current. This also happens in the interior of our planet and it is called the reversal of the magnetic field when the poles change polarity from north to south. We do not know why the field changes its polarity. The last switch of polarity occurred about 780,000 years ago and it is not known when the next one will take place. The flip of the magnetic field could be due to the change in direction of the convection currents moving the molten Fe–Ni-enriched matter in the Earth's core (Olsen and Mandea 2008).

The first striking evidence of sea floor spreading was observed when analyzing the magnetic data obtained by magnetic sensors (called magnetometers) towed by the a research vessel in order to measure the fields of magnetic intensity on the sea floor. Vine and Matthews published this finding in 1963. Their paper talked about the *existence of long stripes* of normal and reversed magnetization running parallel to each other on each side of the spreading ridge, like a mirror image. Molten material containing magnetite and titanomagnetite had magnetized and upon solidification these elements pointed towards the direction of the Earth's magnetic field. This observation on the magnetism that was seen on the mirror image stripes is a way of showing that the solidified material is moving away from the axis of the ridge as more magma is injected to take its place. The magnetized formations reported by Vine and Matthews consist essentially of dykes and basaltic flows. It is believed that the Earth's magnetic field reverses itself (the North Pole becomes the South Pole and vice versa) about once every one million years. Therefore, the width of the stripes of magnetized anomalies will depend on the extent of the magmatic material injected into the crust and will be comparable to the rate of lithospheric opening (Harrison 1968).

References

Bell DR, Rossman GR (1992) Water in earth's mantle: the role of nominally anhydrous minerals. Science 255(5050):1391–1397

Ben-Avraham Z, Nur A, Jones D, Cox A (1981) Continental accretion: from oceanic plateau to alloctonous terrains. Science 213:47–54

Boyet M, Carlson RW (2005) 142Nd evidence for early (>4.53 billion year) global differentiation of the silicate earth. Science 309:576–581

Broecker WS (1985) How to build a habitable planet. ELGEO Press, New York, 291 p

Brownlee D (2008) Comets and the early solar system. Phys Today 61(6):30–35
Convay EJ (1943) The chemical evolution of the ocean. Proc Roy Irish Acad B48:161–212
Dalrymple GB (2001) The age of the Earth in the twentieth century: a problem (mostly) solved. J Geol Soc London Spec Publ 190(1):205–221
Dhuime B, Hawjesworks CJ, Cawood PAA, Storey CD (2012) A change in geodynamics of continental growth 3 billion years ago. Science 335:1334–1336
Gastil G (1960) The distribution of mineral dates in time and space. Am J Sci 258:1–35
Engel AEJ, Engel CG (1964) Continental accretion and the evolution of North America. In Advancing Frontiers in Geology and geophysics, Volume in honour of M.S. Kishnam, Hyderabad, Indian Geophysical Union 17–37
Green DH, Ringwood AE (1969) The origin of basalt magmas. In: Hart PJ (ed) Geophysical monograph No. 13, The Earth's Crust and Upper mantle, American Geophys Union, pp 489–495
Green DH, Ringwood AE (1970) Mineralogy of peridotite under upper mantle conditions. Phys Earth Planet Inter 3:359–371
Hazen RM (2012) The story of earth. The first 4.5 billion years, from stardust to living planet. Viking, Published by the Penguin Group Inc., New York, 306 p
Hanika R, Boyet M, Blichert-Toft J, O'Neil J, Rosing M, Paquette J-L (2012) The elusive Hadean enriched reservoir revealed by 142Nd deficits in Isua Archean rocks. Nature 491:96–100
Harrison CGA (1968) Formation of magnetic anomalies. J Geophys Res 73:2137–2142
Hekinian R, Binard N (2008) Le feu des abysses. Editions QUAE, France, 175 p
Jacobs JA (1967) The Earth's core and geomagnetism. Pergamon Press, Headinggton Hill Hall 137 p
Labrosse S, Poirier J-P, le Mouël J-L (2001) The age of the inner core. Earth Planet Sci Lett 190:111–123
Le Pichon X (1968) Sea floor continental drift. J Geophy Res 73(12):3661–3696
Melbom A (2008) The rise and fall of great ideas. Science 319:418–419
Morgan WJ (1968) Rises, trenches, great faults and crustal blocks. J Geophys Res 73:1959–1982
Morgan WJ (1972) Plate motion and deep mantle convection. Am Assoc Petrol Geol Bull 56(2):203–213
Mukhopadhyay S (2012) Early differentiation and volatile accretion recorded in deep-mantle neon and xenon. Nature 486:101–104
Myers JS, Crowley JL (2000) Vestiges of life in the oldest Greenland rocks? A review of early Archean geology in the Godthabsfjord region, and reappraisal of field evidence for >3850 Ma life on Akilia. Precambian Res 103(2–4):101–124
Olsen N, Mandea M (2008) Rapidly changed flow in the Earth's core. Nat Geosci 1:390–394
Pappalardo RT, Head JW, Greeley R, Sullivan RJ, Pilcher C, Schubert G, Moore WB, Carr MH, Moore JM, Belton JS, Goldsby DL (1998) Geological evidence for solid-state convection in Europa's ice shell. Nature 391:365–368
Pearson DG, Parman SW, Nowell GM (2008) A link between large mantle melting events and continent growth seen in osmium isotopes. Nature 449:202–205
Schidlowski M (2001) Carbon isotopes as biogeochemical recorders of life over 3.8 Ga. The age and accretion of the earth. Earth Sci Rev 59:235–263
Turikian KK (1968) Oceans Foundation of earth science series. Prentice-Hall Englewood-Cliffs, New Jersey 120 p
Vine FH, Matthews DH (1963) Magnetic anomalies over the spreading ridges. Nature 199:947–949
von Huenen R, Scholl DW (1991) Observation at convergent margins concerning sediment subduction, subduction erosion, and growth of the continental crust. Rev Geophy 29:279–316
Wilde SA, Walker JW, Peck WH, Graham CM (2001) Evidence from detrital zircon for the existence of continental crust and oceans on the Earth 4.4 Gyr ago. Nature 409:175–178

Wilson M (2005) Radioisotope tracers reveal extensive melting in Earth's distant past. Physics Today, September issue 19–21
Wilson JT (1972) Continents adrift. Scientific American W.H. Freeman and company, California, 172 p
White WM (2013) Geochemistry. Wiley, London 672 p

Chapter 3
Diving into the Abysses

Abstract Most of the Earth's surface is under water. The hostile environment of the underwater world makes it difficult for direct exploration by human beings. Underwater visibility is limited, it is impossible for humans to breathe in water, its salty nature makes seawater denser than air (about 1300 times heavier than air), and underwater currents tend to displace any man-made engines at depth. Despite these difficulties, humans have always been attracted to the underwater world. Since antiquity men and women have tried to penetrate the sea and elucidate its mysteries. In 1934, exploration with manned submersibles first became a reality. Even if technology has advanced our capabilities, the spirit of underwater exploration has remained the same. Each dive into the unknown is a unique and exciting experience, bringing both joy and fear.

Why Do We Explore the Sea Floor?

It is important to investigate the sea floor and its environment because this will increase our knowledge of the hidden parts of our planet's interior. Below the sea floor's surface, the rigid lithosphere is thinner and therefore closer to the Earth's more ductile Mantle where most of the transfer of matter and energy takes place (See Chap. 2). By exploring the sea floor, we have a better chance to comprehend the internal structure and evolution of our Earth.

Seawater covers 71 % of our planet. Five oceans cover the Earth. The largest is the Pacific Ocean, which equals 46 % of all the World's oceans. The Atlantic equals 26 %, the Indian Ocean is equivalent to 21 %, the Antarctic is 6 % and the Arctic represents 4 % of our planet's total ocean coverage. As of today (2013) we have explored less than 30 % of the sea floor.

Our reasons for exploring the ocean's depths could be purely scientific, since we want to know more about the Earth and its history, but there is also an economic incentive for increasing our knowledge about the sea floor because our mineral resources on emerged landmasses are limited. However, the most probable

R. Hekinian, *Sea Floor Exploration*, Springer Oceanography,
DOI: 10.1007/978-3-319-03203-0_3,

reason for pursuing undersea exploration is our basic curiosity, which drives many of us to want to penetrate what could be considered as being the last frontier of planet Earth.

Early oceanographic expeditions, before and at the turn of the twentieth century, were more concerned about the water mass rather than the sea floor itself. Biological and physical oceanographic investigations were the main concern of the scientific community. Modern oceanographic expeditions started at the end of the Second World War and have continued because the world's industrialized nations have considerably invested in ocean floor exploration.

What triggers Earthquakes? Why do volcanoes erupt? It is by understanding the processes taking place on the sea floor that we will have more knowledge of our planet and this will also help us better understand the extraterrestrial bodies of our Universe. Since volcanic activity is far greater on the ocean floor (see Chap. 2), this makes it an ideal place for studying and understanding the mechanisms of creation and subsequent destruction of the lithosphere-crust. This is primordial information if we want to learn about the history of our planet.

Bathymetric Mapping and Sea Floor Imagery

Modern Oceanography began with the British deep-sea expedition on board the H.M.S *Challenger*, when, under the leadership of Sir Wyville Thomson, men first scientifically explored the Pacific, Atlantic and Indian Oceans. The earliest attempts to "sound the depths" (i.e. measure the depth of the oceans) occurred during the H.M.S *Challenger* expedition between 1872 and 1876 (Linklater 1972). The crew used a hemp rope about 2 cm thick, which they dropped over the side as a depth-gauge. The first bathymetric charts were made through the initiative of Prince Albert the 1st of Monaco in 1905 and Georg Wüst (1964) published the first bathymetric chart of the Atlantic Ocean in 1935.

Any manned-submersible exploration must begin with precise mapping of the sea floor. For today's mapping of the sea floor, modern technology uses underwater acoustic sound propagation. Sound is generated by explosive devices such as air or water guns as well as by real explosives (TNT) giving rise to sound impulsion received by electric amplifiers (hydrophones). The "sonar" (acronym for **So**und **Na**vigation and **R**anging) is an instrument that emits ultrasound waves (pulses of inaudible sounds) and then listens to the echoes of these waves as they hit any targets on the ocean floor or in the water column. This is a technique that was developed after determining the effects of sound propagation within water for determining the depth of the sea floor.

Sound waves differ from radio waves and light waves because they have a lower frequency and they can propagate in the water column along great distances, at a speed of about 1500 m/s. Sound waves are then reflected by any obstacles encountered and are received on the surface ship by means of hydrophones, which are receivers used to hear the sound wave's signal. Knowing the amount of time necessary for sound to propagate in both directions (i.e. for sound to travel to the sea floor and back again to the surface), it is possible to measure the ocean floor's depth.

A multi-beam sonar system was first developed by the U.S. Navy for military purposes. However, in the 1970s, this device was commercialized by an American company and called *Sea-Beam.* The signals received by the Sea-beam's hydrophones are registered in real time and can also retrace the ocean bottom's topographic contour lines on a drawing table. It was around the middle of the 1970's (1976) that multi-beam swath mapping started to be regularly used by civilian oceanographic vessels. In May 1977, the first multi-beam echo-sounder (manufactured by General Instrument Corporation in the USA) was installed by CNEXO on the N.O. *Jean Charcot.* The equipment was tested in the spring of 1977. The capability of this multi-beam echo-sounding equipment was to cover a corridor as wide as ¾ of the sea's depth during one single passage of the ship. A bathymetric contour map was instantly obtained during the ship's transit (Renard and Allenou 1979).

The second generation equipment for swath mapping started to be used in the 90's and considerably reduced the time required for mapping while increasing the amount of coverage of the sea floor. For example, the multi-beam bathymetric system called Simrad EM12, which is installed on the N.O. *L'Atalante,* is a low frequency (13 kHz) device located on the ship's hull with a capacity for measuring the sea floor's depth from 100 to 11,000 m. The multi-beam is constructed with sets of devices called transducer units (sound projectors and hydrophones) so there can be as many as 162 beams emitting sound waves and covering a swath 7.4 times the water's depth. The sound waves are directed towards the front of the ship with a 2° opening of the sound beam which is horizontally oriented to the heading of the ship and there are lateral openings which can vary between 3° and 15°. Narrowing the sound wave's propagation beam will increase the detail of the sea floor structures that are recorded. The main advantage of this system is to produce an instantaneous bathymetric map with precise topographic contour lines at the same time as the ship navigates at a normal speed of 12 kn.

The first detailed map of the ocean floor produced for non-military, scientific purposes was obtained during the FAMOUS project using a satellite navigation system (Global Positioning System, GPS) in the North Atlantic. However, it was in 1980 that the N.O. *Jean Charcot* used the SeaBeam swath bathymetry system in order to map, for the first time, a portion of the East Pacific Rise during the *Searise project* (IFREMER former CNEXO). By 1990, with the arrival of the new generation of swath mapping systems, we were able to reduce the time of acquisition by a factor of ten. In addition to conventional contour line mapping, the new system was also able to produce a sonar image resembling a negative black and white photo of the ocean floor's structures and relief.

Global Satellite Positioning (GPS) systems introduced in 1968 have been used to enable more precise navigation on the sea surface. When the system is coupled with multi-beam data acquisition, this will provide a very realistic bathymetric map in real time on board a ship. In 1997, using formerly classified satellite data sets and a specific modeling algorithm, David Sandwell and Walter Smith produced a new sea-floor physiographic map of the World's oceans. The de-classification of Geosat images by the US Navy and the European Space Agency's (ERS-1) altimetry data allowed scientists to further generate computer models that inferred the features of the sea floor. This satellite image declassification started in the mid 80's and full declassification was achieved by 1995. The satellite data covers the entire World's oceans but there are some limitations to its resolution. Satellite altimetry data is based on measurements of the Earth's geoids that approximately follow the average surface of the sea. The fact that the satellite tracks are spaced between 1 and 5 km and because the sampling rates along each satellite track are on the order of 4-6 km, it is true that we do not have a totally accurate resolution for recognizing small scale topographic features.

The wavelengths for satellite observation are on the order of 15–30 km hence the topography of structures less than 30 km in length will be difficult to interpret. For this reason, a diving program cannot be based simply on satellite altimetry data because of the inaccuracy in defining the details the topography of the ocean floor. In fact, satellite altimetry observation of the ocean floor is unable to replace the bathymetric data obtained from surface ships. However, satellite images will certainly help to decipher large structures hiding under water and they can provide very useful information for preparing ocean-going expeditions.

The discovery of inactive plate boundaries such as paleo-plates (ancient microplates), fracture zones, pseudo-faults, linear ridges, and seamount chains has become more frequent since the 1980s. This is due to the results from new developments in satellite altimetry and multi-beam bathymetric surveys conducted over the various regions of the World's Oceans. In addition, the interpretation of satellite altimetry data enables us to observe the crustal structures created at ridge axes, which are often preserved during spreading so they could be detected at a distance from their original source. This information is very helpful for retracing the history of oceanic basins and their evolution.

If we wish to reconstruct the history of oceanic basins, it is important to be able to recognize and differentiate ancient structures formed on divergent plate boundary regions (such as oceanic spreading centers) from younger intraplate constructions (i.e. of hotspot origin). Although some of the linear structures representing fracture zones are well delineated when consulting general satellite altimetry data, abyssal hill provinces with their numerous volcanic cones forming discontinuous structures of unknown origin are more difficult to identify. For this reason, it is best to combine satellite observations with swath bathymetry data and even include on-site (manned or unmanned submersible) observations of the sea floor.

Navigation Systems and Other Vital Equipment

Prior to carrying out any detailed sea floor investigations with deep towed instruments, remote controlled Vehicles (ROV) and/or manned submersibles, it is wise to have a general overview of the landscape in order to choose the most adequate targets.

In order to navigate towed instruments and manned submersibles on the sea floor, it is necessary to have a reference frame with respect to the on-site **topography**. Acoustic transponders, which float on an anchor at only about 200 m high above the sea floor and have battery-powered sonar and pinger-sets sending an acoustic signal at different frequencies, are deployed in a triangular network several miles wide (Fig. 3.1). The transponders will emit and receive sonic pulses at different frequencies and when coordinated with signals from the submersible and its surface ship, they will enable precise navigation within a given geographic network. Any instrument or vessel deployed within such a network can be located on a geographic grid by the surface ship, which is, in turn, related to the satellite grid.

Undersea vehicles can carry different kinds of instruments depending on the goal of exploration. However, navigation systems, sonar, a compass, a load pressure-gauge and some electronic devices for detecting water infiltration, as well as devices for measuring oxygen and ambient humidity are critical instruments which are always carried on board. Devices that enable communication with the surface ship are also vital instruments that are carried on board manned and unmanned submersibles.

Early Underwater Exploration

Since antiquity, the sea has always intrigued humankind. When *Alexander the Great* was returning from India in the year 325 BC, he camped on a beach on the coast of Persian Gulf. While there, he decided he wanted to dive into the blue water to look at the fish. He requested his men to build a wooden barrel so he could descend about 10 m, from where he was able to see the underwater world through a window made from the skin of a donkey. The Greek philosopher, *Aristotle,* who was a contemporary and teacher of Alexander, described the primitive diving bell using the Greek term "*corinpha*" meaning cauldron (Riffaud 1988).

The USA was the first nation to build submersibles to be used for military purposes during the American Civil War in the 1860's (Jarry 2003). In 1776, David Bushnell built a one-man submersible called the *Turtle*, which look like an oval egg, or like the two carapaces of giant turtles that had been put together. This construction was made of wood and capable of diving as deep as 15 m. The *Turtle* was powered by the diver's own hand which turned a tendril, and the *Turtle* was also able to carry a charge of explosives. During the War of Independence (1776),

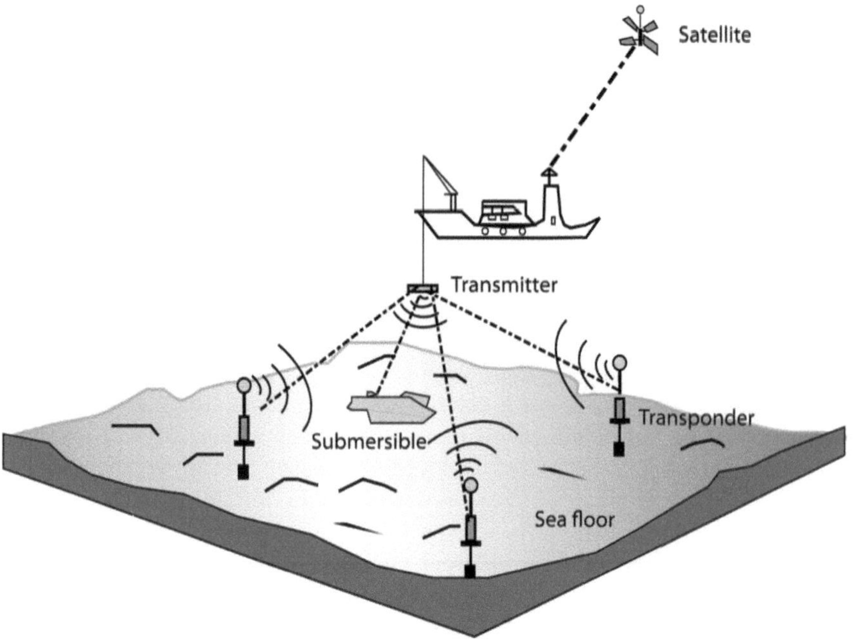

Fig. 3.1 Sea floor navigation system uses bottom-moored transponders disposed on a triangular grid. The transponders send a signal at different frequencies to the ship. The ship is positioned by a satellite navigation system

when the colonies of North America were fighting for their independence from the British, the English had concentrated their fleet in a bay of the Hudson River in New York. Among the vessels there was the admiral ship, *Eagle,* which was the target that the Colonial General, George Washington, wanted to destroy. The *Turtle* submersible, piloted by Sergeant Ezra Lee, dropped its charge of explosives into the water before it could reach its target, however this made so much noise that the British fleet hastily left the harbor (Jarry 2003).

In June 1930, the scientist Charles W. Beebe and an engineer, Otis Barton, both from New York, dove in a bathysphere (a term coined from the Greek words meaning "deep" and "ball") which descended to a depth of about 250 m near the Bahama Islands. The bathysphere was made of cast iron and had two portholes. The sphere hung on the end of a cable connected to a ship. More details on the history of humankind's penetration into the sea can be found in a book entitled "*Les Grandes Aventures des Hommes sous la Mer*" written in 1988 by Claude Riffaud, one of the early directors of the CNEXO (now IFREMER) center in Brest.

Today, human penetration in the deep-sea environment has become a necessity. To explore and understand the formation of our planet requires that scientists be able to observe what is taking place on the sea floor. This involves the preparation of an extensive diving program whose first task is to have detailed bathymetric charts, which are like road maps for the journey. Another important requirement is to be able to send human beings to the sea floor in order that they have the possibility to observe and sample. This can be achieved with the use of modern technology, which has the capacity of allowing human eyes to observe and record data from the sea floor and so we can then place this information in a global, geographical context.

Bathyscaphes

During the first half of the last century, bathyscaphes were used for exploration of the great deeps and were the only engines that could attain all the depths of the oceans without imploding under the pressure found at the bottom of the seas. A bathyscaphe descends because of the pull of gravity and rises by releasing weights. In 1934, the architect Guillaume DeVos drafted a project for a "sort of bathyscaphe" that he called "Aquastat". However it was in 1946 that Auguste Piccard, working in Belgium, designed the first bathyscaphe called FNRS 2 after the name of the funding agency which paid for its construction: "Fonds National de la Recherche Scientifique" (the National Science Research Foundation of Belgium). The vessel was finally built and ready to use in 1948. In 1950, Auguste Piccard and his son Jacques constructed another bathescaphe named "Trieste" with cooperation from the Italian Navy. The Trieste went to sea in 1953 and made several dives at relatively shallow depths (<4,000 m). With the idea of attaining the deepest parts of the ocean floor, the US Navy purchased the *Trieste* in 1958 and then made it available to the scientific community. On January 29, 1960, the *Trieste,* with Don Walsh and Jacques Piccard on board, broke a diving record when they descended to 10,912 m deep in the Mariana Trough located in the western Pacific. For the first time at that depth, divers observed a sedimented bottom and the presence of fish. This was proof that Life exists at all depths in the sea.

Archimède

In 1955, the French Navy, the CNRS (*Centre National de Recherche Scientifique* of France) and the FNRS (the National Science Research Foundation of Belgium) built a bathyscaph first called the "*B 11000*", and later renamed "*Archimède*" (Photo 3.2). Because of its weight and the difficulties encountered in lifting this engine, *Archimède* could only be transported by being towed by a support ship. The French navy ship *Marcel Le Bihan* (1200 tons, 72 m long) named after a French pilot of World War II was used to tow the *Archimède* (Jarry 2003).

The sphere of the French bathyscaphe was built in Belgium. *Archimède* weighs 200 tons in water and uses 162,000 l of kerosene (having a density 0.67) as its liquid for floatability. *Achimède* is propelled by 20 HP electric engines with 110-volt batteries that are protected from seawater in storage tanks filled with oil.

If necessary, the batteries' storage tanks could eventually be released to make the submersible lighter in case of emergency. *Archimède* has one vertical propeller, which allows it to more upward or down in the water column, plus two lateral propellers and one at the stern, which enable the bathysphere to move horizontally. Twelve light projectors providing a total of 1000 W are disposed around the submersible.

The sphere has a diameter of 2.1 m and is composed of two hemispheres forged in steel alloyed with nickel-chromium-molybdenum. When I dove inside the *Archmède*, I had to enter the sphere by climbing down a ladder inside a narrow access tube within the bathyscaphe's hull. The sphere is able to host three persons and is built to withstand 11,000 bars pressure, which is probably greater than the pressure on the sea floor at any depth. The shell of *Archimède's* sphere is 15 cm thick and only the sphere itself weighs 19 tons. Once they have entered the sphere, the three passengers, called divers, will breathe oxygen and release carbon oxide and water vapor. Oxygen bottles and chalk (which can absorb water vapor) are placed into containers in order to provide dry air. *Archimède* has the capability of diving to all ocean depths. In order to see outside, the 2 crewmen and invited scientific observer were obliged to look through individual binocular eyepieces. The distortion of the image was such that things outside the bathyscaphe were about 1.6 times larger than reality. This type of system was not very satisfactory for an observer who needed to clearly identify various structural features or life forms.

Archimède became operational in 1961. After a few tests in the Mediterranean Sea near Toulon, in July 1962 *Archimède* made a historical dive in the Kuril Trench (Japan) and reached a depth of 9,545 m. In 1963, Archimède was chartered by the US Office of Naval Oceanographic Research and sent to San Juan (Porto-Rico) to dive into the Porto-Rico Trench. Nine dives were made between depths of 3,000 and 8,300 m in the Trench (Jarry 2003). The goal was to test the operational aspect of the *Archimède* with some limited scientific prospecting. Sediment cores and a few rock samples were collected. Other dives were made in the Mattapan trench in Greece (1965), and again near Japan in the Kuril trench (1967). However, the first comprehensive and coordinated scientific program with the submersible *Archimède* occurred during the study of the North Atlantic spreading ridge system in a project called FAMOUS (an acronym for French-American Mid-Ocean Undersea Survey) which started in 1972 and terminated in 1974 (see Chap. 7) (Riffaud and Le Pichon 1976).

Although bathyscaphes such as *Trieste* and *Achimède* have demonstrated their capability for working at great depths and by making some scientific contributions, their use is not really efficient. They are heavy, slow moving on the sea floor, and also they are potentially dangerous since they use kerosene to maintain their floatability. Furthermore, the extensive logistics of technical maintenance at sea does not permit research missions to schedule dives on a daily basis. Also, the fact that most instruments used on the sea floor are located underneath the bathyscaphe means that checking and repairing equipment must be done by scuba-divers at sea. This is often difficult and time consuming. All these constraints limit the efficiency for the use of bathyscaphes for today's scientific research.

Manned and Unmanned Submersibles

The deep-diving submersibles used for today's oceanographic research consist of manned and unmanned vehicles. Most modern manned deep-diving submersibles are constructed around a sphere made of an alloy of steel and titanium about 2 m in diameter, which can resist the high pressure found at ocean depths. The pressure at 6 km (6,000 m) depth is about 0.75 tons/cm^2. A sphere is the most adapted form for resisting such high pressure. The speed of descent for a submersible is roughly 30 m/min. Thus, it will take about 90 min to reach a depth of 2,000 m. Inside the sphere, the atmospheric pressure is at one and the passengers breathe oxygen. Their released CO_2 and water vapor are absorbed in the sphere by means of a system using chalk and CO_2 filters. The average autonomy for research diving vessels is 120 h, but most dives are aimed at a total duration of only 8–10 h.

Several industrial nations have deep-diving submersibles that are used for research : Japan has the *Shinkai* (Jamstec organisation), Russia has the *Mir (2)*, the USA has *Alvin* (WHOI) and *Pisces IV and V* (HURL = Hawaii Undersea Research Laboratory), France has *Archimède, Nautile* and *Cyana,* Germany uses Yago, (IFM-GEOMAR Leibniz-Institüt fur Meereswissenschaften) (Fig. 3.2). China has recently entered in the game of deep-sea exploration by building its first submersible called the "*Jiaolong*" that made a test dive at 7,072 m depths in the Mariana Trench in June 2012.

Alvin

Alvin is operated by Woods Hole Oceanographic Institution (WHOI) but is owned by the US navy and is contracted through US government funding (Fig. 3.2). It is the submersible that has the highest record for diving, since it usually makes between 100 and 175 dives each year. The support ship of *Alvin* is now the R.V. *ATLANTIS*. For several years, in the 60's and 70's, a catamaran called *Lulu* was used as *Alvin*'s support ship (Ballard and McConnell 1995). The submersible *Alvin* has a maximum operating capability of 4,500 m deep and it weighs 16 tons. *Alvin* carries one pilot and two scientists. The vessel has two lateral portholes oriented at an angle of 45° and a frontal porthole for the pilot. *Alvin* is not as comfortable as the French submersible *Nautile,* because the observers must sit behind the pilot who has the best view through the front porthole. Sitting on each side of the pilot

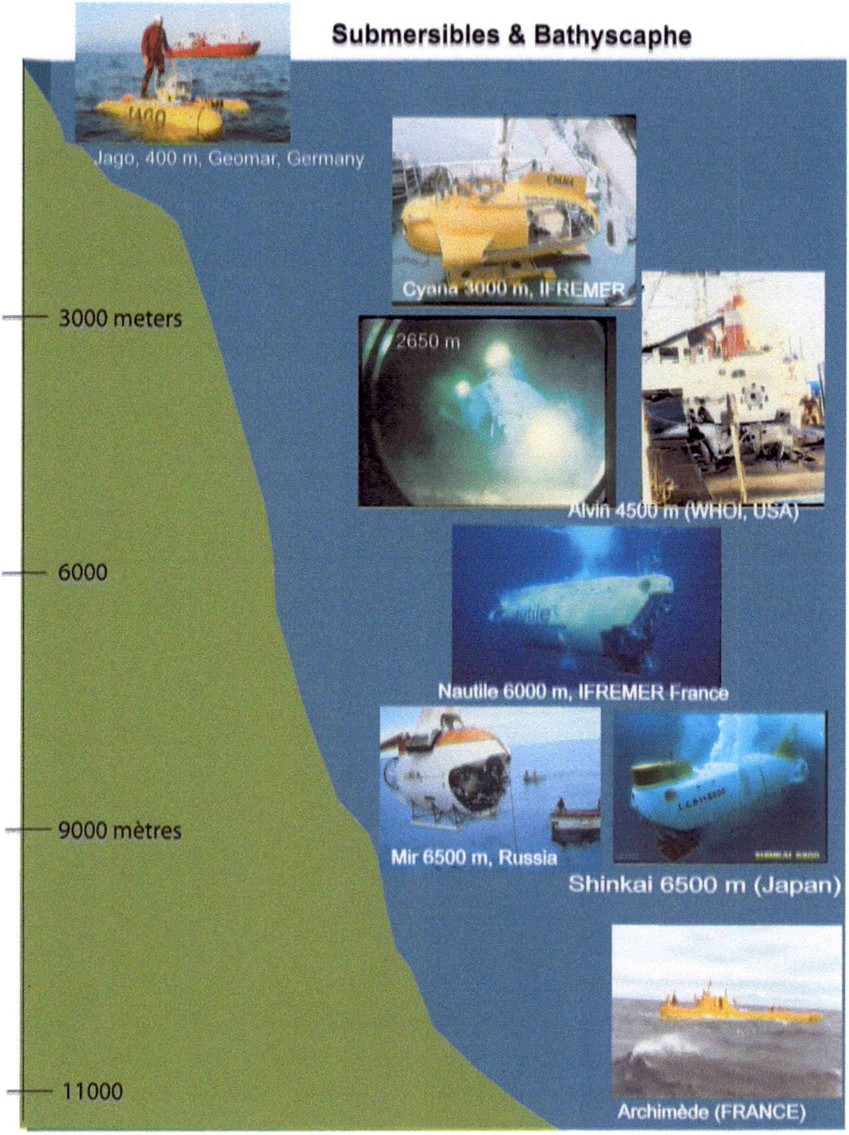

Fig. 3.2 Manned submersibles used for scientific exploration and their depth capabilities

our legs must be bent and we have to look through the side-window portholes, which do not have the same view as the pilot. This was sometimes a handicap because when we wanted to sample rocks, or ask the pilot to make a stop for a picture or a video run, we had to move to his porthole to point out the target.

Cyana

Cyana was built by Jacques Cousteau and was later transferred to CNEXO (now IFREMER). Raymond Kientzy, who was one of the early chief pilots on board the submersible, proposed the name *Cyana* for this small vessel. It is able to dive to depths of about 3,500 m and was first operational during the FAMOUS project in 1974 (Fig. 3.2). *Cyana* is smaller than *Alvin*, since it weighs only 8.5 tons and it is 5.1 m long and 2.7 m high. The submersible carries one observer, plus a pilot and a co-pilot. All the French submersibles carry two pilots and only one observer. The French submersible group has always been hesitant about carrying two scientists during their dives because they want to use dive time in order to train new pilots who could also help the observer and the pilot inside the sphere. This policy is based on the inexperience of most scientific observers who would probably be unable to handle the various pieces of equipment found on board, in the event of an emergency, or even during normal diving missions.

Both, *Alvin* and *Cyana* have two manipulating arms. The portholes of *Cyana* are slightly inclined which make observations easier and this engine is therefore more efficient for conducting geological exploration than the *Alvin*. In addition, *Cyana* is the best engine for observation because it has the advantage of taking a tilt of about 45°. On a vertical slope the observer can request that the submersible be inclined so he can get closer to the target than in other manned submersibles. After about 30 years of diving operations, *Cyana* is no longer being used, however this submersible participated in numerous projects in the Atlantic, such as FAMOUS, and in the Pacific, discovering hydrothermal deposits on the sea floor. Because of the high cost of maintaining and running two submersibles, *Cyana* was increasingly less assigned to scientific missions until it was declassified and retired in the year 2000.

Nautile

The *Nautile* was constructed by the French Navy in Toulon and became operational in 1982. It is operated by IFREMER and GENAVIR (GEstion NAVIR). In December 1983, a letter was sent to ask members of the French research community about their ideas for a name to be given to the new submersible that was simply called SM 97 when first constructed. Various people submitted about 51 propositions for a name for the new submersible. The name "*Nautile*" was suggested by Claude Riffaud (and it refers both to Jules Verne's book "20,000 Leagues Under the Sea" as well as to a shellfish called a nautilus). However the name Nautile is also an acronym for "***N**avire **A**byssal d'**U**tilisation **T**echnologique d'**I**ntervention **L**égère et d'**E**xploration*". The submersible can dive up to 6,000 m deep, which enables it to descend to 97 % of the ocean floor. 6,000 m depth corresponds to a pressure of 600 bars equal to about 0.7 tons of weight

(1 bar = 1 kg/cm^2). The *Nautile* is 8 m long, 2.7 m wide, 3.81 m high and weighs 18 tons. In order to be lighter in weight, the *Nautile* was constructed with a titanium-steel alloy and its floatability is due to the use of synthetic foam. It is equipped with two vertical and two lateral thrusters, and one main axial motor with a propeller, which can be oriented in different directions. Nickel-cadmium batteries power the engines. The sphere of the *Nautile* has a diameter of 2.10 m and is equipped with 3 portholes about 12 cm in diameter. The maximum speed at the bottom is usually less than 2 kn (1.7 kn). It is equipped with 7 projectors, which can provide up to 2000 W of light. The *Nautile* is similar to the Russian-designed submersible *MIR,* built in Finland, and the Japanese submersible, *Shinkai.* In a good year, *Nautile* will be able to log about 170 dives (see Figs. 3.2 and 7.27).

The first scientific expedition of the *Nautile* took place in the **Japan Trench** in 1985 with Xavier Le Pichon as chief scientist. The support ships of the *Nautile* are the *N.O. LE NADIR*, N.O. *L'ATALANTE* and N.O.“*LE POURQUOI PAS?*”. The research vessel “*Le Pourquoi Pas?*” is a very suitable ship for multidisciplinary purposes and sea-going capability. It is 107 m in length, 20 m wide, 6600 tons, and is the largest French oceanographic vessel sponsored by both IFREMER and the French Navy. “*Le Pourquoi Pas?*” is able to carry both the *Nautile* and the ROV “*Victor 6000*”.

Nautile has participated in many different types of expeditions including locating old shipwrecks. The “*Titanic*” was one of the major diving targets of the *Nautile*, which performed about seventy dives in depths of 3,970 m after the *Titantic*'s discovery in 1985. So far, no other scientific objectives have attracted more attention, used more ship time or had more dives than what has been done on the *Titanic* site off the eastern coast of North America.

Pisces

The *Pisces V* and *Pisces IV* are three-person, battery-powered submersibles with a maximum operating depth of 2,000 m (Fig. 3.2). Both submersibles are operated by HURL (Hawaiian Undersea Research Laboratory) affiliated with the University of Hawaii, USA. They were designed and built by Hyco International Hydrodynamics in Vancouver British Columbia in 1968–1970. The pressurized hull has an inside diameter of 2.1 m and is made of a steel alloy with 3 forward-looking acrylic windows, 15 cm in diameter. The two *Pisces* vehicles enable us to make video recordings, to place instruments on the sea floor, and to conduct sampling. They have two remote-controlled arms, which can be manipulated. These arms are useful for placing thermometers in strategic places, or for picking up biological and geological samples and placing them in a collection box, and they are even helpful in maneuvering the submersible around areas that are otherwise difficult to navigate. In *Pisces V,* a typical dive lasts up to 10 h. However, the vessel has an emergency life support system that is capable of keeping the three-person crew alive for as long as 140 h in case of emergency.

Unmanned Submersibles and Robots

The evolution of mankind's penetration into the deep is moving towards the increased construction and use of robots, which are more practical and less expensive than the use of manned submersibles.

Deep-towed bottom camera stations were the early vehicles first used to explore the sea floor landscape. For example the "troika" sledge, which is similar to a snow sledge that could be towed along the sea floor, was extensively used to prepare the FAMOUS diving operation in 1972 (Fig. 3.3). Jacques Cousteau designed this instrument during the early sixties.

Other video camera and photographic devices are mounted on large grabs and metallic frames as well as on coring and drilling pipe devices. Frames built of a titanium alloy were the most commonly used for transporting cameras and electronic devices prior the use of more sophisticated robots. One of these deep towed camera engines which has often been used by IFREMER during our expeditions was called the *Scampi* (acronym for **S**ystème de **Cam**era **P**onctuel **I**nteractive). *Scampi* is an Interactive Camera System towed at 2 kn behind the ship. It is 2 m × 0.8 m × 1.2 m in size, and designed to operate at 5–10 m above the sea floor, but it can descend up to 6,000 m deep. Usually these instruments carry a total of 2000 W of lamps and 35 mm color film for slide photographs. Each device contains 800 frames. The pictures should be taken as close as possible to the bottom, at <15 m from the target.

Remotely **O**perated **V**ehicles (**ROV**) are not only able to do video and photographic coverage but they are more suitable for preliminary exploration such as mapping and bottom sonar imaging for detecting small structures which cannot be seen from surface ships. ROVs are able to detect small fissures, make detailed magnetic surveys and can be sent out to locate hydrothermal vents. They could also carry instruments to be deployed on the sea floor. The ROV are attached to a coaxial electrical cable and/or a fiber optic cable that transmits energy to the vehicle while it is performing on the sea floor. The commands and feedback data to the surface are controlled by the instrument's operators, directly from the ship (Figs. 3.4, 3.5). The engine navigates on the sea floor while tethered by an "umbilical cord" that helps the device to move with its own electrical propulsions over an area with a radius of about 200–300 m in diameter. Usually ROVs are easier to transport, therefore they can be used from many support ships as long as they have sufficient deck space and feasible launching and recovery equipment. An ROV and its support ship will form an inseparable couple since they are dependent on each other and it is necessary that there be excellent coordination between the pilots of the ROV and the ship's officers on the bridge in order to avoid any difficulties. The bridge must anticipate the motion of the ROV on the bottom so it will not overshoot the limits of distance that the vehicle can travel without being retained or pulled by the ship. The ROV's tether is not strong enough to resist breaking in the event of strong pulls by the vessel. The tether has the same neutral floatability as the ROV and must be allowed to be loose and free at all times,

Fig. 3.3 One of the first deep towed camera sledges called "Troïka" used during preparation of project FAMOUS in 1972, in the rift valley of the Mid-Atlantic Ridge. (Copyright IFREMER, R. Hekinian)

although the tether can also be easily tangled and this will limit the maneuverability of the vehicle. Such difficulties are related to the cable and to the positioning system of the ship that has to follow the movement of the vehicle moving about at a few meters above the sea floor. Usually the ships using ROV have a dynamic positioning system controlled by a computer that can correct any shifts in position of the ship with respect to the ROV and its target. This helps the ship to stay close to the right spot within the assigned field of exploration.

The US ROV most used at the present time is the *Jason II-Medea* system managed by Woods Hole (WHOI, Woods Hole Oceanographic Institute in Massachusetts, USA). This ROV has high precision acoustic and imaging sensors, a sampling platform for any recovered material, and has a maximum depth capability of 6,500 m. The French ROV *Victor 6000* (Fig. 3.4) is a modular remote controlled vehicle, which can descend up to 6,000 m depth. It was first used on a scientific expedition in 1999 on board the RV. L'ATALANTE. The *Victor 6000* is composed of two systems: (1) The vehicle itself with its service equipment, propulsion, video survey equipment, lighting and navigation services, (2) The scientific module composed of sampling tools and scientific equipment which are specific to a given mission. *Victor 6000* is relatively heavy with respect to its being handled and to the space requirements for its four 20 meter-long containers carrying all the material which could be required for operational purposes.

The Canadian ROV called *Ropos* has a 5,000 m depth capability and has been successfully deployed from many oceanographic vessels. *Ropos* is connected to a cage by a 300-meter long tether (Fig. 3.5). It has two manipulating arms, two main video cameras and a payload for samples of about 36 kg. As opposed to the French ROV *Victor 6000* and because of its compact size and small amount of accompanying equipment, *Ropos* it is the most suitable choice of an ROV to be used on normal-size ocean going vessels. The German ROV "*Quest*" acquired in 2004 and with a diving capability of 4,000 m is also equipped for modern exploration. Both *Ropos* and *Quest* are easy to transport and handle on board normal sized ocean going vessels and their cost of handling is less expensive than that required for

Fig. 3.4 ROV (Remotely Operated Vehicle) "Victor" has a 6000 m depth capability. Dimensions L = 3.1 m × 1.8 × 2 m. Mass in air = 4000 kg. (Copyright IFREMER, V. Renard)

Fig. 3.5 Canadian remotely operated vehicle (ROV) called *Ropos* (Photo courtesy of S. Scott)

"*Victor*". For example, only two containers are needed on board the support ships using the Canadian or German vehicles.

ROV and AUV are generally more manageable than the submersibles and require a limited crew. About three to four people for 8–9 h performances and only a total of six people are necessary for 24-hour operations. Autonomous Underwater Vehicles (AUV) or Underwater Untethered Vehicle (UUV) are still at their early stage of development. They are unmanned and unattached vehicles that are free to move about in deep water thanks to a remote controlled navigating system. Their power is supplied by on-board batteries.

Future advances in deep-sea exploration would be to develop an engine capable of communicating data as well as transferring material to an underwater platform that could serve as a storage area until the data and samples are recovered. Another variant for such a data storage platform would be to enable the transfer of equipment and data to underwater stations which could then relay this information and/or material to the surface either mechanically, by means of a cable, or electronically via a surface buoy equipped with a solar battery. Such an underwater station would be comparable to space stations receiving material and vehicles from Earth.

Since the discovery of hydrothermal deposits and their associated biological community, scientists have been eager to undertake sustained time-series investigations in order to understand the time scale variation of oceanic processes. Indeed, it has become common to conduct some of our research by repeatedly visiting the same sites. This started in the early 80 s′ at 13°N on the EPR, when biologists wanted to observe the changes in animal distribution and specimen types around hot hydrothermal vents and when geologists wanted to sample and carry out various geophysical measurements.

The increasing capabilities of sending robots underwater makes deep-sea operations much less expensive than when using manned submersibles. Also, the time scientists can spend on the sea floor is limited with manned submersibles, averaging 7–9 h total. With a robot, the time on the bottom is unlimited as long as the ship can maintain its station and supply power to the fiber optic cable.

However, the remote controlled system does not provide the same full concentration of the human intellect as when scientists are at direct contact with the sea floor. One of the advantages of manned submersibles is to have a human being, with his eyes and his brain, at almost direct contact with the target so he can react instantaneously with the environment. This permits the observer to change the plans of the dive rapidly in response to a given situation. Such a change in tactics is most relevant to specific targets in biology with quick moving life forms and is also relevant to eruptive events related to hydrothermal volcanic sources where rapid response is also required.

The eyes of an experienced marine geologist are necessary in order to discern freshly erupted flows and their changing morphology. Although a larger overview might also be made from data recovered by deep-towed sonar and television camera stations, the final proof of correct data interpretation will come with the in situ (on-site) observation obtained from manned submersibles.

The advantages of manned or unmanned observations and data recovery are related to the cost of maintaining HOVs (human occupied vehicles) compared to adjusting to the distortion of 2-D images on a screen. It is a fact that the dimensions and perspectives of a Television image are not the same as on-site observations.

We need to take into consideration the liberty of an independent submersible when compared to using a system involving a tether and fiber-optic cable, which are major disadvantages for an ROV. In fact, both systems are complementary to each other and could be used in various combinations for conducting detailed observations and sampling. It would be a shame to give preference to one technology at the expense of the other because later, if it appears that manned submersibles and/or ROV techniques would be needed, it might be too late to return.

Are we ready to dismiss the possibility of unmanned flight in outer space? I do not believe so, because man is eager to explore new frontiers and confront new elements and I believe that no machines can replace human beings. Furthermore, if we dismiss the possibility of using manned undersea vehicles, we may loose our skills in this area. Indeed, the technology of building such instruments and training pilots and engineers has taken considerable time, so perhaps we would be wise not to allow this heritage of the twentieth century to disappear.

Diving in a Manned Submersible

Most people who have dived with submersibles at great depths have the sensation of succeeding in a significant accomplishment. We travel into the unknown to look at another aspect of our Earth, which has been hidden by the sea. The sun's light propagates down to only about 100–150 m in the water, after which there is only darkness for human eyes. The artificial lights used on the submersibles consist of several projectors, which can create about 2000 W. The light projectors are located on top of the portholes and on the mechanical arms used for sampling (See description of *Nautile*). In order to obtain a fairly good color picture, the object to be recorded should be located less than 3–4 m from the camera.

Outside a submersible, light projectors allow limited visibility, less than 10–15 m distance, but to economize energy, they are not turned on all the time, so sometimes we navigate in the dark. Also, visibility for human eyes does exceed 45–50 m and an object will only become recognizable at <15 m depending on the clarity of the water column.

A submersible has to be able to move freely. Because of its weight, it sinks like a stone, descending in the water at a speed of about 30 m/min. Submersibles are able to adjust their rate of descent by releasing weights. Also, when the submersibles have almost landed on the ocean floor, they will release weights to obtain neutral buoyancy, so they don't merely sit on the ocean floor but are a few meters just above the bottom. Generally, iron pellets are stored in appropriate reservoirs and they can be released as needed. In order to move along the sea floor, it is important to be in a state of neutral buoyancy. The weight release mechanism is electromagnetically and mechanically controlled from the interior of the sphere (Fig. 3.6).

Although, I must have participated in more than 50 dives during my career, each dive remains a unique and exciting experience. In order to retain information from each dive, television videos and a tape recorder (which saves all the comments made inside the submersible) are also part of the data recovered during every dive.

A Normal Diving Day

Even if all the dive objectives have been carefully prepared prior to the expedition, the actual moment of diving remains incredibly intense and exhilarating. Scientific and technical meetings are programmed each evening before every dive. The

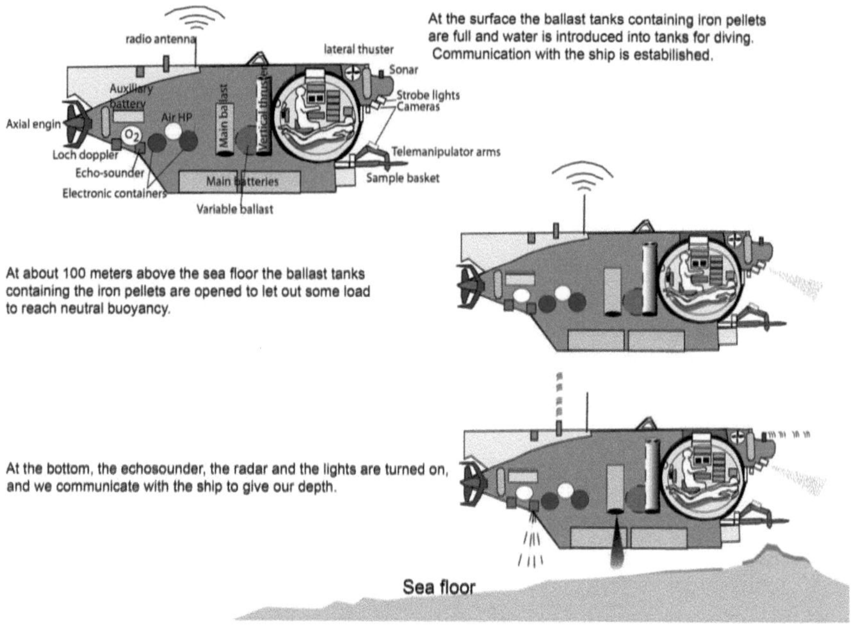

Fig. 3.6 A dive in a manned submersible on the sea floor lasts about 8–9 h. The submersible sinks when it takes on water and rises to the surface when it releases its ballast (iron pellets). Several tasks can be achieved such as in situ observations, continuous video-camera recording, sampling, plus deploying and using various pieces of equipment

diving objectives, launching the submersible, the state of the bottom transponder network as well as navigation on the sea floor are discussed in great detail. The meeting and discussion requires the presence of the diving operation manager, the pilots and engineers from the submersible group, scientists, plus the ship's captain and the officer of the bridge. The targets are clearly defined by the chief scientist and by the scientific observer assigned to the dive. Three people are usually involved during each dive: a scientist (or an observer), a pilot and an engineer, who is also the co-pilot. The scientific observer and the pilots and co-pilots are the main people concerned with a given dive and it is critical to understand their opinions and objectives if the surface operating team is going to follow the activities of the next day's dive.

It is also during the evening before a dive that the submersible group starts to prepare the submarine itself. In the morning of the dive, all the groups (i.e. the scientists, ship-board personnel on watch, and submersible team) get an early start at 6 o'clock so they can perform their respective routine checklists for the equipment involved. At 7 AM, after a quick breakfast, weights consisting of 150–300 kg of iron pellets contained in 27 bags of 25 kilos each are poured into special compartments on the side of the submersible. In addition, four ballast bags of iron pellets are attached to the side of the submersible, and they will be released before reaching the bottom in order for the vehicle to attain a state of neutral buoyancy when approaching the sea floor. The weights of the three passengers inside the sphere of the submersible are also taken into account just prior to diving for adjusting the total weight of the submersible.

The choice of the diving scientist is usually made according to his special interest and his qualifications for the target and tasks of a given dive. The person chosen will be well acquainted with the objectives of the dive and a day or two earlier, he will begin to prepare himself mentally and physically for this great adventure. Usually it is recommended to avoid drinking too much liquid the day before, although there are facilities within the sphere should we need to relieve ourselves. It is also recommended that the scientific observer get a good night's sleep.

Early in the morning, the ship goes on station and positions itself with respect to the bottom transponder network, which has been deployed on the sea floor earlier that day. The transponder beacons are laid on the sea floor and mounted with acoustic devices that send signals of different frequencies to the submersible. The person responsible for the surface ship's navigation has the ship pass over each one of the bottom transponders that were launched a couple hours before diving. The calibration of the transponder network with respect to satellite data is made, and the positions are recorded on the computer screen by the navigation team.

Before diving, it is important to check the surface and bottom currents, if any information is available, and this is another of the roles of the surface ship's navigator. Because of underwater or surface water currents, the submersible could be shunted off from its original target, and sometimes it is necessary to adjust the trajectory of the submersible while diving. When the bottom navigation system is ready and calibrated, the captain will inform the chief pilot and the head of the submarine group that all systems are ready to go and that they will reach the diving site in 10–20 min.

The submersible's checklist is done on the deck, with the pilot inside the sphere and the engineer outside. They communicate with each other about all the vital points of the submarine. It takes about 40 min to complete this checklist. At this point, the co-pilot informs the scientific observer that it's time to go inside the sphere. Meanwhile, the chief steward has handed the lunch meal to the co-pilot who is already inside the sphere. Now it is the scientist's turn to enter the small sphere in his special red or yellow fireproof jumpsuit. It is also recommended to wear warm clothes underneath the suit because the outside temperature at the bottom is only about 1–2 °C at 3,000 m deep in all the world's oceans. The ambient temperature inside the sphere will drop progressively as time passes. As the interior becomes colder, water will start to condense on the interior walls.

When the ship has reached the target area, the captain or the officer on the bridge will send this information to the submersible group leader, who will tell the pilot of the submarine to get ready. The pilot will climb up the ladder inside the sphere and sit on the edge of the hatch. He then verifies the hatch's seal and cleans the grooves around the hatch with a rag. This is a common procedure to avoid that any dust or dirt will prevent proper closing the airtight hatch. The head of the submarine group stands on the back deck and orders the man at the winch to start to bring the submersible into position at the stern of the ship. The submarine is pulled towards the rear of the ship so it is in position, ready to be hoisted off by an "A" frame. Inside the submersible, we can hear the creaking of the winch as it pulls the submarine standing on its "cradle" towards the back of the ship so it will be in position to be lifted. The pilot is still sitting on the edge of the open hatch with a walkie-talkie in his hand. Then he takes a last look at the sky, waves with his hands to the people on deck and says, "We'll see you this evening". The answer is always the same, "Have a good dive," and he disappears inside the submarine after closing the hatch behind himself.

Once the sphere is closed and secured, we are now under oxygen and the reading and ambient lights are switched on to make a last check of the instruments. The rope hanging on the "A" frame comes down to click on and lift the submarine, which rises into the air prior to be lowered overboard. A scuba diver with his feet flippers and eye mask stands on top of the hatch and accompanies the submarine into the water. It is only when the engine has been placed into the water that the diver releases the submarine's cord from the ship's "A" frame. Now the submarine is on its own, totally free from the support ship. The submersible rolls on the waves like a bobbing cork and we feel a bit uncomfortable for a few minutes. One last surface communication is made with the ship to let them know that all systems are "go" and then the pilot asks permission to dive.

As soon as permission is given from the bridge, the pilot opens the valves to let water enter into the water ballast tanks, which makes the submarine heavier until it starts to sink. We can still see the three scuba divers through the port holes, and the three divers will accompany the submersible a few meters down in order to make a final check of the ballast compartments and the equipment. When they think that all is fine, they give us an OK sign with their hands and off we sink (Fig. 3.6).

The submarine drops like a stone at a speed of about 30 m/min and while it sinks it rotates like a spinning top. Complete darkness comes very quickly depending on the visibility conditions of the seawater in the diving area. After 5 min, below about 150 m depth, we are in complete blackness and no sunlight is noticeable. It is at about 1,000 m deep that we usually hear a "bang" indicating that the outside pressure is compressing the sphere of our submarine. By the time the submersible has reached the bottom, the sphere where we three are staying will have shrunk about 1–2 cm in size, depending on the depth we have reached. Looking through the porthole, sometimes we are able to discern phosphorescent plankton with flashing green lights.

The adventure of discovering the sea floor in an area never before seen is extremely exciting and each dive is full of surprises. Let me share the experience of a mission inside *Cyana* when we went to visit an off-axis seamount called Clipperton about 16 km west of the East Pacific Rise in 1982 (see Chap. 7).

When we became aware that we were approaching the bottom on the basis of the sounds of the sonar, which had picked up some wave reflection from surrounding mountain ranges, it was time to be more vigilant. The pilot (Nivaggioli, or "Niva", for short) looked at the pressure gauge and watched the depth readings. At 100 m from the bottom we should also be able to pick up the altitude readings of the submarine based on instruments placed on its bottom. Niva read out "seventy, sixty meters, fifty…"

That's it, we were very near the bottom, so Niva had to flip a switch on an electromagnet to liberate 50 kilos of steel ballasted-bags hanging from each side of the submarine. We were now lying flat on our bellies, our noses pressed to the portholes and we were excited to think about who was going to be the first to actually see the sea floor. In the meantime, the noise of the pinging sonar waves bouncing sound off the surrounding cliffs made a squealing noise every time they hit an obstacle. We were surrounded by cliffs about 150 m high. When the submersible reached an altitude of about 30 m above the sea floor, we were able to guess at the shapes on the rocky bottom and soon we found ourselves among giant lava flows at about 2,650 m deep. We didn't actually touch bottom but remained about 30 m above the sea floor surface. The pilot turned his light projectors on full force and using its motor, the submersible started to turn around in a circle, just to see what the surroundings looked like. In the meantime, a message was sent to the surface ship telling them that we arrived on the bottom and requesting that they give us our exact position before we started to move.

The pilot usually took a short break before starting to move the submersible in order to stabilize and re-verify the equipment. Then we would proceed on our itinerary as planned. The first reflex prior to getting underway was to take a rock sample and a few photographs of the landing site.

After working about 5 h on the sea floor, the submersible's batteries would be getting low and it would soon be time to request permission to return to the surface. *Time has flown by and suddenly I realized my dive was nearly over. I've been so engrossed in my work, I've forgotten to take time to eat, but this is my usual way of operating.*

Generally a dive involves a lunch break, although this will depend on the target as well as the divers. Some people prefer to take a brief break while others, like me, are eager to continue. If this is the case, the pilot usually moves to the rear of the vessel and allows the co-pilot to take over. This helps the pilot to relax and gives an opportunity to the co-pilot to be trained in handling the submarine. This kind of pilot/co-pilot approach does not exist on *Alvin* since there is only one pilot on board, therefore this submersible's pilots must be trained during special training dives without any scientists on board.

When it was time to leave the ocean floor, Niva pushed the electro-magnet control button to release the steel pellets stored in the ballast compartments on the side of the submersible. Slowly, our little yellow submarine rose back to the sea's surface.

Our return trip was generally more relaxed because we felt that we had accomplished our task. This was the time I used in order to write up some final notes before I forgot the details, and to make an inventory of the videotapes and the number of pictures taken with the outside camera.

Arriving near the surface, the sunlight penetrated into the sea giving a bluish-gray coloration to the water. When we were about 15–20 m below the surface we could start to feel the waves that made the submarine roll gently. Looking from the porthole, I was able to see the round ball of the sun, and I felt happy to be back on the surface. Now I could look forward to warming myself up after a tiring but fruitful day of adventures at the bottom of the sea.

References

Ballard RD, McConnell M (1995) Explorations: a life of underwater adventure. Hyperion, New York, 407 p

Jarry J (2003) L'Aventure des bathyscaphes. Editions du Gerfault 26 Rue Jacob, Paris 303 p

Linklater E (1972) The voyage of the challenger. Published by Doubleday and company Inc, Garden City, 288 p

Renard R, Allenou J-P (1979) Sea beam, multi-beam echo-sounding in "Jean Charcot" description, evaluation and first results. International Hydrographic Review Monaco LVI(1):35–67

Riffaud C (1988) Les grandes aventures des hommes sous la mer. PUBLISHER, Albin Michel SA. 22 Rue Huyghebs, 75014, Paris, 456 p

Riffaud C, Le Pichon X (1976) Expédition FAMOUS: A trois milles Mètres sous l'Atlantique. Publisher Albin Michel 22 Rue Huyghens, 75014 Paris: 267 p

Wüst G (1964) The major deep-sea expeditions and research vessels 1873–1960 progress in Oceanography (2) Pergamon Press, Oxford, 1–52

Chapter 4
Sea Floor Rocks

Abstract Sea floor rocks can reveal Earth's history. Based on the various sampling operations conducted by ocean going scientists, it is inferred that the relative distribution of oceanic rocks includes about 50 % basalts/dolerites, 20 % gabbros and 30 % peridotites. Basalt and dolerite are the main volcanic rocks of the upper crust, gabbros are formed during the solidification of a magma chamber or magma conduit, and peridotites are either the heavy mineral residues left within the reservoirs after magma solidification or they could be the remains after a partial melting of mantle material. Basalt is the most common type of volcanic rock found on the sea floor. Basalt and other related rocks have been extruded after partial melting of the Earth's mantle material. Rocks from the sea floor differ from those encountered on land due to their shape and their chemical composition. The effect of seawater and the pressure it exercises on hot, outpouring lava will fashion the shape of deep-sea volcanic rocks giving the sea floor a different appearance than what we see in subaerial environments. Curved and spherical-shaped pillow lavas are only found on the sea floor due to the fact that seawater pressure is equally applied on all directions of the lava flows and their cooling surfaces. Basalts consist of silicates of magnesium, plus iron and calcium oxides. Less common silica-enriched rocks ($SiO_2 > 53$ %) such as andesites ($SiO_2 =$ 53–59 %), rhyolites (>70 %) and trachytes ($SiO_2 = 59$–64 %) are also found on some undersea structures such as domes and seamounts.

The history of rocks is what the field of *petrology,* my specialization, is all about. The word "petrology" comes from the Greek words: "petros" meaning stone and "logos" meaning "study of" or "learning about". The description and classification of volcanic rocks helps us to understand their origin, their subsequent evolution, and the subsurface circulation of the magmas responsible for their formation. Mineralogical and chemical composition will influence the rate of lava extrusion, as well as the lava's cooling rate and the shape of various lava flows. The magmatic history of the various geological provinces under the sea is affected by a process of partial melting of deep-seated mantle as well as by crystal-liquid fractionation in shallow depths within a magma reservoir.

R. Hekinian, *Sea Floor Exploration*, Springer Oceanography,
DOI: 10.1007/978-3-319-03203-0_4,

Sea floor rock samples will reveal their origin so we can understand how they have been extruded after the partial or total melting of material rising from depths >30 km. Upon decompression, during their ascent to shallower depths, the minerals forming the mantle will be fused. Since each mineral forming the rocks is sensitive to changes in temperature and pressure, we can study these minerals to infer their original depths and the extent of their melting before they were erupted on the sea floor.

The relative abundance of different rock types encountered on the ocean floor may not reflect the absolute quantity of any particular type. The fact that we do not have easy access to deep-seated formations means we are limited to observations of just the thin veneer of the Earth's crust, with the exception of areas such as fracture zones where the crust has been sliced and uplifted to expose more deep-seated rock formations. Regions where this type of material is most likely to reach the sea floor are magma-starved areas where volcanic activity is lacking or limited. Such areas are found on slow spreading ridge systems, which are preferential zones for exposing the deep-seated rocks from the Earth's mantle. Ocean drilling operations are also used to probe the Earth's interior in order to sample different, deeper formations.

Rocks: A Combination of Several Basic Minerals

A mineral is a natural, inorganic compound with an orderly internal chemical structure and physical properties that cause it to assume a specific crystalline form. The fascinating aspect of the minerals forming the Earth's rocks is that they are basically composed of atoms of just a few different elements, all of which could be considered as being the building blocks of our planet. The different elements are bonded together to form a mineral lattice. Each individual crystalline lattice has a regular arrangement of elements constructing a network or a space lattice. Each element in the lattice is composed of atoms with their specific weight, electric charges and radii. An atom is the smallest portion of an element, which can join with another atom or several atoms to form compounds and minerals. For example the mineral known as quartz (or silica dioxide, SiO_2) is formed with one atom of silica (Si) and two atoms of oxygen (O_2).

The atoms and elements are bonded together due to a sharing of their electric charges. The space lattice between the atoms has the capacity of absorbing or releasing electrons, thus creating energy during the formation of a new compound. Whether the space lattice releases or absorbs an electron will depend on whether it is charged positively (so it will release an electron) or negatively (so it can absorb an electron). The transformation or growth of a mineral lattice will occur during partial melting and magma solidification and/or during the alteration of a rock. The factors that govern the behavior of all the elements are their ionic radius and their electric charges, which characterize the atomic configuration of the elements within a lattice.

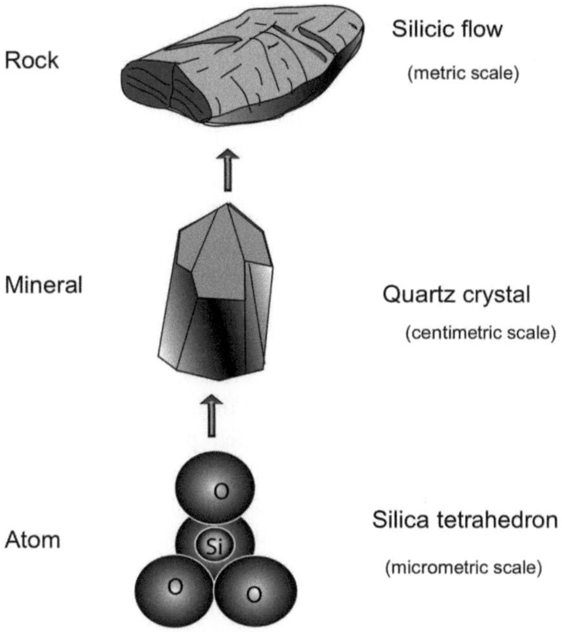

Fig. 4.1 Most minerals forming rocks are silicates, and their basic lattice is defined by the silicon-oxygen tetrahedron. The dominant elements of the Earth's lithosphere-crust are oxygen and silicon

The atomic configuration of the elements provides a guide for understanding the degree of compatibility between the various elements composing a mineral's crystalline lattice. The size of each atom's radius and the electric bonding of each element will govern the architecture of the oxides and silicate minerals, which compose both crystalline and glassy (aphyric) rocks. A decrease in ionic size will be proportional to the degree of element affinity forming the atomic configuration. A list of the most commonly occurring elements based on the size of their ions (from large ions to small) and related to an increase in the number of their electrons is as follows: K, Na, Ca, Mg, Fe, Al, Si.

The most common minerals forming Earth's rocks are composed of various silicate minerals, which are essentially made up of silica (Si) and oxygen (O) (Fig. 4.1). The oxygen atoms in oxides as well as in silicate minerals are found surrounding the other elements, such as silica.

The main elements considered as being the "building blocks" of the most common minerals found on Earth are: oxygen, magnesium, silica, and iron. Potassium, sodium, aluminum, titanium and calcium are other lighter (less dense) elements that are also important in the formation of the common minerals found in basaltic rocks.

Elements that are bonded to oxygen atoms (or surrounded by oxygen atoms) are expressed in a number expressing their weight percent (Wt %) of the bulk rock's composition. Other "trace elements" (measured in parts per million = ppm) do not contribute to the formation of the most common mineral phases that are usually found in most rocks. However, trace elements are often bonded and/or have become partially substituted for the major elements in a crystalline lattice. Some of these elements could either replace the major elements or even form other individual minerals. For example, the most common metallic trace elements forming minerals and exploited by industry are gold, iron, silver, nickel, platinum, copper, lead, cobalt, zinc, antimony and palladium.

In the 1960's, before the invention of Microprobe Analyses, which could determine the chemical composition of individual crystals, we performed chemical analyses on rock samples by using a mass-spectrometer. Our samples needed to be ground into a fine dust before inserting them into a spectrometer, which could indicate their chemical composition. By analyzing glass, one could have a better idea about the bulk-melt composition than when studying a crystalline sample where the individual minerals forming a particular rock might significantly vary in their chemical contents. Analytical uncertainty could also result if the whole rock powdered-sample had not been properly homogenized. Mineralogy was done using onionskin thin slices of rocks (called "thin sections") that were placed under a microscope so we could see the mineral crystals and recognize them on the basis of their particular shapes and colors. Therefore, crystallized samples were best for mineralogy studies, however glassy samples were the best for mass spectrometry since they did not have any crystals that might unduly influence the analyses' final results.

Basaltic Rocks

The top layer of the sea floor is primarily composed of about 50 % basalt, which is the most common type of volcanic rock to be erupted. Basaltic rocks have solidified after the eruption of lava composed of several minerals such as *plagioclase* (silicates of aluminum, calcium and sodium), *clinopyroxene* (silicates of aluminum, magnesium, calcium and iron), and *olivine* (silicates of iron and magnesium), which could be enriched in forsterite (Fe-rich olivine = $Fo_{70\text{-}88}$). Less than 5 % of basalt's mineral content is composed of opaque metallic-oxides such as spinel, titanomagnetite or ilmenite.

The mineral composition of basalt reflects the composition of the upper portion of the Earth's oceanic lithosphere. Partial melting starts at the borders of mineral grains where interstitial liquids are formed and then escape towards shallower depths. If a basaltic melt circulating in the lithosphere solidifies prior to its extrusion at the surface, it will be called a dolerite. The mineral composition of dolerite is close to that of basalt, but since dolerite rocks have solidified more slowly within the crust, they are therefore more crystalline than are the erupted basaltic rocks.

The name "basalt" has an uncertain origin according to Albert Johannsen (1931). It could have come from the Greek word meaning "touchstone", due to its shiny appearance and its use in engraving coins. A historical naturalist, Pliny the Younger, has written that most statues from Egypt were carved from basalt, and the word “beechen” means “hard” in Egyptian. The name basalt also could have originated from the Hebrew word “barzel” meaning iron, since basalt is hard and heavy, just like iron.

It was in the 1960’s that several scientists started to study sea floor rocks in relation to their geological settings. For example, Engel and Engel in 1963, 1964, and later Hekinian (1968) and Kay et al. (1970) emphasized the chemical and mineralogical differences between the various basaltic lavas from the sea floor. In addition to their differences from most subaerial rocks, basalt samples also show compositional variation in relation to their geological locations. For example, samples of basalt samples collected from seamounts and islands in oceanic basin areas are seen to differ from the basalt lava erupted on accreting ridge systems. These authors coined the term *“alkali basalt”* for the volcanic rocks from islands and elevated seamounts and *Mid-Ocean Ridge Basalt* (MORB) to define the rocks formed on spreading ridge systems.

One of the first comprehensive papers dealing with the terminology of basalt and the inferred origin of these newly discovered oceanic rocks was published by Engel and Engel (1963) in “Science”. At about the same time, in 1964, I was a student and also working on oceanic basalt samples at Colombia University. My adviser and professor, Arie Poldervaart, was excited about some preserved glassy samples that I had found in sediment cores from the Indian Ocean when I was looking at the samples in the core-laboratory of the Lamont-Doherty Geological Observatory. The pebble-sized glassy fragments, which had been quenched at contact with seawater, represented a melt issued directly from the partial melting of mantle material. It was on the basis of chemical analyses as well as on the basis of mineralogical studies that the two major types of sea floor-generated basalt, called MORBs (Mid-Ocean Ridge Basalt) and alkali basalts or Intraplate Oceanic Basalt (IOB) were identified on a global scale.

Mid-Ocean Ridge Basalt (MORB)

Based on their mineralogy, chemistry and geological environments, a simplified term, “Mid-Ocean Ridge Basalt” (MORB), was adopted by marine scientists to define the exceptional character of these volcanic rocks and to distinguish them from their land-based counterparts. The MORB is also called a *tholeiite*, named after the German town of Tholey, where they were first collected and studied by land geologists.

MORBs are subdivided into three types: the **N**ormal (N-MORB), **T**ransitional (T-MORB) and **E**nriched (E-MORB). These classifications are based on their bulk rock’s chemical composition in relation to their degree of enrichment in “incompatible compounds”, such as the Large Ion Lithophile Elements (LILE), for example, potassium (K), sodium (Na), zirconium (Zr), niobium (Nb) and yttrium (Y), which do not easily enter into a mineral lattice.

The normal Mid Ocean Ridge Basalt (N-MORB) is deprived in calcium and potassium oxide ($Na_2O + K_2O = < 3$ %), and has various chemical ratios such as potassium to titanium ($K/Ti = < 0.15$), zircon to yttrium ($Zr/Y = < 3$), as well as chemical contents such as Zr (50–100 ppm), Y (20–40 ppm), strontium (Sr = 50–120 ppm) and niobium (Nb = < 4 ppm) when compared to Transitional MORB (T-MORB). The Transitional MORB has intermediate values of $Na_2O + K_2O$ (<3–3.5 %), K/Ti (<0.14–0.25), Zr/Y (<2–4) in comparison to the Enriched rocks (E-MORB), which contain higher $Na_2O + K_2O$ (>3.5 %), K/Ti (>0.25), Zr/Y (1–2), Zr (120–200 ppm), Y (30–60 ppm) and Nb (7–10 ppm).

Alkali Basalts/Intraplate Oceanic Basalts (IOBs)

Alkali basalts are also called Intraplate Oceanic Basalts (IOBs) and these types of basalts are found on islands and hotspot generated seamounts along volcanic chains or in ocean basins. They differ from MORBs since the IOB (also known as "hotspot basalts") have a different chemical composition for their clinopyroxene and plagioclase minerals.

The normal Mid-Ocean Ridge Basalt (MORB) contains calcium enriched and sodium–potassium depleted plagioclase and clinopyroxene and is less enriched in incompatible elements than an IOB. In comparison, the alkali basalts from intraplate (IOB) regions and hotspots are enriched in incompatible elements and have higher contents of LILE (Large Ion Lithophile Elements) such as $Na_2O + K_2O$ (3.5–5 %), K/Ti (>0.45), Zr/Y (5–16) Zr (300–400 ppm), Y (25–30) ppm, Sr (600–1000 ppm) and Nb (15–140 ppm) than do the MORBs.

The IOB rocks, which are enriched in LILE, could be the result of the partial melting of a different mantle source than that of the MORBs. Hoffman and White (1982) have suggested that enriched melts are due to the mixing of sediment and mantle material as a result of recycling the subducted oceanic lithosphere within the mantle (See Chap. 2). Since the continental crust is made up essentially of silica and alkali enriched material, the processes of selective partial melting and fractionation of mantle material producing the IOB in today's modern-ocean could be related to the events that once gave rise to ancient continents during the geological time-span of our planet's formation.

Different origins for the MORBs and IOBs could be a valid theory for explaining the chemical differences observed between the erupted volcanics found in diverse geological provinces such as spreading ridges or ocean basins. However, this idea is more controversial for explaining small-scale geochemical variations. For example, when MORBs and alkali basalts are extruded on the same site and/or close to each other in the vicinity of the same volcanic structure (<100 km apart) during cyclic volcanic eruptions, it is unlikely that separate melt sources could account for the differences observed. In this case, the differences between alkali basalts and MORBS erupted in the same area are probably due to the partial melting of different source components (minerals).

Model for Explaining Small-Scale Magma Heterogeneities in Sea Floor Basalt

Most models suggested that the compositional variations in basalt, such as the difference between N, T and E- MORBs, are due to the partial melting of a heterogeneous mantle source and the passive upwelling of melt in the lithosphere (Yoder 1976; McKenzie and Bickle 1988; Plank and Langmuir 1992). It was also suggested that compositional differences could result from fractional crystallization and a subsequent mixing in shallow level magma chambers (e.g. Christie and Sinton 1981). Although these models are valid for explaining some large-scale (spatial and temporal) compositional variability, they do not provide a satisfactory answer for the production of the different types of basalt extruded in a limited area (only a few tens of meters distance between sampling sites) nor can these models explain the differences noted in basalt extruded in small batches underneath the ridge axis.

From 1981 to 1984, several French cruises (*SeaRise*, *Clipperton*, *Cyatherm*, *Geocyarise*) carried out on the EPR near 13 and 11°N revealed the diversities in composition of volcanic products erupted at relatively short distances from each other (See Chap. 7). Basalts collected on the ridge axis as well as on off axis provinces located at less than 20 km from each other were found to differ in their chemical composition. The extrusion of the diverse types of MORBs as well as alkali basalts, all of which were found on the same edifice as well as at short distances (<2 km) between sampling sites, was very puzzling. It was difficult to explain the reasons for these differences simply by means of the classical views referring to partial melting of a heterogeneous mantle or on the basis of the fractional crystallization processes taking place inside magma chambers. Indeed, based on the chemistry of their incompatible elements, each individual type of basalt seemed to indicate a different parental lineage.

Therefore, in order to explain the diversity of MORBs and alkali basalts found in the same area, Hekinian et al. (1989) and Bideau and Hekinian (1995), proposed an alternative hypothesis to the one elaborated by Hoffman and White (1982) and others. The hypothesis of Bideau and Hekinian (1995) is based on a multi-stage melt extraction model, which involves several steps of magma extraction and accumulation above the melting region of a heterogeneous mantle source. The model assumes that magma supply is discontinuous, or in other words, that there are periods with a lack of magma supply, which agrees with the observed cyclic nature of undersea volcanism. The *multistage melting and extraction theory*

(Bideau and Hekinian 1995) also involves a cyclical production of basaltic magma, when similar processes would be repeated several times under similar conditions. In the field, this is observed by compositional zoning, which reflects cycles of magmatic activity. Such cyclic magmatic events have been observed during the detailed morphological and systematic sampling of volcanic outcrops on volcanoes from the Pitcairn hotspot (Hekinian et al. 2002), on the MAR spreading center near 34°55′N (Hekinian et al. 2000) and elsewhere on the EPR near 13°N (Hekinian et al. 1989).

Thus, the diversity of basalts found in a given area can be explained by the presence of small-scale mantle heterogeneities. The small-scale variability includes the mineralogical diversity of the mantle material. This newer multi-stage mixing theory involves the partial melting of a heterogeneous composite-mantle made up of spinel-lherzolite containing olivine (49–55 %), orthopyroxene (25–29 %), clinopyroxene (18–21 %) and spinel (1–2 %) (Bideau and Hekinian 1995). This composite mantle, with lenses of clinopyroxenite (>80 % clinopyroxene) in a lherzolite matrix, will first give rise to an alkali enriched melt (having the composition of enriched MORBs and alkali basalts), which accumulates in the warm lithosphere. As melting continues, the higher temperature phases will be extracted giving rise to a depleted melt of residual MORBs (N-MORB and T-MORB types). The mixing of the two types of melts (alkali-rich and residual) inside the magma chamber will subsequently give rise to the various types of MORBs as well as to the alkali basalts mentioned above. This was also called the "3-M Theory", which stands for Mantle Melting and Mixing.

Silica-Rich Lava and Obsidian

It has been thought that silica-rich rocks (such as rhyodacites or felsites) are found mainly on continents and ancient subaerial regions such as islands and continental landmasses. In fact, silica-rich lavas are close in composition to granitic rocks forming continental crust but they are also erupted on the sea floor associated with various types of geological environments. They have a relatively high viscosity and thus have greater tendency to erupt explosively than their mafic counterparts such as basaltic flows. Hence, the emplacement of silica-rich lava is often associated with island arcs in subduction zones and with mantle plumes forming hotspot volcanoes. Silica-rich lavas are rarely found on accreting ridge systems except when associated with areas influenced by mantle plume upwelling. Examples of silica-rich lavas associated with mantle plumes giving rise to hotspot volcanism are found in Iceland, in the Azores (Faial Island), in the Galapagos islands, as well as on the "Axial volcano" of the Juan de Fuca Ridge (Wanless et al. 2010). Because of their highly viscous nature when compared to basaltic flows, silica-rich lavas are more limited in the extent of their dispersion on the ocean floor (see Chap. 7).

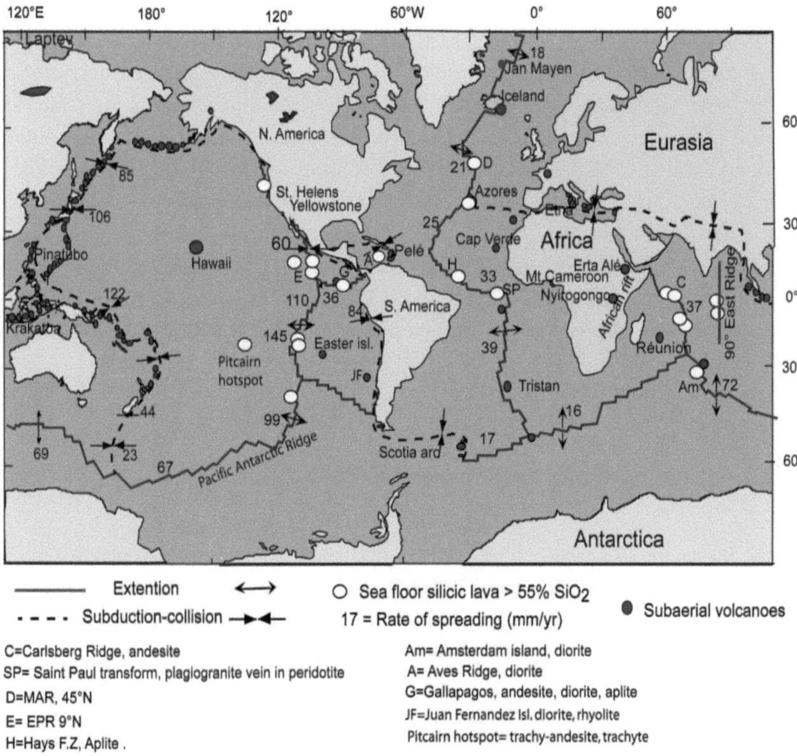

Fig. 4.2 World distribution of silicic lava (*white circles*) collected from the sea floor and from subaerial volcanic sites (*filled dots*). Most subaerial volcanism contains silica-enriched flows

Silica-rich flows are less abundant than basaltic lavas, but it is often difficult to distinguish them from other flows on the ocean floor because silica-enriched lava forms a relatively small fraction of any volcanic structure (only a few tens of cubic meters up to 1 km^3) (Fig. 4.2). Today, with more detailed sampling of our ocean floor, it has become obvious that these types of rocks occur in all the world's modern oceanic environments. Thus, it is important to find out how such types of rocks are formed in relation to our present day oceanic crust.

Oceanic silica-rich lavas differ from subaerial lavas by their thick glassy crust and their form or surface morphology. Thus, field observation helps to determine some morphological criteria in order to distinguish them from basaltic flows. Because of their high viscosity, they are more elongated with smoother and/or a more scoriaceous surface than N- MORB flows. They rarely show corrugated ridges on their cooling surfaces and are mainly characterized by thick (5–20 cm thick) glassy, aphyric margins with large cavities (1–2 cm long) caused by gas release during volcanism. Some surfaces appear scoriaceous due to the coagulation (welded glassy material) of the glassy crust.

To my knowledge, the only visual observations of silica-rich lava reported from the ocean floor are from the Pitcairn and the Society hotspots and from the south

Pacific-Antarctic Ridge (PAR) spreading axis (Hekinian et al. 1999; Hekinian et al. 2002; Stoffers et al. 2003). Most of these flows are blocky and/or form giant lobated tubes with flattened tops. They have conchoidal fractures and elongated vesicles (see Chap. 8). Due to their high viscosity and consequent low rate of diffusion of ions to the potential sites of their crystal lattice, highly siliceous lavas commonly solidify as black obsidian rather than as a crystalline rock. The viscosity of ascending magma will increase as solidification or crystallization takes place. Therefore, near the surface of the rising lava, the flow will decrease its velocity of ascension and form domed structures, usually enriched in silica (in the form of obsidian). Silica-rich lavas are responsible for the formation of volcanic "pipe" intrusions similar to what have been observed on Mont Pelée on the Island of Martinique.

Composition of Silica-Enriched Rocks

Several types of silica-rich rocks consisting of trachytes, trachy-andesites, trachybasalt, andesites, dacites and rhyolites occur in the different oceanic environments such as hotspot volcanoes and spreading ridges. Some are classified as being silica and incompatible-element enriched rocks, or *Trachytic rocks*.

T*rachytic rocks* have the highest SiO_2 (59–64 %) and total alkali content ($Na_2O + K_2O$ = 8–11 %), followed by *trachy-andesites* (SiO_2 = 53–59 %, $Na_2O + K_2O$ = 4–8 %), and *trachybasalts* (SiO_2 = 49–53 %, $Na_2O + K_2O$ = 4–7 %). Other silica-enriched rocks are classified as being *rhyolitic rocks* because they are silica enriched and incompatible-element depleted rocks. They have a SiO_2 content of 53–59 % and $Na_2O + K_2O$ contents of 3–5.5 %. They could be dacite (SiO_2 = 59–66 %, $Na_2O + K_2O$ = 3–5 %), andesite (SiO_2 = 52–59 %, $Na_2O + K_2O$ = 3–5 %) or rhyolites (SiO_2 = 70–73 %, $Na_2O + K_2O$ = 5–6 %) which are most commonly found on spreading ridges, island arcs and back-arc basins.

The major differences between the two types of silica-enriched rocks (trachytic and rhyolitic) are in their incompatible element contents for approximately the same amount of silica. However, the andesite-dacite suites from back arc basins are known to be different from common MORBs and IOBs. Previous detailed studies (Turner and Verhoogen 1960; Myashiro 1974) have claimed that the island arc andesites consist essentially of calc-alkaline suites enriched in augite (a calcium-alumina-silicate mineral) and hypersthene (an iron alumina silicate mineral), pyroxenes, plagioclase phenocrysts and with a high alkali content when compared to the tholeiitic basalt suites found on Hawaiian and Icelandic volcanoes.

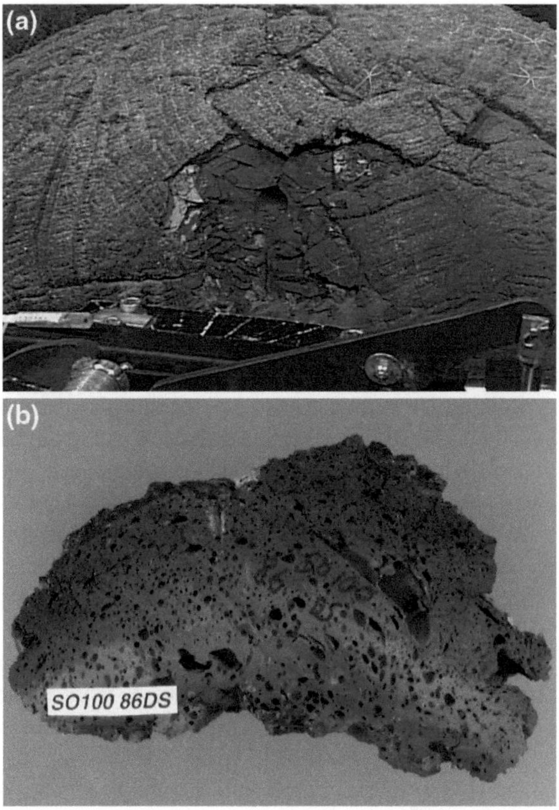

Fig. 4.3 Photographs of silica-enriched lava flows. **a** Bottom photograph taken by the submersible Alvin on the axial graben of the East Pacific Rise near 9°N (Courtesy of M. Perfit) shows a rounded (lobated) silica-rich lava flow covered with a thick (>2 cm) glassy crust. **b** Sample of silica-rich lava dredged (SO100-86-2 and SO100-91-01) from the axial graben of the PAR (Pacific Antarctic Ridge) near 38°04.94′S–110°59.83′W at 2193–2200 m depth shows large gas cavities

While alkali-enriched volcanics having a hotspot origin and back-arc volcanics are both well documented in their silica-enriched rocks, very little is known about similar lava erupted along the spreading ridge systems of the World. In the East Pacific, only a few cases have been documented, such as on the flank of the North East Pacific Rise (NEPR) at 9°30′N and 10°30′N (Thompson et al. 1989; Batiza et al. 1996) and on the South East Pacific Rise (SEPR) between 17 and 21°S (Fig. 4.3a, b). Silica-rich lavas were also recovered from the Pacific-Antarctic Ridge (PAR) near 37°35′S (Stoffers et al. 2003; Hekinian et al. 2000) (see Chap. 7). Other andesites were recovered from the 95°W propagator of the Galapagos Spreading Center (Clague et al. 1981) and from the Juan de Fuca region (Wanless et al. 2010).

Origin of Silica-Rich Lavas

The origin of viscous silica-rich lava is not well understood. These rocks could be the result of crystal-liquid fractionation during partial melting of the lithosphere and/or crystal-liquid fractionation processes in a magma reservoir. Another alternative hypothesis is that partial melting of hydrated (altered) oceanic crust produced low temperature (<1000 °C) mineral assemblages.

The bimodal basalt plus silica-enriched lava association during volcanic eruption could either be due to a common origin and/or to separate melting source components. Marsh et al. (1991) suggested that while small portions of Hawaiian silica-rich lava found in association with the basalt could be the result of fractional differentiation, in comparison, considerable amounts of rhyolites have been extruded in association with other Icelandic flows. The origin of these silica-rich flows might be tied to that of the associated, less evolved lavas of basaltic composition. The heat required to produce silica-enriched lava is lower than what is necessary for basalt. Hence another alternative hypothesis is that the silica-rich lavas are derived from an independent source with a lower degree of partial melting.

Support for both of these interpretations is given by the isotopic composition of basalt plus silica-rich lava suites. Indeed, the neodymium (Nd) and strontium ratios of some basalt-silica-enriched lava associations (such the suites found on the SEPR Pacific-Antarctic Ridge and in some intraplate regions) agree with the hypothesis of a common parent for the basalts and the silica-rich lavas. Other cases, such as the Icelandic suite, might not be due to a common origin but rather each one of the associated pairs could have been derived from its own parental melt.

The criteria for recognizing lithospheric melting are: (1) the presence of disequilibrium minerals derived from previous periods of solidification and remelting (anatexis) and (2) the inclusion of solids from the melt source. The solidus temperature of silica-rich lava is lower than the liquidus for basalt, which could supply heat during crystallization. During partial melting and melt segregation, small amounts of solidification during magmatic ascent could provide heat to the shallower lithosphere and thereby lower the temperature of the solidus. This will be more significant where the oceanic crust has thickened during intense episodes of volcanism. Also, it is important to keep in mind that the most evolved silica-rich lavas of dacitic and rhyolitic composition are too low in their compatible element concentrations to have been originated directly from mantle-derived melts.

Experimental work on Fo (the Forsterite content of peridotite)-Di (Diopside content of clinopyroxene)-Qz (Quartz) and H_2O at pressure = 20 kilobars (Kushiro 1969) has shown that if the liquid is saturated with an excess of H_2O, it will give rise to a rhyolitic melt at 950 °C. As soon as a liquid of rhyolitic composition is removed and if melting continues, eventually the remaining melt will move towards the Enstatite-Diopside field on a Kushiro (1969) diagram and give rise to basaltic liquid. This is an elegant way of explaining the formation of both silica-rich and basaltic liquid from the same parental melt source. This source could be a mantle lherzolite, which has undergone several degrees of partial melting. If we use this hypothesis, it is likely that the strontium isotopic ratios of both extruded basalt and silica-rich lavas should be comparable.

Based on the positive correlation of the ratios of radioactive isotopes of $^{87}Sr/^{86}Sr$ and $^{206}Pb/^{204}Pb$ in an area of the Foundation Seamount Chain, Haase et al. (2005) have speculated that mixing between a plume that is enriched in radioactive components and some non-radioactive upper mantle material could also have occurred. The plume has carried deep-seated material made up of products enriched in LILE elements from recycled sediment found in subduction zones. The basalt and the silica-rich lava of the Foundation Seamount Chain is more enriched in the $^{206}Pb/^{204}Pb$ ratio than that of the magma erupted on the PAR (Pacific Antarctic Ridge). This suggests that the lava in these two provinces originated from different sources. In addition, it was shown that the basalt from both areas lies on a linear mixing trend.

On the contrary, in the PAR, the andesite and dacite (silica-rich lava) have a higher $^{87}Sr/^{86}Sr$ ratio for a given $^{206}Pb/^{204}Pb$, which might indicate the assimilation or a contamination of material enriched in the $^{87}Sr/^{86}Sr$ ratio. Seawater has a high $^{87}Sr/^{86}Sr$ ratio (0.709), hence it is not excluded that the assimilation of altered oceanic crust (metamorphosed to the greenschist and/or amphibolite facies) is responsible for this type of enrichment. Haase et al. (2005) estimated that about 25 % metamorphic crust would be necessary and would need to be assimilated in order to produce silica-rich lava in the PAR.

Metamorphosed amphibolite is often found in oceanic crust as well as in ophiolite complexes, and this product has a high distribution coefficient for Nb (niobium) relative to La (lanthanum). A melt of amphibolite composition would yield a relatively low Nb/La ratio. Haase et al. (2005) have shown that two separate fields of Nb/La ratios can exist for the basalt-silica-enriched lava (andesite-dacite) association: one with low values (Nb/La < 0.9) and the second with higher values (Nb/La $= 1.0$–1.2). These different fields also coincide with the locations of the two basalt-silica-enriched flows. Those with lower Nb/La values are found on the ridge segments south of latitude 39–40°S and the others, with higher Nb/La values, occur at 37–38°S.

It is also speculated that the partial melting of heterogeneous mantle sources having variable Nb/La ratios influenced the two different ridge segments. Furthermore, the relatively high chlorine content, and Cl/K ratio is another indication supporting the assimilation of altered (metamorphosed) oceanic crust for producing these lavas. In conclusion, silica-rich lavas have probably originated from a combination of partial melting at the source, plus crystal-liquid fractionation and the assimilation of previously altered oceanic crust. It is also likely that silica-rich lavas do not have the same parent melt as do other basalts found on the ocean floor.

Gabbroic Rocks

Gabbroic rocks constitute another important component of the oceanic lithosphere and probably comprise about 20 % of the exposed outcrops on the ocean floor. They represent the coarse grained equivalent of a basalt which has solidified in a magma reservoir and/or conduit. Along with the dolerite-dyke components of the oceanic crust, gabbroic rocks are very significant because they suggest the presence of a magma chamber. Gabbros differ from the dolerite formed in dykes due their grain size (larger crystals) and texture. Because gabbros are generally formed in a warmer and more homogeneous temperature environment than are the basalt and dolerite, they tend to develop larger mineral grains with well-defined crystal outlines and they generally lack a matrix.

The importance of studying gabbroic rocks is that their texture and mineral composition give a great deal of information on the dynamics of crystal-liquid fractionation. The gabbroic complexes show layered structures due to the sequential cooling of a magma reservoir. The heavy and first-formed minerals such as olivine and spinel drop to the bottom of the magma chamber while the lighter minerals such as plagioclase and clinopyroxene will be crystallized near the top. Many gabbroic complexes show stratigraphic sedimentation due to crystal-liquid fractionation. Indeed, when ophiolite complexes are exposed in subaerial environments, the sequential variation of layered gabbroic units is best exposed. The base of the complex is marked by the presence of an olivine-spinel enriched layer (dunites), covered by wehrlite, then massive olivine gabbros, gabbronorite and ferrogabbros.

On the ocean floor, it is difficult to have access to all these sequences because there is a lack of continuous observations. However, on the basis of drilling and on-site observations in submersibles, it has been possible to draw an analogy to subaerial observations. The lighter and most evolved gabbros are called "anisotropic gabbros" and consist of Fe–Ti enriched ferrogabbros as well as silica (plagio-granite) gabbroic rocks. The most obvious cumulates are the dunites, wehrlite and pyroxene-olivine enriched gabbros.

Most gabbroic rocks are exposed in fracture zones and are rarely found along spreading ridge segments (Dixon et al. 1986; Davis and Clague 1990). Occasionally, gabbros occur on the uplifted walls of magma-starved ridge segments such as on the slow spreading ridge of the south Atlantic and in the Indian ocean in an area where the crust is thin and close to a ridge-transform (fracture zone) intersection. On the ocean floor, many gabbroic rocks are found associated with serpentinized peridotites. Often veins and veinlets of gabbroic rocks are found as intrusive bodies running sub-parallel to the foliation of peridotites. Usually these intrusive bodies have a heterogeneous and wide range in their mineral composition when compared to the more massive gabbroic complexes forming layered bodies issued from normal crystal liquid-fractionation. This was inferred to be due to continued compositional change as the percolating liquid reacts with the surrounding solid rocks. Sometimes gabbros are metamorphosed and often show streaky layers of oriented hydrated minerals around lenticular preserved minerals. This type of gabbro is also called "flaser gabbro", since the word "*flaser*" in German means "streaks". These streaky gabbros are formed during tectonic-stress generated activities.

It is not always clear whether the gabbro-peridotite-basalt complexes that are found together are interrelated due to the crystal-liquid fractionation process. The gabbro-peridotite could also be emplaced *a posteriori* during a forceful injection or during thrust-folding tectonic processes. Detailed mineralogical and geochemical analyses are often required in order to determine their mode of emplacement.

Massive and thick-layered gabbroic units, which are comparable to subaerial ophiolite complexes, are found in deep-seated troughs such as the Terevaka transform fault, the Pito Deep and the Hess Deep in the Eastern Pacific as well as in the Atlantic (Gorringe Bank, Vema transform fault). These gabbros are tabular blocks with angular contour lines and show a granular texture. Usually these massive gabbros are olivine gabbros, which are found in the lower part of the gabbroic complex. These massive gabbros are more homogeneous in composition than the gabbros observed intruding into the peridotites. The massive gabbros form layered complexes going from evolved ferrogabbros and/or isotropic gabbros to olivine gabbros and plagioclase dunite cumulates. Gabbroic cumulates are formed by crystal fractionation which accumulates in the lower part of a solidifying magma pool. The cumulate crystals have well-defined separate crystalline outlines. The crystals are large in size, varying from 2–3 cm in diameter to a few millimeters across. The more fractionated types of gabbros, such as ferrogabbros and plagioclase-enriched gabbros, abound in the upper part of the gabbroic complex as well as forming intrusions within peridotites.

The order of mineral crystallization for gabbroic rocks is spinel, followed by olivine, plagioclase, clinopyroxene and ilmenite. This is in agreement with the order of crystallization observed in basaltic rocks.

In addition to land-based ophiolite complexes, deep sea drilling has provided other direct evidence on the stratigraphic relationship of gabbros with other units. It was during the *GLOMAR CHALLENGER* leg 37 on the Mid-Atlantic Ridge near 37°N–34°W at about 30–40 km west of the Mid-Atlantic Ridge rift valley (Aumento and Loubat 1971; Melson et al. 1977) that a basalt-gabbro-peridotite complex was drilled at 2619 m depth. It is at this site (# 334) that 50 m of basalt overlaying 68 m of gabbro and peridotite were drilled (Clarke and Loubat 1997). The drilled core penetrated 376 m into the crust and recovered 254 m of sediment and 118 m of hard basement rock.

Based on textural and compositional variations, the major gabbroic sequences observed in the oceanic environment consist of Fe-bearing gabbros (ferrogabbro), isotropic gabbros, gabbronorite, olivine gabbro, troctolite, werhlite and dunite. Ferrogabbros consist of plagioclase (an $_{30\text{-}55}$), clinopyroxene and Fe–Ti magnetite—ilmenite. Isotropic gabbros are crystalline rocks with a homogenous crystal orientation. Gabbronorites consist of olivine, calcic-pyroxene and plagioclase (anorthite content of $an_{55\text{-}75}$). They could contain orthopyroxene but some gabbronorites do not have any of this mineral. Olivine gabbros are coarse-grained rocks made-up essentially of olivine, plagioclase, clinopyroxene and othopyroxene. They have a plagioclase composition of $an_{65\text{-}80}$, calcic-clinopyroxene and olivine with $Fo_{75\text{-}87}$. Troctolites have a patchy green and white appearance and are a variety of gabbroic rocks composed of plagioclase (anorthite-bythonite) and olivine-pyroxene. They are either cumulates from a gabbroic complex and/or late impregnated melts. Wehrlites differ from dunites by their higher content in pyroxene, diopside and clinopyroxene. The olivine is commonly serpentinized. Dunites are cumulates with an equigranular texture (same-sized crystals) consisting essentially of magnesian-olivine ($Fo_{89\text{-}94}$) altered into serpentine and orthopyroxene. Chromites, which are dark colored minerals of iron, magnesium and aluminum oxydes, are disseminated throughout the dunite cumulates or form independent units.

Peridotites

Peridotites are green colored rocks that have the closest composition to some meteorites (achondrites) coming from space, probably due to the disintegration of older planets. Peridotites have the same composition as the most abundant compounds on Earth and consist of silicates of magnesium, iron and calcium, forming the minerals of olivine, pyroxene, spinel and garnet. Peridotites probably make up about 50 % of the Earth's mantle and they are found exposed both on land and under the sea.

The first extensive dredging operation, to collect peridotites on a spreading ridge was carried out in the summers of 1947 and 1948 with the RV *ATLANTIS* under the leadership of Maurice Ewing. Several hundred pounds of rocks from various parts of the Atlantis fracture zone on the MAR at 500–2500 fathoms were collected and later studied by Professor Shand from Columbia University (Shand 1949). The importance of these dredged rocks became obvious when scientists noticed the presence of some deep-seated peridotite, which was thought to form the Earth's mantle and was believed to be the parent melt as well as a residue of the basalt that covers the major part of the modern sea floor. Peridotite, found in the lower crust and upper part of the Earth's lithosphere, was formed from a combination of melting and crystallization processes during the ascent of solidified minerals (mainly olivine) and melt beneath spreading ridge segments.

Two types of peridotite are found on the ocean floor: lherzolites and harzburgites, which take their names from the locations in Europe where they were first studied (See glossary). The harzburgites are considered to be the residue left after the partial melting of a mantle peridotite. They remained after the melt separated from the mantle and circulated upward before giving rise to basaltic volcanism. Residual harzburgite is quite close in composition to the original melt prior to melting and consists essentially of olivine ($Fo_{85\text{-}95}$), orthopyroxene, spinel and variable amounts of clinopyroxene (diopside-augite). During partial melting, clinopyroxene (with the lowest melting temperature) is the first mineral to melt and rise towards the upper part of the lithosphere where it could supply a magmatic reservoir. Unaltered peridotite, called lherzolite, consists essentially of olivine, which is a dense silicate containing iron and magnesium, plus orthopyroxene (a similar mineral but less dense) and clinopyroxene containing aluminum and calcium. The minor mineral constituents of all peridotite rocks are spinel and/or garnet, which consist of chromium, magnesium, aluminum and iron oxide.

The importance of peridotite as a main constituent for the source of basaltic volcanism was revealed during experimental laboratory work on the melting of a synthetic peridotite, equivalent to an eclogite, a granular rock composed essentially

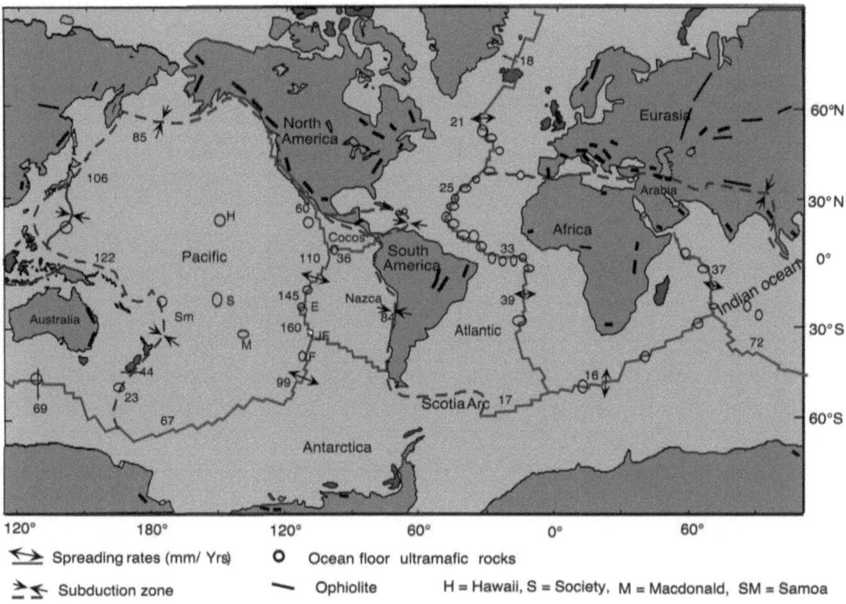

Fig. 4.4 World map distribution of ultramafic rocks from the sea floor shown by empty circles and the land based ophiolite deposits containing peridotite are indicated by heavy lines. The numbers indicate the total spreading rates of the ridge segments

of pyroxene and garnet, such as spinel-peridotite and garnet-peridotite. The rock was melted at temperatures near the solidus up to 2000 °C above and within the pressure range of 0–15 kilobars (55 km depth) during experiments carried out by Green and Ringwood (1967). The results of these studies have shown that with decreasing pressure, rock melting can and will take place, and the portion of the melt obtained experimentally was similar to basaltic lava erupted on the sea floor. This melting experiment yielded tholeiite basalt (similar in composition to a MORB), at a pressure less than 15 kilobars corresponding to about 47 km depth in the lithosphere, and alkali-olivine basalt, at pressure greater than 15 kilobars. Garnet-peridotite is stable only in the upper mantle, and it could be yet another source for alkali-basalt melts (Fig. 4.4).

Any peridotite exposed on the sea floor is unstable and alters into serpentinite (a hydrated aluminum silicate) during interaction with seawater. The most common peridotites (harzburgites) to be found are the residue of the partial melting of mantle material. This residue is the result of different amounts of partial melting going from 2 % for the most enriched melts in incompatible elements up to about 20 % for the least-enriched melts. Serpentinization (alteration of peridotite) essentially gives rise to two types of minerals: lizardite and antigorite (Aumento and Loubat 1971).

The alteration of peridotite and other ultra-mafic rocks takes place under low temperature hydration of the mafic minerals, at probably less than 500 °C. For example, the clay mineral sepiolite (Bonatti et al. 1983) was formed under experimental conditions from the alteration of peridotite at a temperature of about 174 °C. Peridotite containing olivine fosterite ($2Mg_2SiO_4$) is transformed into serpentinite in the presence of water through the following two reactions:

1) Peridotite $+2H_2O$ = Mg2 + (release) $+SO_2$ (release) $+H_2O$
2) $2Mg_2SiO^{4+}$ (forsterite) $+3H_2O$ = $Mg_3Si_2O_5\ (OH)_4$ (antigorite) +MgO (OH) (brucite).

During the serpentinization process, the fluid becomes enriched in silica (SiO_2) as well hydrogen (H_2). The acidity (low pH) of the fluid will further increase during the oxidation of the ferrous iron-bearing components in the minerals. This is illustrated for the Fe-rich component (ferrosilite) of the pyroxene, which gives rise to magnetite through the following reaction: $3FeSiO_3$ (ferrosilite) $+H_2$ = Fe_3O_4 (magnetite) $+H_2$ (aqueous) $+3SiO_2$ (aqueous).

Hydrogen and silica release during the alteration of peridotite also increases the acidity of the fluid so it becomes more corrosive and facilitates the leaching of metals from the rock (see Chap. 6). The density (3.30 g/cm^3) of fresh peridotite is higher than that of the basalt, however when it is altered (serpentinized), its density (2.550 g/cm^3) decreases below that of basalt (Christensen 1978; Miller and Christensen 1997). Thus, serpentinization is accompanied by a decrease in the weight of peridotite.

Emplacement and Distribution of Peridotite

The emplacement of residual and/or the un-fractionated mantle peridotite has been found in different environments such as on the rift walls and rift-mountains of magma-starved spreading ridge segments. It is also emplaced in fracture zones and their transform zones where deep slices of the crust-upper mantle have been exposed (Fig. 4.5). The classic example of oceanic peridotite exposure is when tectonically governed processes lead to the formation of a sea floor basement composed of mantle-derived peridotites and associated mafic intrusions. In diverging plate boundaries at ridge spreading, which involves tectonic stretching, the upper mantle peridotite is drawn upwards where it will undergo partial melting. The emplacement of peridotite is also facilitated by the degree of serpentinization due to magmatic fluid and seawater circulation in the lithosphere.

Hess (1955) postulated the idea that seismic layer 3 (located underneath the basalt-dyke-gabbroic complex of the oceanic crust, which is in the lower crust) consists of serpentinized peridotite. As mentioned before, serpentinization is accompanied by a decrease in density of the serpentinized peridotite so it becomes lighter. The upper mantle will rise during convection, but when cooled below 500 °C, it will become hydrated and give rise to the serpentinization of peridotite.

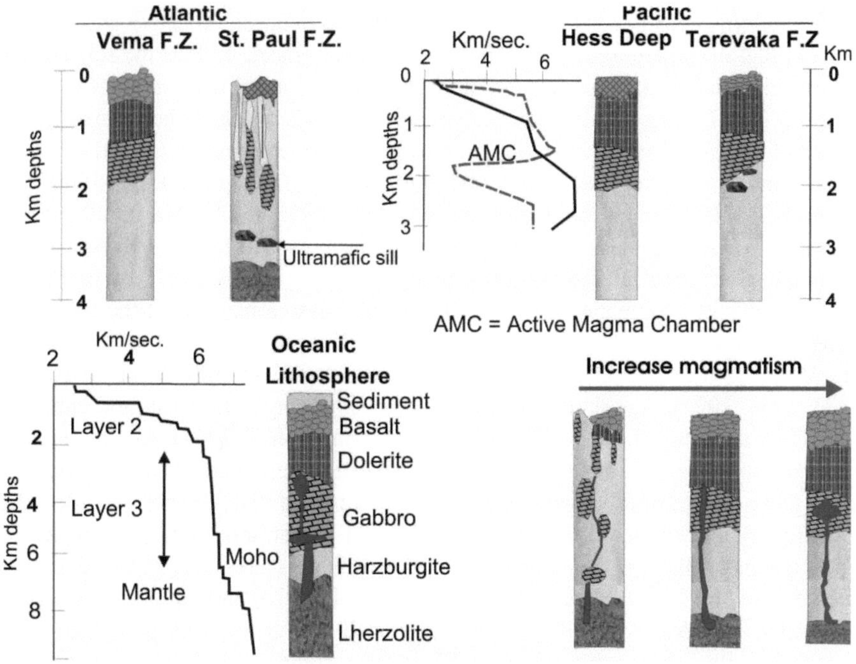

Fig. 4.5 Stratigraphic section showing the various types of lithology of the Earth's upper lithosphere exposed in deep troughs of the Atlantic and Pacific oceans. The Vema fracture data is reported from Auzende et al. (1989). The St. Paul F.Z. (see Chap. 8 on the St. Peter and St. Paul Rocks Transform Fault) section is reconstructed after data obtained on a small Intra-Transform-Ridge (ITR) representing a magma-starved spreading center (Hekinian et al. 2000). The Hess Deep represents a section of 1 million year old East Pacific Rise exposed north of the Galapagos Islands and the Cocos-Nazca spreading ridge in the northeastern equatorial Pacific (Francheteau et al. 1992; Hekinian et al. 1993)

In addition to being lighter when compared to basalts, serpentinized peridotites also have low frictional strength and low permeability. The presence of such relatively soft complexes enhances the deformation of the material through which they intrude. Serpentinization facilitates detachment faulting, often accompanied by rock fragmentation and avalanches along slopes. It is believed that serpentinization occurs with a change in volume. The volume changes will depend on the degree of serpentinization and might be as much as 20–50 % of the bulk rock.

The exposure of serpentinized peridotites will provide a unique opportunity to study mantle-basalt relationships and the magma genesis of the most common rock types found on the sea floor. The tectonic constraints related to the mechanics of peridotite emplacement are important for understanding the lithosphere's deformation and its influence on melt circulation underneath spreading ridges. Escartin et al. (1997) suggested that serpentinization weakens the oceanic lithosphere and enhances faulting.

In modern oceans, ultramafic intrusions are also exposed on active and passive margin regions where two plates of different density meet, such as the seismically active margins of the circum-Pacific subduction zone's deep trenches where faulting and folding of drifting oceanic plates takes place. Other examples of ultramafic rocks such as peridotites and dunites were reported from the Tonga Trench in the southwest Pacific at 9,000–9,400 m deep (Fisher and Engel 1969) and in the Porto Rico trench in the western Atlantic (Shido et al. 1974). In subduction zones, down-sliding slabs of serpentinized peridotites will undergo large volume reduction due to the release of their volatile contents (mainly water, up to 25 %).

Examples of ultramafic complexes found on passive margins at the transition zone between oceanic and continental lithosphere were reported from the Galicia Bank during the ODP Ocean Drilling Program, legs 103, 149 and 173 (Boillot et al. 1988, 1995). Age dating on a drill-core containing both basalt and gabbroic rocks was determined to be 122–100 Ma (Charpentier et al. 1998). Another area in the most southern part of the Galicia Bank near the Iberian coast where peridotite and gabbroic complexes were sampled is found on the Gorringe Bank. The Gorringe Bank consists of two shallow seamount-like structures called Gettysburg and Ormond. The first indication of peridotite in this area was reported during dredging by the R.V. *R.D. CONRAD* (Gavashi et al. 1973) and the N.O. *J. Charcot* (Hekinian and Aumento 1973). The *Cyagor* diving cruise was organized with *Cyana* and its support ship *LE SUROIT* in May–June 1981. During the *Cyagor* cruise (Auzende et al. 1982), geologists observed a lithological sequence of the peridotite-gabbro-basalt association. The gabbroic succession above the peridotites was dated as being 143 Ma (Ar/Ar age data, Feraud et al. 1986). On the basis of the scientists' observations, it was suggested that an eastern tilt of the oceanic basement followed by a vertical uplift gave rise to the shallow topography of the Gorringe Bank. In a larger context, it was believed that the Gorringe Bank was uplifted during a major tectonic motion when the African plate under-thrusted the European plate in a southeast-northwesterly direction (Fukao 1972).

The presence of mantle peridotite on the sea floor will depend on the spreading rates, which are influenced by the extent of volcanic activity.

Slow Spreading Ridges. Peridotites associated with slow spreading centers (<3 cm/yr. total rate) along the northern MAR are best documented at 45–46°N (Aumento and Loubat 1971), at 36°15′N in the "Rainbow field" (German et al. 1996), at 37°N (DSDP Leg 37, Melson et al. 1977), 35°40′N (Gracia et al. 1997), 30°N (Blackman et al. 1998; Blackman et al. 2004), and 24°–24°40′N (Allerton et al. 1996) (Fig. 4.5). Several Russian-French, US, Japanese and German investigations have been carried out in the Cap Verde fracture zone (14°25′–14°45′N) and at 12°58′N–12°10′N in the rift valley (Matsumoto et al. 1998) during the last 15 years. These areas where nearly continuous outcrops of peridotite were observed are comparable to the Saint Peter's and St. Paul's Intra-transform spreading ridges (see Chap. 8) because they also contain mylonitized peridotite, pyroxenite and gabbro which suggest that the emplacement of peridotite and associated rocks must have occurred due to the forceful injection and squeezing of the upper mantle material during differential compressive and strike-slip plate motions. The short intra-transform spreading ridge segments located within the Saint Paul's multiple transform fault investigated by the *Nautile* in 1997–1998 showed the presence of a thin crust (probably less than 300 m thick) associated with large peridotite outcrops (Hekinian et al. 2000).

The presence of topographic highs at the intersection of the MAR and Atlantis transform fault near 30°N indicated that localized peridotite outcrops could be exposed during detachment faulting. This topographic high of peridotite is called the "Atlantis Massif" and is found at 1,800 m depth, 15–20 km west of the MAR axis on crust that is 1–2 million years old. Gravity anomaly measurements indicate that the lower crust and the upper mantle forming this outcrop are not symmetrically distributed (Blackman et al. 1998). This asymmetrical distribution could be the result of a detachment faulting of the serpentinized massif. From the deep tow survey (TOBI) it was shown that the massif is about 1.5 km in diameter (Blackman et al. 2004).

Geophysical studies such as magnetic, gravimetric and bathymetric surveys were undertaken during Russian cruises during the "Equaridge Project" by the R.V. *AKADEMIK NIKOLAI STRAKHOV* in 1988–1991 (Udintsev et al. 1996) in the equatorial Atlantic. The results of these cruises have shown that where peridotites are exposed, there is very little sign of volcanism. For example, at the MAR bordering the axial valley near 14°45′N, the rift mountain region shows almost continuous outcrops of serpentinized peridotite which suggests the presence of a thin volcanic crust and abundant mafic and ultramafic intrusions. This is also suggested by the presence of ultramafic rocks found within the rift valley, which are also associated with localized volcanic constructions. In this case, it is believed that localized ("puntiform") volcanic activities have created part of the oceanic crust exposed in the rift valley floor deeper than 3850 m, while other parts have been affected by the intrusion of ultramafics. Also, the presence of basaltic constructions on the rift mountain area of the MAR at various depths (3,000–1,600 m depths) indicates that alternate volcanic activities have taken place with the intrusion of ultramafic-gabbro associations during at least the past 2 million years (when assuming a half spreading rate of about 10 mm/yr). In addition, mass-wasted material consisting essentially of ultramafics and basaltic debris were derived from localized volcanic constructions bounding the rift valley to the east between 3,800 and 3,000 m at 14°45′N on the MAR. This area shows a thin sediment cover as well as fresh un-sedimented talus suggesting that tectonic activity is an important ongoing process with respect to the more ephemeral volcanic events.

Serpentinized peridotites are found along the wall of the axial graben as well as on the inside corner of the individual ridge segments, either associated with a transform or a non-transform discontinuity in slow spreading ridge segments. The fact that the rift mountain region bordering the axial valley shows almost continuous outcrops of serpentinized peridotite suggests that these rocks have been intruded at 12°40′N on the MAR. This intrusion is related to the cold lithosphere and its lack of magmatism. Although the exposure of peridotite is unrelated to depth, nevertheless it would make sense to look for more peridotite outcrops on magma starved ridge segments that are deeper than 3,500 m along the MAR.

Ultra-slow Spreading Ridge. The Gakkel Ridge in the Arctic Ocean is an example of an ultra-slow spreading ridge segment (11 mm/yr half rate) (Michael et al. 2003). Studies on other ultra-slow spreading ridges such as the southwest Indian Ridge (SWIR) (Chu and Gordon 1999) led to the suggestion that volcanism decreases as the spreading rate decreases (Karasik 1974). The SWIR near 41°S–49°W and 34°S–69°W has a mean depth of 4,700 m and a thin crust of 4–5 km (Muller and Jokat 2000; Meyzen et al. 2003). In 2001, scientists aboard the *Healy*, a US Coast Guard icebreaker, and on a German research icebreaker, the R.V. POLARSTERN, started the study of the ultra-slow spreading ridge located in the Arctic Ocean. The Arctic Mid-Ocean Ridge Expedition (AMORE) expected the Gakkel Ridge, where the spreading rate is only 1 cm (0.39 inches) per year, to exhibit little, if any, volcanic activity (Michael et al. 2003). The spreading rate on the Gakkel is about 20 times slower than that of the more frequently studied lower latitude ocean-ridge systems. This ridge extends 1770 km (1100 miles) from north of Greenland to Siberia. It is the deepest and most remote portion of the global mid-ocean ridge system. Because the spreading rate decreases progressively towards Siberia, it is expected that the amount of melting and magma production would also decrease away from Greenland towards the east (Michael et al. 2003). In fact, the central portions of the ridge showed virtually no volcanism and large faults have exposed pieces of the Earth's mantle directly on the sea floor.

Fast and ultra-fast Spreading Ridge. Very few areas of ultra fast spreading (half spreading (>60 mm/year) ridge segments were found to be associated with outcropping ultramafic rocks. This is because fast-spreading ocean ridges are volcanically very active with a large magmatic budget and with more steady state magmatic upwelling conditions than other slower spreading centers. However, even in the Pacific, a colder lithosphere occurs near equatorial regions and continues south to 13°30′S (Garrett transform) where larger transform faults are predominant (Fig. 4.5). Indeed, in the Equatorial Pacific the ultra-fast spreading ridge segments located between 3 and 9°S reach a maximum depth of 3,200–4,700 m near 3–5°S (Lonsdale 1989). This area is compatible to a probable exposition of lower crust and upper mantle peridotite.

A cold lithosphere in the south Pacific is also observed at the boundaries of the Easter and Juan de Fuca microplates (25 and 30°S respectively), which are characterized by transforms and magma-starved structures located at the ends of the ridge propagators (Pito and Endeavor deeps) (Naar and Hey 1986; Francheteau et al. 1988; Hekinian et al. 1995). The Terevaka transform fault, displacing the West Rift branch of a spreading ridge near 24°N–116°W, and the Hess Deep bordering the Cocos-Nazca Ridge in the Galapagos region near 2°14′N–101°33′W are other areas with exposed peridotites. The difference between the Garrett transform and the Hess Deep is that the Hess Deep peridotite is formed underneath the EPR spreading center while the peridotites from Garrett were formed on the EPR and then transported into the transform during the tectonics of spreading (see Chap. 8).

Some of the conditions for the emplacement of exposed peridotite are listed below. We may need:

1. A slow-spreading segment with a small magmatic budget where the lithosphere is more heterogeneous than on fast spreading ridges (Hess 1962; Dick and Natland 1996). This implies a long-lived low-magma budget where the mantle undergoes serpentinization, uplift and steady state lateral spreading. Such processes will characterize the entire ridge segment.
2. The presence of deep (>15 km) earthquake epicenters, which are an indication that fracturing of the rigid lithosphere will enable seawater circulation and the formation of a low-density hydrated peridotite (Aumento and Loubat 1971; Bonatti 1976).
3. Low-angle faulting during tectonic extension beneath the rift valley during a period of low magmatism will enable the alteration and exposure of peridotite on the ridge axis.
4. A denudation of the oceanic crust, which can take place during a stretching of the lithospheric plates (Tapponier and Francheteau 1978) or by the viscous drive of the asthenosphere (Sleep 1969; Lachenbruch 1973) leading to exposure of peridotite. The mechanism responsible for the tectonic denudation at segment tips is related to an asymmetrical extension (Karson and Dick 1983; Karson 1990) during spreading. Denudation could influence the temperature gradient in the lithosphere, which is sensitive to mantle melting processes.
5. The dehydration of down-moving slabs in fore-arc regions is also responsible for serpentinization (Haggerty 1987; Mottl 1989). For example, the serpentinite mud found in the Mariana fore-arc in the western Pacific's mound is due to the subduction of a plate (Fryer 1992).

The hypothesis of peridotite emplacement on the sea floor has long been debated and it is most likely that several processes were involved in the emplacement of these kinds of rocks. Thermal conductivity is sufficient to inhibit peridotite melting and trigger a local diapir. Diapiric upwelling as well as low-angle faulting mechanisms are both attractive models to explain the emplacement of mantle peridotites. The occurrence of peridotites in spreading centers fits with the diapiric model. When sheared, serpentinized peridotite undergoes plastic deformation under stress and it could incorporate fragments of country rock. This mixture is an indication of diapiric motion or shearing during low angle faulting. Also, during spreading in a magma-starved area, serpentinized peridotite upwelling will be facilitated. A forceful injection of mantle peridotites could take place in a brittle lithosphere accompanied by tectonic deformation and melt circulation. However when the serpentinized peridotites occur in transform faults and are associated with only a strike-slip motion, it is more difficult to explain their presence by simple diapirism, unless a component of crustal opening occurs during lithospheric readjustment.

The failure to find fresh peridotite outcrops on the ocean floor even during drilling indicates that serpentinization has been very intense because of olivine's instability at shallow depths, under low temperatures and in the presence of water. Independently from the process of serpentinization, there are two alternative processes that could drive mantle peridotite to rise within the lithosphere where it will partially melt. These two processes are: (1) the spreading and opening of the crust-lithosphere at ridge axes, which could facilitate the upwelling of passive mantle material which will then melt due to decompression, and/or (2) the dynamic flow model which involves a forceful uplifting of the mantle during partial melting as a result of localized heat conduction. This will require a region of lower viscosity in the mantle due to thermal, compositional and density changes caused by melting. The localized or focused mantle upwelling is associated with thermal pulses generated by the circulation of hot mantle material in the asthenosphere, when it enters the lithosphere. Such localized, thermal pulses were detected by Bonatti et al. (2003) during a study of a 20 million-year-old record of exposed lithosphere underneath the Mid-Atlantic Ridge.

Melt Impregnation of Peridotites

The heterogeneous nature and the structural and compositional diversities of exposed oceanic crust are a reflection of the asthenospheric circulation interacting with the rocks that make up the rigid lithosphere. The asthenosphere-lithosphere interactions are suggested by the exposed mantle residues derived from the partial melting of mantle material, which was impregnated with basaltic liquids that did not reach the surface but which had previously solidified within the lithosphere.

When a melt rises after the partial melting of mantle peridotite, it will travel through the lithosphere's openings, breaks, faults and fissures to create new oceanic crust. Evidence for melt impregnation dispersed within the peridotite and other formations (i.e. dunites, gabbros) forming the lithosphere is observed by the presence of small millimeter-sized dykelets and larger veins (centimeter to meter size) of solidified melt in the form of minerals. An increase in partial melting will depend on a region's spreading mechanism as well as on mantle plume upwelling, which is able to bring up hot material during lithospheric decompression.

The composition of an impregnated melt does not necessarily reflect the composition of the original parental mantle source. Indeed, chemical reactions could have taken place between the impregnated liquid and the surrounding formation during the liquid's ascent as well as during crystal-liquid fractionation, which could alter the original mineral assemblages. The study of impregnated melts that have solidified within the lithosphere can give us an indication about the way magma arrives at the surface. It can also give us an idea about the extent and the amount of melt circulating within the lithosphere. This impregnation phenomenon takes place during an intermediate stage between when a melt has been extracted from the mantle until the moment of its extrusion on the surface during a volcanic event.

Section of the Oceanic Lithosphere

The association of dolerite (dykes)-gabbro-peridotite is not easily found on the ocean floor. For many years, geologists have used land-based ophiolite sections to infer the lithology of the oceanic lithosphere. In subaerial regions, tectonic events such as folding, faulting and diapiric ascent are responsible for exposing deep sections of the lithosphere-crust. Deep-seated formations on the sea floor are essentially exposed in transform faults (fracture zones), on spreading ridges and in subduction zones when a slice of the lithosphere has been uplifted and overrides other structures.

A geophysical approach for defining the crust, lithosphere and mantle boundaries is based on the physical characteristics of the rocks related to the transmission of sound waves. Geophysical information about crustal thickness is directly related to rock density and will reveal variability with depth, which suggests a layering of the lithosphere. The concept of a layered lithosphere model based on seismic experimental data suggests that the axial asthenosphere—lithosphere boundary coincides with the roof of axial magma reservoirs under spreading ridge axes, or we may also be seeing a mass of solidified magma, which percolated from the asthenosphere towards the sea floor.

If the lithosphere has been affected by mantle magma injection and alteration (serpentinization), this will hide any original layering of the lithosphere. Thus, it is important to reconsider our view of seismic layering and take into account that the various suggested geological models based on seismic observations may simply be a general approximation. In order to obtain a stratigraphic section of the Earth's outer shell, which is as close as possible to reality, we need to redefine the rigid lithosphere-crust boundary for each specific area of the world's oceans.

Sea Floor Exposed Through Time

Prior to the 1960's, our knowledge about deep-layer sections which sliced into Earth's outer shell leading to exposed peridotites, gabbros and dykes (dolerite) was mainly inferred from our understanding of the land-based geology concerning peridotites and associated volcanic rocks exposed on mountain chains and called ophiolites. The term ophiolite means "snake rock" (from the Greek *ophis* = snake and *lithos* = rock). Also commonly called "greenstones", ophiolite complexes are composed of greenish and dark colored veined rocks made up of serpentinized peridotite overlaid by dykes and basaltic lava, which are usually associated with folded mountain belts and therefore deformed and altered. Ophiolite complexes were exposed during the collision and subduction of the oceanic lithosphere at contact with continental landmasses.

As spreading and plate motion takes place, peridotites and associated formations are transported across the oceanic basins and are eventually pushed against the active and passive margin regions (subduction trenches) where they are thrusted, folded and exposed on a subaerial environment and/or subducted underneath continental margins and volcanic arcs (Nicolas 1990). The model of ophiolite emplacement over time covers the period from 570 Ma up to the Tertiary and Cretaceous (50–40 Ma) belts in the Alps-Carpathians-Caucasus-and Himalaya chains. This model suggests that ophiolite complexes have been formed episodically since the beginning of the Pangaea land break-up more than 200 million years ago, when continents began colliding. Thus, several types of sections of ophiolite provinces in the World (in Oman, Cyprus, Anatolia, the Alps, the Apennines and in Vourinos, Greece) were exposed during the closure of the Tethys Ocean.

Present Day Exposed Sea Floor

Cold areas with a thin or non-existent basaltic crust are ideal places for observing the geological rock sequences. Such regions are magma-starved, and are located in major depressions or deeps (>4000 m depths) associated with spreading ridge segments and fracture zones.

In the North Atlantic, the crustal thickness varies considerably since under Iceland it has a maximum depth of about 40 km (Darbyshire et al. 2000). On the southern tip of the Reykjanes Ridge along the MAR, it was found that the crust is about 9 km thick. The same crust thickness (9–10 km) was also found in the Famous area near 37°N on the MAR (Whitmarsh 1973) where the intrusive zone is about 3.5–1.6 km wide. In the equatorial Atlantic, it has been observed that there is a deepening of the average spreading ridge's axial bathymetric depth, which is due to a decrease in magmatism and a less thick crust. This area of the equatorial Atlantic corresponds to a region with a high concentration of fracture zones. The 4°N fracture zone, along with the St. Peter and Paul's Rocks, the Romanche and the Chain fracture zones are among the most prominent (see Chap. 8).

In the Pacific Ocean, similar situations are found in the equatorial region where the Wilkes, Quebrada and Gofar transform faults occur. Other major fracture zones in the Pacific include the Garrett and Eltanin FZ, which displace the East Pacific Rise segments, and the Terevaka transform in the Easter Microplate (see Chap. 8) (Fig. 4.6). If we hope to see the geological sequences that exist under the ocean and within the crust, these are the areas where we should be looking.

Stratigraphic Columns

Although geologists have been able to observe and study land-based stratigraphic sequences, very few oceanic sections of the crust and upper mantle have been identified up to this day. In oceanic environments, we are obliged to use

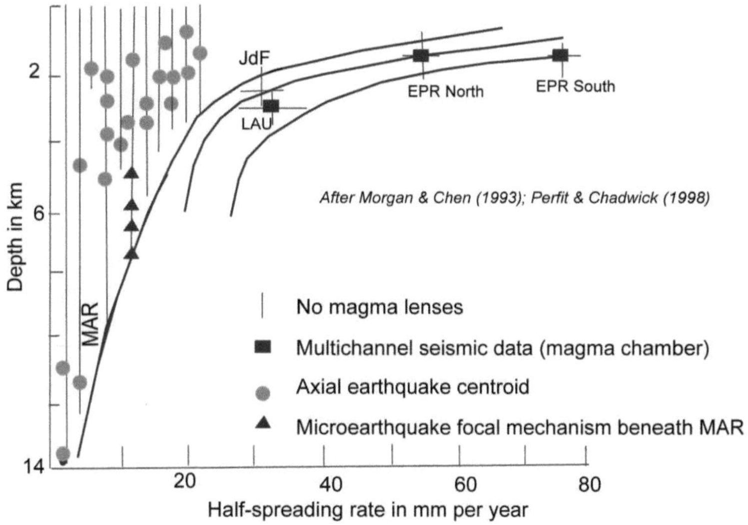

Fig. 4.6 Seismic depths versus spreading rates show the differences in the storage of magma lenses or the brittle nature of the lithosphere diagram inspired after Perfit and Chadwick (1998). The Juan de Fuca (JdF), the Lau back-arc basin (Lau) and segments of the East Pacific Rise (EPR) are indicated. The two lower curves are the results of numerical experiments by Phipps Morgan and Chen (1993). The model results suggest that a steady-state magma lens exists only at half-spreading rates greater than 20–30 mm/yr. The upper curve shows the modeled 750 °C isotherm, which is in agreement with the depth of the seismically observed brittle-ductile transition on slow spreading ridges

geophysical methods such as seismic refraction, or we must rely on deep-sea drilling operations or on direct in situ sea floor observations using submersibles or remote controlled vehicles (ROV). A classic example of stratigraphic sequences of the outer layers has been extrapolated. This stratigraphy consists of a thin (less than 2 km thick) basalt-sheeted dyke complex overlying less than 5 km of gabbros and ultramafic cumulates.

The Summary of Stratigraphic Columns represented on Fig. 4.6 was constructed on the basis of the few available field observations and sampling carried out by submersible. The geology of these areas will be further described in subsequent chapters dealing with the results obtained in fast spreading ridge segments, in the Terevaka Transform fault, the Garrett transform in the southeastern Pacific, the Hess Deep region in the central eastern Pacific, on Mid-Atlantic slow spreading ridges, the Saint Peter's and Saint Paul's Rocks and the Vema fracture zones in the equatorial Atlantic and northern Atlantic respectively.

The reconstruction of the oceanic stratigraphy is based on the assumption that the rate of spreading taking place during step faulting remains constant over the surveyed region (Karson 1998).

The Brittle and Ductile Transition Zone

The brittle/ductile transition zone marks the boundary of material instability and material exchange between the upper mantle (asthenosphere) and the lithosphere-crust. This boundary is related more to the physical nature of the oceanic lithosphere than to compositional changes in the material (Fig. 4.6).

Earthquake analyses indicate that knowing about the centroid depth (focal mechanism and depth source of an earthquake) has helped to constrain the thickness of the ridge axis and the thickness of the lithosphere. Earthquake epicenters, micro-earthquake observations, and focal depth mechanisms best retrace this boundary in rupture-depth (brittle nature), which is 7–8 km below the axial rift valley in slow spreading segments (Toomey et al. 1990; Perfit and Chadwick Jr 1998) (Fig. 4.6). The depth of brittle lithosphere as a function of spreading rate shows that faulting of the oceanic lithosphere reaches up to about 20 km deep, which also corresponds to the vertical limit of faulting (Wolfe et al. 1995) in amagmatic regions associated with slow spreading ridge segments (Fig. 4.6). Intermediate (>25 mm/yr half spreading rate) and fast spreading centers (50–80 mm/yr half rate) do not show prominent seismic activity but instead they are associated with steady-state magma chambers located at less than 4 km underneath the ridge axis (Perfit and Chadwick Jr 1998). Also, the calculated (750–800 °C temperature (Perfit and Chadwick Jr 1998) derived from the standard plate-cooling model of Parsons and Sclater (1977) corresponds to the brittle-ductile boundary in the lithosphere. Similarly, the temperature conditions associated with the brittle-ductile transition zone were also estimated to coincide with the formation and stability of key minerals such as olivine at 700–1000 °C (Kirby 1985) and to be above the temperature of formation of metamorphic chrysotile (Evans et al. 1976).

A model of brittle/ductile transition for the sub-solidus temperature of spinel-peridotite was also proposed to occur at 750 °C based on both the pyroxene and the olivine-spinel equilibrium. Other temperature calculations for the brittle/ductile boundary were obtained from the amphibole-chlorite geothermometry (determination of the temperature of a given mineral) and based on observations of the metamorphic mineral assemblages in peridotites. The average temperature of metamorphism was calculated to be at 550–600 °C (Bazylev and Silantiev 2000). The brittle/ductile transition zone is attributed to the high to low re-equilibration of the temperature conditions of minerals forming rocks during tectonic uplift, which could affect the physical behavior of the lithosphere.

Examples of deformation are observed in the minerals, which compose intruded crystalline rocks such as gabbros and peridotites.

In gabbroic rocks, such as those studied near 15°N on the MAR, for example, there are indications that successive magmatic and tectonic events took place in the brittle-ductile transition zone. This ductile deformation is marked by the recrystallization of magmatic minerals including orthopyroxene, which is seen as large (0.2–0.3 mm) neoblast (newly formed) polygonal minerals in the granulite, found in the upper amphibolite facies (Cannat and Casey 1995). Brittle-ductile deformation is also observed when plagioclase has recrystallized so hornblende has been transformed into actinolite, which is seen as small (<0.04 mm) irregular-shaped neoblast crystals in the lower amphibolite-greenschist facies. The decrease of the neoblast sizes suggests an increase in the flow stress, which is consistent with strength increase as temperature decreases in the brittle domain.

Fractures and micro-fractures are often filled with lower amphibolite to greenschist facies indicating that the plutonic sequences have undergone brittle deformation. These veins and veinlets are also filled with hydrothermal precipitates suggesting that circulating fluids were present after deformation. The temperature of mineral transformation is obtained from some critical mineral compounds such chlorite, which equilibrates at 150–299 °C (Chatelineua and Niva 1985). Also the presence of actinolite in gabbroic rocks suggests a fluid temperature of about 300 °C during brittle deformation accompanied by vein development (Liou et al. 1974).

In peridotite, the preferential orientation for the flattened orthopyroxene surfaces gives an indication of crystal-plastic deformation, while new generations of olivine crystals display a well-developed euhedral shape. Crystal-plastic deformation is also seen during the formation of the fine-grained matrix and in the closely spaced sub-grain boundary between olivine and pyroxene (i.e. porphyroclastic texture). The term "porphyroclast" is used to indicate a pseudoporphyritic (coarse grained) texture showing foliation and the term "granoblast" is used when the minerals are sub-equigranular without any foliation.

Often the peridotite rocks are called tectonites when the plastic deformation and mineral recrystallization that took place at high temperature is recognized. Kinky structures (deformed lattices showing foliation) and small-recrystallized minerals are observed as replacement products of orthopyroxene, clinopyroxene and olivine. The most frequent textures are porphyroclast and granoblast. Porphyroclasts are set in a mosaic-like matrix of the recrystallized minerals. The most extreme case of plastic deformation is that observed in mylonite (metamorphic gabbro) where peridotite has been transformed into a fine-grained granulated mosaic-like texture. These types of deformation are believed to take place in the upper mantle (Mercier and Nicolas 1975), probably in the lithosphere. More insight into the plastic deformation of the Earth's lithosphere can be found in a publication by A. Nicolas (1990).

The rigidity of the plate thickness increases with the age of the lithosphere. Thus a particular depth above which the upper mantle has negligible strength will determine the location of the elastic layer. Extrapolating from the cooling plate model of Parsons and Sclater (1977), the isotherm (thermal gradient) will also correspond to the lower limit of the elastic thickness. On average, the rigid lithosphere underneath the sea floor is less than 100 km thick (see Chap. 2).

References

Allerton SR, Searle RC, Burton B (1996) Bathymetric segmentation and faulting on the Mid-Atlantic Ridge, 24°00′N to 24°40′N: In tectonic magmatic, hydrothermal and Biological segmentation of Mid Ocean Ridges. In:MacLeod CJ, Tyler TA, Walker CL (eds) Geological Society Special Publication, London 118:29–48

Aumento F, Loubat M (1971) The Mid-Atlantic Ridge near 45°N: serpentinized ultramafic intrusions. Can J Earth Scie 8:633–663

Auzende J-M, Cornen G, Juteau T, Lagabrielle Y, Lensch G, Mevel C, Nicolas A, Prichard H, Ribeiro A, Vanney JR (group CYAGOR 11) (1982) The Gorringe Bank: first results of submersible expedition. CYAGOR 11, Terra Cognita 2:123–130

Auzende J-M, Bideau D, Bonatti E, Cannat M, Honnorez J, Lagabrielle Y, Malavieille J, Mamaloukas-Frangoulis V, Mevel C (1989) Direct observation of a section through slow-spreading oceanic crust. Nature 337:726–729

Batiza R, Niu Y, Karsten JL, Boger W, Potts E, Norby L, Butler R (1996) Steady and non-steady state magma chambers below the East Pacific Rise. Geophys Res Lett 23(3):221–224

Bazylev BA, Silantiev SA (2000) Geodynamic interpretation of the subsolidus recrystallization of mantle spinel peridotite 1. Mid-ocean ridges. Petrology 8(3) 2:91–213

Bideau D, Hekinian R (1995) A dynamic model for generating small-scale heterogeneities in ocean floor basalt. Jour Geophys Res 100:10141–10162

Blackman DK, Cann JR, Janssen B, Smith DK (1998) Origin of extensional core complexes: evidence from the MAR at Atlantis fracture zone. J Geophys Res 103:21315–21334

Blackman DK, Karson JA, Kelley DS, Cann JR, Früh-Green GL, Gee JS, Hurst SD, John BE, Morgan J, Nooner SL, Ross DK, Schroeder TJ, Williams EA (2004) Geology of the Atlantis Massif (Mid-Atlantic Ridge, 30°N): implications for the evolution of an ultramafic oceanic core complex. Mar Geophys Res 23(5–6):443–469

Bonatti E (1976) Serpentinized protrusions in the oceanic crust. Earth Planet Sci Lett 32:107–113

Bonatti E, Clocchiatti E, Bonatti R, Clocchiatti P, Colantoni R, Gelmini G, Marinelli G, Ottonello R, Santacroce M, Taviani AA, Abdel-Meguid HS, Assaf, El Tahi MA (1983) Zabargad (St. John's) Island: an uplifted fragment of sub-Red Sea lithosphere. J Geol Soc 140(4):677–690

Bonatti E, Ligi M, Brunetti D, Cipriani A, Fabretti P, Vlentina Ferrante V, Gasparini L, Ottolini L (2003) Mantle thermal pulses below the Mid-Atlantic Ridge, and temporal variations in the formation of oceanic lithosphere. Nature 423:499–505

Boillot G, Giradeau S, Kornpobst J (1988) The rifting of galicia margin: crustal thinning and emplacement of mantle rocks on the sea floor. Proc Ocean Drill Program Sci Results 103:741–756

Boillot G, Besler MO, Krawczyk CM, Rappin D, Reston TJ (1995) The formation of passive margins: constraints from the crustal structure and segmentation of the deep galicia margin (Spain). Geol Soc Spec publ N°90:71–91

Cannat M, Casey JF (1995) An ultramafic lift at the Mid-Atlantic ridge: successive stages of magmatism in serpentinized peridotites from the 15°N region. In: Vissers RLM, Nicholas A (Eds) Mantle and lower crust exposed in oceanic ridges and ophiolites, Kluwer Academic Publications, p 5–34

Charpentier S, Kornprobst J, Chazot G, Cornen G, Boillot G (1998) Interaction entre lithosphère et asthénosphère au cours de l'ouverture océanique: données isotopiques préliminaires sur la marge passive de Galice (Atlantique-Nord). Comptes Rendus de l'Académie des sciences, Sciences de la Terre et des Planètes 326:757–762

Chatelineua M, Niva D (1985) A chlorite solution geothermometer. The loz azufres mexico geothermal system. Contribution mineral Petrol 91:235–244

Christie DM, Sinton JM (1981) Evolution of abyssal lavas along propagating segments of the Galapagos spreading center. Earth Planet Sci Lett v56:321–335

Christensen NI (1978) Ophiolites, seismic velocities and oceanic crustal structure. Tectonophysics 47:131–157

Chu D, Gordon RG (1999) Evidence for motion between Nubia and Somalia along the Southwest Indian Ridge. Nature 398:64–67

Clague DA, Frey FA, Thompson G, Rindge S (1981) Minor and trace element geochemistry of volcanic rocks dredged from the galapagos spreading center: role of crystal fractionation and mantle heterogeneity. J Geophys Res 86(B10):9469–9482

Clarke DB, Loubat H (1997) Mineral analyses from the peridotite-gabbro-basalt complex at site 334, DSDP Leg 37. In: Aumento, Melson et al. 1997, Initial Report of Deep sea Drilling Project vol. 37 Washington (US Government printing Office) P1008:847–855

Davis A, Clague D (1990) Gabbroic xenoliths from the northern Gorda Ridge: Implications for Magma chamber processes under slow spreading centers. J Geophys Res 95:10885–10905

Darbyshire FA, White RS, Pristleir KF (2000) Structure of the crust and upper mantle of Iceland from combined seismic and gravity study. Earth Planet Scie Let 181(3):409–428

Dick HJB, Natland JH (1996) Late-stage melt evolution and transport in the shallow mantle beneath the East Pacific Rise. In:Leg 147 Mével C, Gillis K., Allan JF, Meye PS (eds) Proceedings Scientific Results ODP , pp 103–133, Ocean Drilling Program College station Texas

Dixon JE, Clague DA, Eissen JP (1986) Gabbroic xenoliths and host ferrobasaltfrom the southern Juan de Fuca Ridge. J Geophys Res 91:3795–3820

Escartin J, Hirth G, Evans B (1997) Nondilatant brittle deformation of serpentinites: Implications for Mohr-Coulomb theory and the strength of faults. J Geophys Res 102(B2):2897–2913

Engel AEJ, Engel CG (1963) Basalts dredged from the northeastern Pacific Ocean. Science 140:1321–1324

Engel AEJ, Engel CG (1964) Continental accretion and the evolution of North America. Advancing frontiers in geology and geophysics, Volume in honour of M.S. Kishnam, Indian Geophysical Union, Hyderabad, pp 17–37

Evans BW, Joannes W, Otterdoom H, Tromsdorff V (1976) Stability of chrysotile and antigorite in serpentine multisystem. Schweiz Miner Petrogr Mitt 56:79–93

Feraud G, York D, Mevel C, Auzende J-M (1986) Additional 40Ar/40Ar dating of the basement and alkali basalt of the Gorringe Bank (Atlantic Ocean). Earth Planet Sc Lett 79:255–269

Fisher RL, Engel C (1969) Ultramafic and basaltic rocks dreged from nearshore flank of tonga trench. Geol Soc Amer Bull 80:1373–1378

Francheteau F, Patriat P, Segoufin J, Armijo R, Doucoure M, Yelles-Chaouche A, Zukin J, Calmant S, Naar DF, Searle RC (1988) Pito and orongo fracture zones: the northern and southern boundary of the easter microplate (South Pacific). J Earth planet Sci Let 89:363–374

Francheteau J, Armijo R, Cheminee J-L, Hekinian R (1992) I MA East Pacific Rise Oceanic crust and uppermost mantle exposed by rifting in hess deep. Earth Planet Sci Lett 100:281–295

Fryer P (1992) A synthesis of Leg 125 drilling of serpentine seamounts on the mariana and Izu-Bonin forearcs. In: Fryer P, Pearce JA, Stokking LB et al. (eds) Proceedings ODP, Scientific Results, 125: College Station, TX (Ocean Drilling Program), pp 593–614

Fukao Y (1972) Thrust faulting at a lithospheric plate boundary the Portugal earthquake of 1969 Earth planet. Sci Lett 18:205

Gavashi AT, Fox PJ, Ryan WBF (1973) Petrography of rocks from the crestal area of the Gorringe Bank. Proc Deep Sea Drill Program Init Rep 13:749–752

Green J, Ringwood D (1967) The genesis of basaltic magmas. Contrib Miner Petrol 15:103–190

German CR, Parsons LM, HEAT Scientific Team (1996) Hydrothermal exploration near the AzoresTriple Jonction. Tectonic control of slow spreading ridge. Earth Plant Sci Lett 138:93–104

Gracia E, Bideau D, LagabrielleY, Parson LM (1997) Along-axis magmatic oscillations and exposure of ultramafic rocks in a second-order segment of the Mid-Atlantic Ridge (35°45′N and 34°07°′N). Geology 25(12):1059–1063

Haase KM, Stroncik N, Stoffers P (2005) Nb-depleted andesites from the Pacifi-Antarctic Rise as analogue for early continetal crust. Geology 33921–924. doi:10.1130/G 21899.1

Haggerty JA (1987) Petrology and geochemistry of neogene sedimentary rocks from mariana forearc seamounts: implications for emplacement of the seamounts. In: Keating B, Fryer P, Batiza R, Boehlert GW (Eds) Seamounts, Islands, and Atolls. Geophys. Monogr. 43:175–185

Hekinian R (1968) Rocks from the Mid-Oceanic ridge in the Indian Ocean. Deep Sea Res 15:195–213

Hekinian R, Aumento F (1973) Rocks from the gibbs Fracture Zone and the minia seamount near 53°N in the Atlantic Ocean. Mar Geol 14:47–72

Hekinian R, Thompson G, Bideau D (1989) Axial and off-axial heterogeneity of basaltic rocks from the East Pacific Rise at 12°38′N–12°51′N and 11°26′N–11°30′N. J Geophys Res V94:17:437–463

Hekinian R, Hoffert M, Larqué Ph, Cheminée J-L, Stoffers P, Bideau D (1993) Hydrothermal Fe- and Si- oxyhydroxide deposits from the South Pacific Intraplate volcanoes and East Pacific Rise axial and off-axial regions. Econ Geol 88(N°8):29

Hekinian R, Stoffers P, Ackermand D, Binard N, Francheteau J, Devey C, Garbe-Shonberg D (1995) Magmatic evolution of the Easter Microplate-crough seamount region (South East Pacific). Mar Geophys Res 17:375–397

Hekinian R, Stoffers P, Ackermand D, Révillon S, Maia M, Bohn M, (1999) Ridge-hotspot interaction: the Pacific-Antarctic Ridge and the foundation seamounts. Marine Geol 160:199–223

Hekinian R, Juteau T, Garcia E, Sichler S, Sichel G, Udintsev G, Apprioual R, Ligi M (2000) Submersible observations of the equatorial atlantic mantle: the St. Paul Fracture Zone, Mar. Geophys. Res 21:529–560

Hekinian R, Cheminée J-L, Dubois J, Stoffers P, Scott S, Guivel C, Garbe-Schönberg D, Devey C, Bourdon B, Lackschewitz K, McMurtry G, Le Drezen E (2002) The pitcairn hotspot in the South Pacific: distribution and composition of submarin volcanic sequences. J Volcanol Geoth Res 121:219–245

Hess HH (1955) Serpentinized orogen. Geol Soc Amerca Spat Pap 62:391–407

Hess HH (1962) History of ocean basins. petrologic studies: a volume in honor of A. F. buddington. In: Engel AEJ, James HL, Leonard BF (eds) Geological Society of America, New York, pp 599–620

Hofmann AW, White WM (1982) Mantle plume from ancient oceanic crust. Earth Planet Sc Lett 57:421–436

Johannsen A (1931) A descriptive petrography of igneous rocks; Volume I, Introduction, textures, classifications. University of Chicago Press Chicago III, Chicago, pp 267

Karasik AM (1974) The Eurasian basin of northern 1rctic Ocean In respect to plate tectonics. Geologiys Arckt Leningrad Nauchno Issled Institut. In Problrmy geologii Plyarnykh Oblastey Zemli Sbornik Statey, pp 23–31

Karson JA, Dick HJB (1983) Tectonics of ridge- transform intersections at the Kane fracture zone. Mar Geophys Res 6:51–98

Karson JA (1990) Seafloor spreading at Mid-Atlantic ridge: implications for the structure of ophiolites and oceanic Lithosphere Produced in slow-spreading environments: In proceedings of the Ymposium TROODOS 1987. In: Malpas J, Moores EM, Panayotou A, Xenophontos C (eds) Geological Survey Dept Nicosia, Cyprus, pp 547–555

Karson JA (1998) Internal structure of the oceanic Lithosphere: a perspective from tectonic window. 177–218. In: Buck WR, Delaney PT, Karson JA, Lagabrielle Y(eds) faulting and magmatism at Mid-Ocean Ridges, Geophysical Monograph 106, Am. Geophys.Union, Washington D.C, p 348

Kay R, Hubbard NJ, Gast PW (1970) Chemical characteristics and origin of oceanic ridge volcanic rocks. J Geophys Res 75:1585–1613

Kirby (1985) Rock mechanics observations pertinent to the rheology of the continental lithosphere and the localization of strain along shear zones. Tectonophysics 119 issues 1–4:1–27

Kushiro I (1969) The system forsterite-diopsite-silica with and without water at high pressure, Amer. J. Sci. Schaider 267:-A 269–294

Lachenbruch A H (1973) A simple model for oceanic spreading centers. J Geophys Res 78:3395–3117

Liou JG, Kuniyoshi S, Ito K (1974) Experimental studies of phase relations between greeshist and amphibolelite in a basaltic system. Am J Sci 274:613–632

Lonsdale P (1989) Segmentation of the Pacific-Nazca Spreading Center, 1°N–20°S. J Geophys Res 94(B9):12197–12225

Marsh BD, Gunnarsson B, Congdon R, Cormdy R (1991) Hawaiin basalt and Icelandic rhyolite: indicator of differentiation and partial melting. Geol Rundsch 80:481–510

Matsumoto T, Kelemen PB, Party OS (1998) Preliminary results of the precise geological and geophysical mapping of the Mid-Atlantic Ridge 14–16°N—tectonic extension along the magma- poor ridge axis. Eos. Trans Am Geophys Union 79:F46

McKenzie D, Bickle MJ (1988) The volume and composition of melt generated by extension of the lithosphere. J Petrol 25(3):623–679

Melson W, Byerly GR, Nelen JA, O'Hearn, Wright TL, Vallier T, (1977) A catalog of the major element chemistry of abyssal volcanic glassas. Smiththsonian Contributionn Earth Sci 19:31–60

Mercier JC, Nicolas A (1975) Textures and fabrics of upper mantle peridotites as illustrated by xenoliths from basalts. J Petrol 16:454–487

Meyzen CM, Toplis MJ, Humler E, Ludden JN, Mével C (2003) A discontinuity in mantle composition beneath the southwest Indian ridge. Nature 421:731–733

Michael PJ, Langmuir CH, Dick HJB, Snow JE, Goldstein SL, Graham DW, Lehnert K, Kurras G, Jokat W, Mühe WR, Edmonds HN (2003) Magmatic and amagmatic seafloor generation at the ultraslow-spreading Gakkel ridge, Arctic Ocean. Nature 423(6943):956–961

Miller DJ, Christensen NI (1997) Seismic velocities of lower crustal and upper mantle rocks from the slow-spreading Mid-Atlantic ridge, south of the kane transformation zone (MARK). In: Karson JA, Cannat M, Miller DJ, Elthon D (Eds) Proceedings ODP Scientific Results 153 College Station, TX (Ocean Drilling Program), pp 437–451

Mottl MJ (1989) Hydrothermal convection, reaction, and diffusion in sediments on the costa rica rift flank: pore-water evidence from ODP sites 677 and 678. In: Becker K, Sakai H et al. Proceedings ODP Scientific Results, 111: College Station, TX (Ocean Drilling Program), pp 195–213

Muller C, Jokat W (2000) Seismic evidence for volcanic activity discovered in central Arctic. Eos 81:265

Myashiro A (1974) Volcanic rock series and tectonic setting. Annu Rev Earth Planet Sci 3:251

Naar DF, Hey RN (1986) East rift propagation along the east Pacific Rise near easter Island. J Geophys Res 91:3425–3438

Nicolas A (1990) Les montagnes Sous la mer. Edition BRGM Avenue de Concyr, B.P. 6009, 45060, Orleans CEDEX 2 France, p 187

Parsons B, Sclater JG (1977) An analysis of the variation of ocean floor bathymetry with age. J Geophys Res 82:803–827

Perfit MR, Chadwick Jr WW (1998) Magmatism at Mid-Ocean Ridges: constraints from volcanological and geothermal investigations. In Faulting and magmatism at Mid-Atlantic Ridges. In: Buck R, Delaney P, Karson JF, Lagabrielle Y, (eds) Gophysical Monogr 106 series II, pp 59–115

Phipps Morgan J, Chen YJ (1993) The genesis of oceanic crust: magma injection, hydrothermal circulation, and crustal flow. J Geophys Res 98:6283–6297

Plank T, Langmuir CH (1992) Effects on the melting regime on the composition of ocean crust. J Geophys Res 97:19749–19770

Shand SJ (1949) Rocks of the Mid-Atlantic ridge. J Geology 57:89–91

Shido F, Miyashiro AM, Ewing M (1974) Basalts and serpentinite from the puerto ricotrench, 1. Petrol. Mar Geol 16(4):191–203

Sleep NH (1969) Sensitivity of heat flow and gravity to the mechanism of sea-floor spreading. J Geophys Res 74:542–549

Stoffers P, Worthington T, Hekinian R, Petersen S, Hannington M, Turkey M et al (2003) Silicic volcanism and hydrothermal activity documented at Pacific-Antarctic Ridge. EOS 83(28):301–304

Tapponnier P, Francheteau J (1978) Necking of the lithosphere and the mechanics of slowly accreting plate boundaries. J Geophys Res 83(B8):3966–3970

Thompson GW, Bryan WB, Humphris SE (1989) Axial volcanism on the eastpacific rise, 10°–12°N:In magmatism in Ocean basins. In: Saunders AD, Norry MJ (eds) Geological Society Special Publication 42, London, pp 181–200

Toomey DR, Purdy GM, Solomon SC, Wilcock WSD (1990) The three-dimensional seismic velocity structure of the East Pacific Rise near latitude 9 30′N. Nature 347:639–645. doi:10.1038/347639a0

Turner FJ, Verhoogen FJ (1960) Igneous and metamorphic petrology. International series in earth science. McGraw Hill Book Comp Inc, New York, p 694

Udintsev G, Beresnev B, Alexander F, Golod VM, Kol'tsova AV, Kurentsova NA, Zakharov MV, Agapova GV, Lyudmila P, Udintsev VG (1996) Geological structure of the strakhov fracture zone (equatorial segment of the Mid-Atlantic Ridge). Oceanology 35(4):544–558

Wanless VD, Perfit MR, Ridley WI, Klein E (2010) Dacite petrogenesis on Mid-Ocean Ridges: evidence for oceanic crustal melting and assimilation. J petrol 51(12):2377–2410

Whitmarsh RB (1973) Median valley refraction line, Mid-Atlantic Ridge at 37∫N. Nature 246:297–299

Wolfe CJ, Purdy GM, Toomey DR, Solomon SC (1995) Microearthquake characteristics and crustal velocity structure at 29°N of the Mid-Atlantic Ridge: the architecture of slow-spreading segment. J Geophys Res 100:24449–24472

Yoder HS Jr (1976) Basic magma generation and aggregation. Bull Volcanol 41:301–316

Chapter 5
Earth's Mantle Melting and Volcanism

Abstract Convection currents inside the Earth's asthenosphere will cause instability at shallow depths in the mantle. Rising material and subsequent decompression melting will form hot, upwelling mantle plumes or diapirs. This phenomenon is more common on slow spreading ridges (total rate <5 cm/yrs), rather than beneath fast (total rate >5 cm/yrs) spreading ridge segments with their extensive fissural magmatism. Magma upwelling after partial melting of the mantle will depend on the force of buoyancy and on the permeability of the lithosphere. The effects of permeability and buoyancy will be modified by tectonic stress-release as well as by compression due to spreading following a cooling of the lithosphere. Instability in the melting zone is related to pressure release during a period without magma extraction. If pressure is released during spreading, the heat supply will generate more melting. With increased tension, the melting zone expands laterally and deepens until enough melt aggregates and accumulates in a confined zone to form a magma chamber. Rapid migration of melt through fissures at shallow depths enables a release of tension in the magmatic zone as it undergoes periods of melting and magma accumulation. When this process is repeated several times, it will trigger successive arrivals of more deep-seated magma, which can replenish the magma reservoir.

Magma Reservoirs Underneath Spreading Ridges

After the partial melting of the mantle, molten material from about 30–60 km deep will rise towards the sea floor and this magma can form shallow sub-crustal reservoirs or "magma chambers" (Figs. 5.1 and 5.2). The location of these reservoirs is found in a transition zone where liquid has been stored prior to its escape on the surface during volcanism. The existence of magma reservoirs has long been inferred from land-based geology records mainly gathered from exposed sections of ophiolite complexes. In the field, the presence of layered gabbros is considered as being the result of crystal-liquid differentiation giving rise to stratiform

R. Hekinian, *Sea Floor Exploration*, Springer Oceanography,
DOI: 10.1007/978-3-319-03203-0_5,

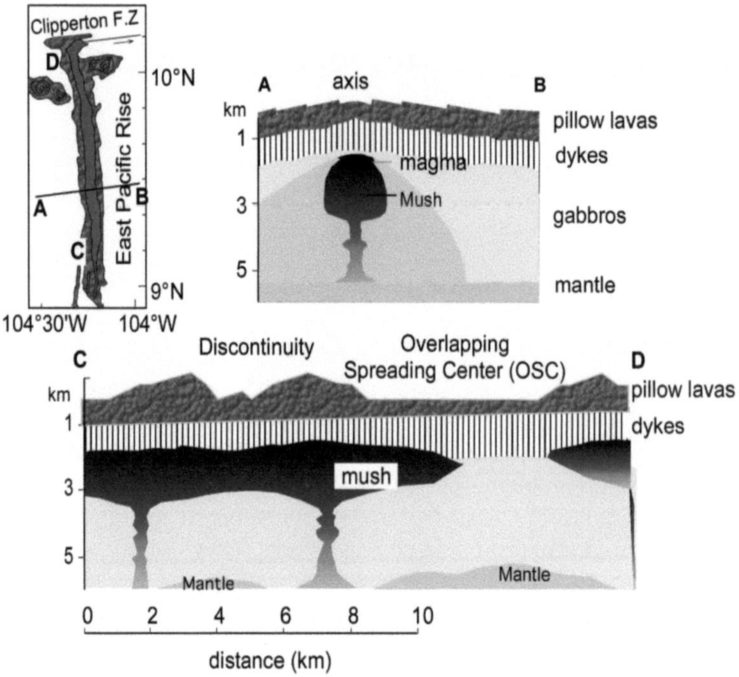

Fig. 5.1 Magma chamber underneath a spreading ridge axis showing the proportion of melt variation with respect to the amount of solids (minerals) underneath accreting ridge segments. A magma lens containing melt is associated with a mixed zone made up of liquid and solid forming a "mush" (After Nicolas et al. 1993). The total size of a magma chamber does not exceed a width of 7–10 km at its base

(layered) units during magma's accumulation in the confines of a reservoir. The volume of melt underneath ridge axes could vary considerably and will depend on the rate of partial melting of upwelling mantle diapirs, as well as on the rate of spreading at the ridge axes.

The presence, shape and size of the magma chambers underneath the ocean floor are inferred essentially from seismic experiments that have been carried out along spreading ridge segments. A magma chamber could be small, short-lived or even non-existent if the volcanic eruptions are simply coming from large conduits. This aspect will be further discussed in a section concerning the magma-starved regions of the planet, such as slow spreading ridge segments (See Chap. 4).

As explained previously, it was in the 1970s that detailed seismic experiments were first conducted using water guns and/or air guns towed behind ships to produce sounds in the water column along with hydrophones to record these sounds. We know that seismic waves travel in different directions and at different speeds within the layered Earth according the type of material encountered. For instance, seismic waves traveling through a liquid are "less deviated" since there are no solid obstacles to deviate their course. A decrease in sound velocity will

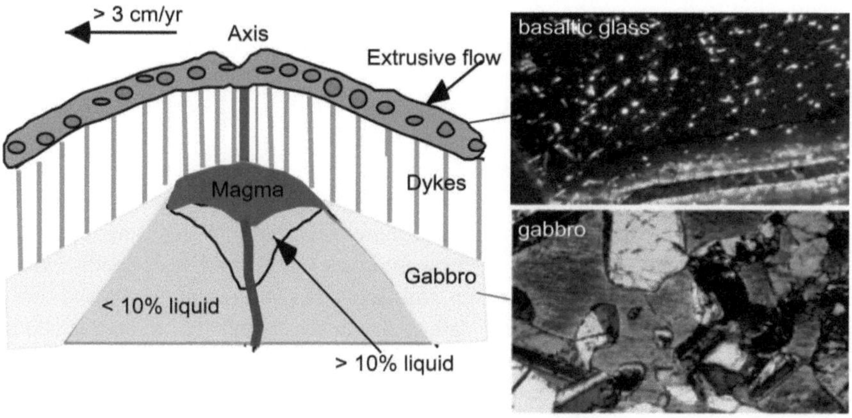

Fig. 5.2 A close-up of the upper lithosphere underneath a fast spreading center shows the various rock formations. Experimental peridotite melting and seismic anomaly measurements suggests that the highest melt concentration (10–35 %) occurs in the upper part (1–2 km) of the magma reservoir and the lower melt (2–12 %) forming a "crystalline mush" is found in the middle and lower crust (1.5–4 km depth) (Sinton and Detrick 1992). The microscopic view is an 80 times greater magnification of solidified glass and of the crystallized minerals forming a gabbro

indicate the presence of softer material, which is probably a partially or completely molten product (See Chap. 2). This type of seismic area is also called a Low Velocity Zone (LVZ). An extensive seismic experiment recorded data along a stretch 30–40 km long on each side of the East Pacific ridge axis across the northern EPR at 12°50′N (McClain et al. 1985) and also at 9°28′–9°35′N (Rosendahl et al. 1976; Detrick et al. 1987). The presence of a narrow Low Velocity Zone (<5 km wide) located in the upper (1.2–2.4 km deep) part of the crust was revealed (Figs. 5.1 and 5.2). The width of this LVZ corresponds to the presence of melted material, usually less than 3–4 km in diameter, across the ridge axis (Detrick et al. 1987).

The first seismic experiment carried out along the axis of a fast spreading ridge segment was made by Felix Avedik during the *Clipperton Cruise* in 1981 (Avedik and Geli 1987) (See Chap. 7 on the East Pacific Rise 12°50′ N) (Fig. 5.2) Further geophysical studies along the EPR and reported in 1992 by Sinton and Detrick have defined the shape of a sub-crustal magma chamber. It consists of a thin (a few hundred meters high) and narrow (1–2 km wide) melt lens overlying a region of crystal mush surrounded by a transition zone of solidified crust.

Seismic data obtained from a variety of geological settings indicates a variation in crustal thickness, which corresponds to changes in the lithosphere's composition. On the EPR, the average crustal thickness is on the order of 6–7 km. There is a Low Velocity Zone at less than 2 km deep underneath the axis of fast spreading centers. In slow spreading centers, the Low Velocity Zone is considerably more variable. For instance, crustal thickness in the Mid-Atlantic Ridge (MAR) is variable due to the presence of islands formed by plumes (see chapter on Azores Islands hotspots). The crust can reach about 9 km in the FAMOUS area (MAR, 37°N (Witmarsh 1975). At 35°N on the MAR, the LVZ is located at 3 km depth and has a bulls-eye shaped negative gravity anomaly suggesting the presence a magma reservoir (Detrick et al. 1987).

Based on seismic data (Perfit and Chadwick 1998) and using thermal calculations, we can conclude that when the sea floor spreading rates are low (with half spreading rates of less than 20 mm/year), there will be a "steady state" magma chamber. This means that most of the upwelling melt will solidify at depths within less than 10 km of the sea floor (Sleep and Barth 1997). On the other hand, only ridge segments with intermediate, fast and ultrafast rates of spreading (total rates >40 mm/year), will sustain a transient, non-steady state magma reservoir since magma is extruded as soon as it rises through the crust. These types of magma chambers are mainly found underneath fast spreading (>60 mm/year) ridges such as the EPR. In slow spreading ridges, magma lenses are absent or ephemeral and there is only sporadic volcanism, with the exception of slow spreading ridges that are associated with mantle plumes (a hotspot), which could result in voluminous but localized volcanism (such as in the region of the Azores).

Volcanic Landscape

The extrusion of lava flows has shaped the sea floor's volcanic landscape. Spreading ridge systems and intraplate volcanoes constitute the primary places where volcanic landscapes are formed and where they vary the most. The general appearance of a spreading ridge segment is that of an elongated dome-shape feature, whose topography becomes lower as volcanism decreases.

Comparisons of the morphological and compositional variability of lava observed on the sea floor along the ridge segments have been made in order to detect magma that has been channeled towards the segments' end in sub-crustal environments as opposed to the magma released on topographic highs. Rift propagation and dyke injection implies that as magma moves away from its main upwelling zone, it will change its physical properties due to a drop of its initial temperature, which will influence the magma's composition, its degree of crystallinity and its viscosity. This is observed by a decrease in crustal thickness, an increase in pillow/sheet flow ratios and an increase in the degrees of magmatic differentiation along the strike of a ridge. For example, the lava composition towards the segment-end, located at some distance away from the source of magma, tends to become more viscous and contains higher contents of incompatible elements (i.e. K, Na, Ba, Sr) and silica.

We must also take into consideration that the rising mantle material is dispersed within the lithosphere and is not simply limited to extrusion on the sea floor surface. A large amount of melt can stagnate in the lithosphere forming separate units such as vertical dykes, horizontal sills or magma ponds. The melt will follow a path of existing crustal weakness and will propagate through irregular fissures and faults (Figs. 5.3 and 5.4), which become conduits of magma that could be released to form volcanic edifices, lava ponds, pillow mounds and other volcanic features.

Lava Morphology

The types of lava morphology encountered on the sea floor vary according to the geological environment, the depth from which the lava has come, its temperature, rate of eruption and its viscosity, which also depends on the lava's composition.

The major parameters involved in modifying the morphology of lava flows are listed below:

(a) ***Magma's viscosity*** *depends on its flow rate, the amount of crystals and its volatile content other than* H_2O *(mainly for silica-rich lava). A magma's viscosity will affect its density. Viscosity is defined as a substance's internal resistance to flow when shear stress is applied. For example, water flows in response to a small amount of stress therefore we say it has a low viscosity. Viscosity is measured in Pascal-seconds* (1 kg/1 m/s = 1 Pa s = 10 poise). *Most lavas including basalt, andesite, rhyolite and others such as trachy-andesite, or trachyte all flow at a rate between* 10^2 *and* 10^5 Pa s. *For comparison, pure water has a viscosity of* 10^{-3} Pa s. *Viscosity decreases with an increase in temperature and an increase in the water content of the magma.* Murase and McBirney (1963, 1973) *have discussed the relationship between temperature, water content and viscosity.*

(b) ***Cooling rate*** *depends on the conditions of thermal diffusivity. Thick crust will insulate the magma flow, and thereby decrease the cooling rate. Large magmatic conduits and magma chambers will cool more slowly than small veins and thin horizontal sheets forming sills. A small conduit such as a lava tube* (<2 m *in diameter) will channel a limited amount of lava for a short distance.*

(c) ***Extrusion rate*** *depends on the size of a conduit, the volume of magma, and the rate of magma delivery. Magma conduits supplied by large magma chambers will take more time to cool (several days or even several tens of years).*

Fig. 5.3 Photograph shows a fissure in the axial graben of the East Pacific Rise segment at 12°50′N, 2606 m depth. The fissure is about 1 m wide and more than 10 m deep. Fissures and fractured crust are pathways for magma extrusion on the sea floor. (Copyright IFREMER, *Cyathem cruise 1982, Cyana dive Cy82-09.*)

Fig. 5.4 Lau Basin showing a columnar dyke complex taken during the *Starmer I* cruise by the submersible Nautile (from Lagabrielle et al. 1994). The dolerite dykes are located in the North Fiji back-arc basin (Lau Basin), at the intersection of a triple-junction between two spreading ridge segments and a fracture zone near 16°40′S–174°W. (Copyright IFREMER, *Starmer I cruise 1989, Nautile dive PL07*)

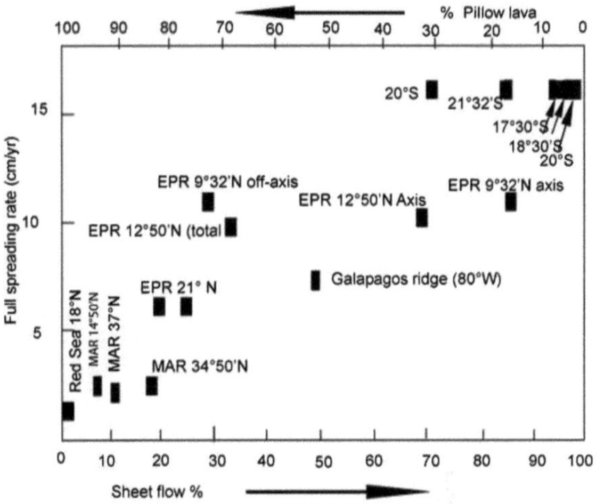

Fig. 5.5 Lava morphology versus spreading rate (modified after Bonatti and Harrison 1988; Perfit and Chadwick 1998) is reported. Pillow lava is more common on off-axial volcanic structures

The lava flows pouring out on the ocean floor will solidify in a variety of shapes, which, in addition to the physical parameters involved such as viscosity, rate of extrusion and cooling, will also depend on the morphology of the landscape where they are released. The advancing front of a lava flow is discontinuous. When lava is extruded on the sea floor a thin surface of the flow (only a few mm thick) is rapidly quenched and will form a solid glassy crust of obsidian that serves to insulate the hot interior of flowing lava. Frequently, the fragile, chilled glassy crust will break and allow hot lava to ooze out in different shapes (Figs. 5.5 and 5.6).

Pillow Lava and Sheet Flows

The surface texture and shape of extrusive flows give us indications about the physical parameters involved during eruption. Pillow lava and sheet flows are the commonest features forming the sea floor landscape. The rounded or tubular form of the "pillows" suggests a diminishing internal strength when compared to the pressure of the seawater on the ocean floor. The fact that pillow lavas are rounded or oval shaped is most likely due to water pressure exerted on the lava flow during extrusion. The presence of sheet flows (both flat flows and ropey lava flows) indicates a hotter and more fluid lava, which flows more quickly over a greater distance than does the more viscous lava forming pillows and tubes. A lava flow will also be affected by the original landscape where it is released. The ratio of pillow lava versus sheet flows is best observed when comparing the spreading rates

Fig. 5.6 Photograph of a pillow lava from the East Pacific Rise axis at 12°50′N shows the recent extrusion of lava seen at 2628 m depth. (Copyright IFREMER, *Geocyatherm cruise 1982, Cyana dive Cy82-09*)

of different ridge segments. (Figs. 5.5 and 5.6). It is observed that as the spreading rate diminishes, the ratio of pillow lava versus sheet flows increases (Batiza 1980). This increased presence of pillow lava is the result of both decreasing magma delivery as well as increased lava viscosity, which coincides with the lower rate of spreading on a Ridge axis.

Small volcanic constructions less than 10–50 m high and about 30–100 m wide are also encountered on the sea floor. They consist of haystacks, pillow lava mounds, hornitos and tumuli. They are formed during short lasting eruptions channeled through lava tubes and generally located in rifted zones. These small structures are rootless and found along lava channels or conduits forming large meandering tubes where molten magma has been transported.

The pillows from slow spreading centers of the MAR and from off-axis seamounts are large (>1 m in diameter), while pillows from fast spreading centers are less prevalent and smaller in size (<1 m in diameter).

As molten lava flows, its surface will rapidly cool and harden at contact with the seawater while the interior of a lava tube will remain at hot temperatures (>900 °C). A dark glassy crust of about 1–2 cm in thickness will form on the cooling surface of the most common basaltic lava flows as a result of the almost instantaneous "chilling" of molten melt at contact with cold seawater (Figs. 5.6, 5.7, 5.8 and 5.9). Hence natural glass (obsidian) is a common product when lava erupts in water. This glassy crust will be rapidly altered by seawater and

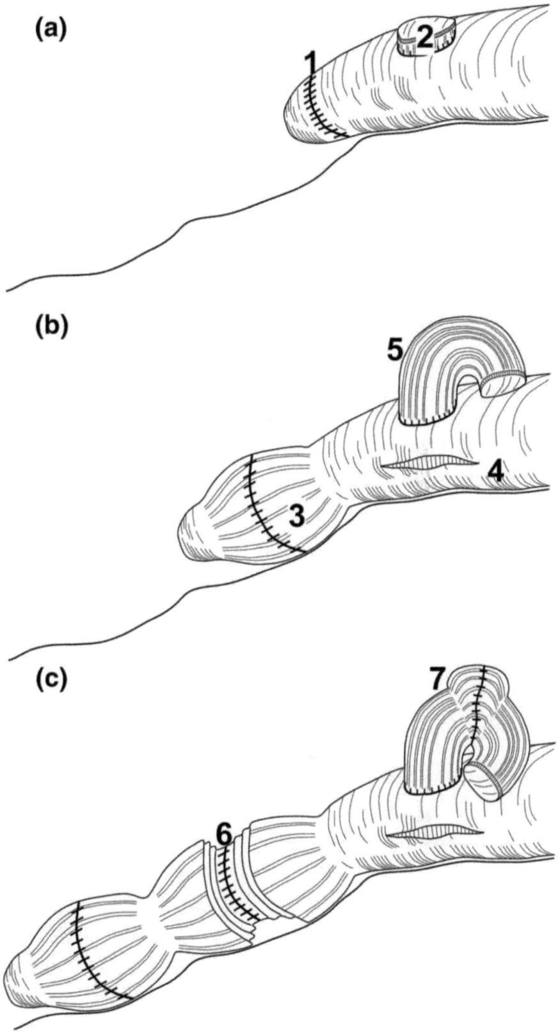

Fig. 5.7 The growth of pillow lava takes place when the surface of the hot flow chills at contact with seawater, while the interior continues flowing (Hekinian and Binard 2008). As the hot melt extrudes various shaped growth takes place. **a** On a single lava tube, a "trap-door" structure (*2*) is formed when lava is extruded on top of a pillow lava, **b** Lateral growths similar to a "tooth paste extrusion" (*5* and *7*) protrude, and **c** Continued sequences of growth with occasional cracks (*1*) and fissures (*4* and *6*) (after Hekinian and Binard 2008)

transformed into a hydrated clay-like material called palagonite (Nayudu 1962). After drainage of the lava flow, when molten material is no longer being fed into a lava tube, if the roof of these channels collapses they will form cavities, which could vary in size from a few centimeters to several meters (up to 3–4 m) in diameter.

Types of Volcanic Flows Erupted During Quiet Eruptions:

Pillow lavas *formed underwater have a variety of shapes. They could be bulbous, or tubular and are often broken by small "Yam shape" protrusions. Pillows show radial jointing and polygonal shaped cooling cracks. The* ***bulbous pillows*** *are sub-spherical and occur on flattish surfaces; vertical pipes channeling lava feed them and they grow upwards and expand laterally on top of the feeding tubes.* ***Elongated tubular pillows*** *are sub-cylindrical in shape because their lava has flowed down-slope.* ***Trapdoor pillows*** *are formed on top of another pillow during the extrusion of lava from a circular crack. They form a cylindrical cap on top of the original pillow. A "****Haystack****" is a volcanic cone with radial draped elongated pillow lava tubes on its sides* (Fig. 5.9). ***Lobated flows*** *are flattened lobes with respect to pillows and their interior shows flat ledges* (<10 cm *thick) representing cooling levels. Such flows fill depressions and fissures. They are formed by the rapid growth and drainage of extruded lava (see also lava pond).* ***Flat flows****, also called* ***sheet flows****, are like lobes without any inflation and they can be ropy, jumbled, and hackly. They are formed during the rapid extrusion of fluid lava. Sheet flows are thin, less than* 30 cm *thick, and they display different shapes according to the landscape encountered.* ***Jumbled flows*** *are a type of sheet flow with broken up and re-cemented glassy crusts.* ***Ropey flows*** *are formed during lava extrusion in a semi-circular and confined pathway therefore the melt laps over itself and cools in "coiled rope" shapes.* ***Blocky and tabular flows*** *consist essentially of silica-enriched lavas that have a surface morphology and texture that is different from basaltic pillow. Depending on their composition, they are blocky (a silica-rich dacite flow) and/or tubular with "corkscrew" shaped tubes and flattened lobes. They can be found on spreading ridges as well as in ocean basins. Their surface morphology is characterized by a spiny and rughose thicker glassy crust* (3–20 cm *thick) than what is found on pillow lava or sheet flows of basaltic composition. The interior of these silica-rich flows is characterized by the presence of large cavities, (up to* 8 cm *in diameter). They are formed during the eruption of viscous lava, which expanded laterally and formed a thicker chilled crust due to the lower cooling rate than that of more fluidal basaltic lava.* ***Scoria-like*** (***Clinker***) *volcanic structure refers to massive or fragmented debris, which have been re-cemented during the advance of a new lava flow. They are similar to the typical "aa" flow found in Hawaii.*

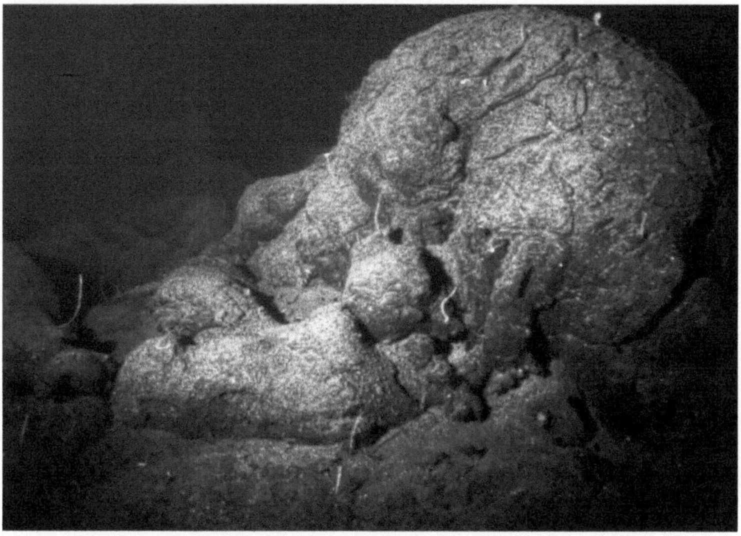

Fig. 5.8 Pillow lava from the East Pacific Rise axis at 12°50′N taken at 2628 m depth showing surface corrugations and lateral small protrusions of "yam" like glassy flows. (Copyright IFREMER, *Geocyarise Cruise 1984, Cyana dive Cy84–24.*)

Fig. 5.9 Accumulation of lava tubes during eruption forms small volcanic structures that resemble a "haystack". Photograph was taken during by the submersible *Cyana* on an off-axial volcano at 2534 m depth called the Southeastern seamount, located at 6 km from the axial graben of the EPR at 12°43′N). Lava has flowed down from a summit orifice ("trap-door") located on top of the cone to form a haystack about 3 m tall. Haystacks are also found on top of a volcanic fissured or rift zone. (Copyright IFREMER, *Geocyatherm cruise 1982, Cyana dive Cy82-14*)

Fig. 5.10 Pillar remains of a collapsed lava pond (2,600 m depth) in the axial graben of the EPR at 21°N taken during the Cyamex cruise (Cyamex scientific team 1978). (Copyright IFREMER, *Cyamex cruise 1978, Cyana Dive Cy78-12*)

Lava Pond

A specific type of volcanic feature called a lava pond or a lava-lake is the result of an extensive extrusion of very fluid lava that flows within a confined depression or graben-like structure. R. Ballard on the Galapagos spreading center in 1977 observed the first "lava ponds" or "lava lakes". However, it was on the EPR axis near 21°N during the Cyamex cruise conducted by Jean Francheteau in 1978 (Francheteau et al. 1979) that the first comprehensive description of these structures was made. Later that year, Jean Francheteau and Pierre Choukroune sat up overnight in a hotel room in Nicosia, Cyprus, in order to write a paper describing how these lava ponds could have been formed and their report was published in 1979.

It appears that lava ponds are the result of collapsed, elongated lava conduits, about 10–200 m wide and up to 200–300 m in length, which are parallel to the ridge axis and form en-echelon arrangements. They were built by a very fast extrusion of hot lava which poured into a depression or fissure formed during crustal extension. These lakes are the collapsed features of submarine flows after magmatic drainage and subsidence took place when lateral spreading out-paced the rate of magma effusion. Lava ponds are usually bounded by normal faults and are filled by successive discharges of very fluid flows. The pillars, which abound in lava ponds, are formed when seawater trapped in the crust underneath the hot lava pool was heated and expanded. The heated seawater would have been forced up through the molten lava in a series of near vertical conduits whose walls were

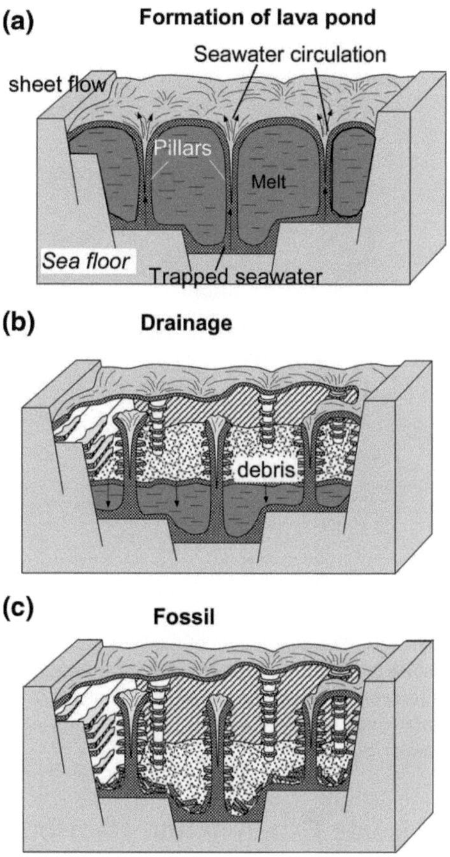

Fig. 5.11 Formation of a lava pond (Hekinian and Binard 2008). **a** Lava is pouring inside a depression forming a "lake" of hot melt with a chilled glassy crust forming the roof. Any trapped seawater underneath the lava pool will rise toward the surface and chill its immediate surroundings thereby forming pillars. **b** The lava lake will drain leaving the solidified rocks forming the roof and the pillars. **c** When the hollow lake is impacted by tectonic events, the roof will collapse inside the pond leaving a ruin of pillars and broken debris

quenched by the rising trapped seawater. Thus, lava ponds contain tall, nearly vertical, hollow pillars with rings of glassy ledges (Figs. 5.10 and 5.11).

As the lava cooled at contact with seawater it would form a succession of sheeted layers, which upon breaking, could leave basaltic ledges. We have observed this chilled-layer type of feature, similar to "bathtub rings", which is left around the cold walls of a depression. When the solidified crust collapsed, it revealed the presence of a depression filled with pillars showing layered columnar features. The appearance of a lava pond could also remind us of a ruined cathedral with a few standing columns after the roof has collapsed. Each layer on the columns, which is 10–20 cm thick, represents the cooling level of a flow. The various cooling levels are thought to reflect the inflation and deflation cycles of underlying magma reservoirs and are expressed as vertical fluctuations on lava pond surfaces (Ballard et al. 1979; Francheteau et al. 1979). Although lava ponds are formed when hot, fluid lava is rapidly extruded on the sea floor, they are often surrounded at their edges by pillow lavas, which are the result of a more viscous lava flow.

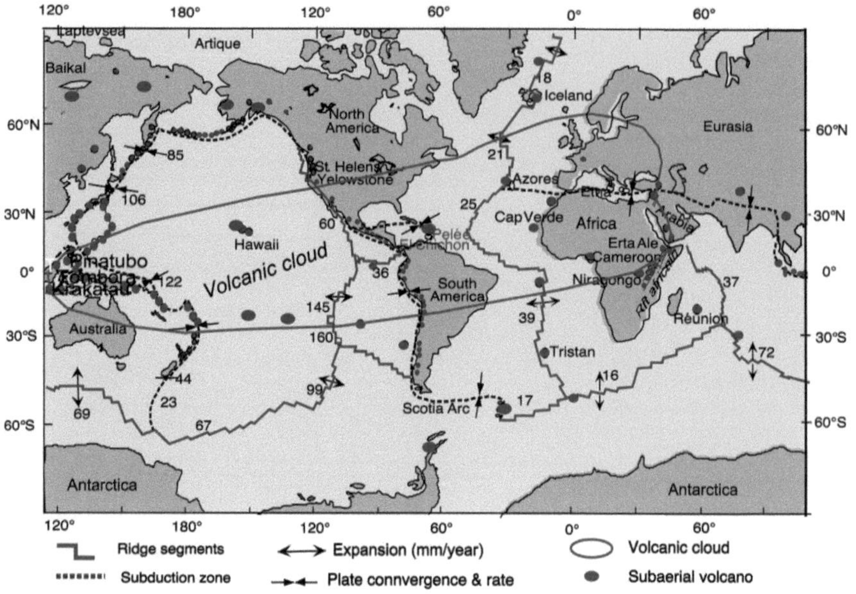

Fig. 5.12 The distribution of the World's subaerial volcanoes and an example of volcanic clouds discharged in the atmosphere during the eruption of the Tambora (1815) and the Krakatoa (1883) volcanoes in Indonesia and Mount Pinatubo in the Philippines in 1991 are shown. The volcanic clouds were detected as going from southwest Asia in the western Pacific Ocean eastward to Europe and Africa

Volcanic Eruption and Distribution

More than 80 % of Earth's volcanic activity takes place on the sea floor. The number of active volcanoes in sub-aerial regions is limited to less than 1,500 active sites when compared to the more than hundreds of thousands of volcanoes existing on the sea floor (Batiza 1981) (Fig. 5.12). The fact that most of the sea floor's upper 2–5 km are covered with volcanic deposits indicates that the main phenomenon building our planet is volcanic activity. The type of volcanic events which take place will depend on the intensity of eruption, or in other words, on the rate at which lava exits from its vent.

During volcanic eruptions when gases and vapor phases are associated with fragmented material they will form pyroclastic flows, and will give rise to catastrophic explosions. Quieter eruptions will occur without the release of large amounts of gases and with limited explosive events. The quieter eruptions are often referred to as fissural events because basaltic lava pours out through linear, fissured terrains such as for Hawaiian volcanoes and on most submarine spreading ridges. Volcanic activities associated with hydrothermalism contribute to the transfer of mantle heat and material between the hydrosphere and the atmosphere.

Type of Volcanic Rocks Formed During Explosive Activity:

Pyroclasts *(from the Greek words "pyros" meaning fire and "klastos" meaning broken) are fragmented pieces of material that were ejected during explosive volcanic eruptions. They refer to the observed scattered blocky or bomb-shaped "clasts" that are fragments of various sizes from a few centimeters up to several meters in diameter (See* Chap. 5). *Pyroclasts clustered during a volcanic eruption take various forms*: (1) *isolated bombs which were blasted from a volcano during explosion or* (2) *when mixed with water, these hot "bombs" form "lahars", which are mixed mud and rock debris that could travel long distances* (100 km) *from their point of origin, or* (3) *avalanche deposits that have rolled down slope.*

Volcanic breccias *refer to accidental explosive rock debris as well as mechanically transported rocks caused by gravity slumping along a slope. Volcanic breccias are cemented with silt and sand size material. This cementing takes place during diagenesis (physical and chemical changes undergone by the material after its deposition) and it is due to volatile exchanges (CO_2, H_2O, hydrated silica solutions) and/or during low-temperature hydrothermal fluid circulation.*

Hyaloclastites *(from the Greek words "hualos" meaning glass and "klastos" meaning broken) usually form flat lying layered deposits or rubble debris of broken up slabs composed of silt to sand size volcanic glassy shards cemented by either a sediment matrix and/or by hydrothermal precipitates. The glass shards are formed during the explosive activity when magma comes into contact with water. After the shards have fallen on the sea floor, low temperature hydrothermal precipitates of Fe and Mn-oxyhydroxides will cement the glassy shards to form slabs of hyaloclastites. The cementing products include hydrated material such as goethite, limonite and palagonite* (Nayudu 1962). *These slabs could be coated with a thin manganese-enriched crust composed of birnessite as well as crusts of the clay minerals. Located on top of lava flows and pelagic sediment, hyaloclastites are mainly encountered on seamounts. The hyaloclastite slabs could extend over hundreds of meters and cover sand-sized loose debris extruded during explosive volcanism (lapilli).*

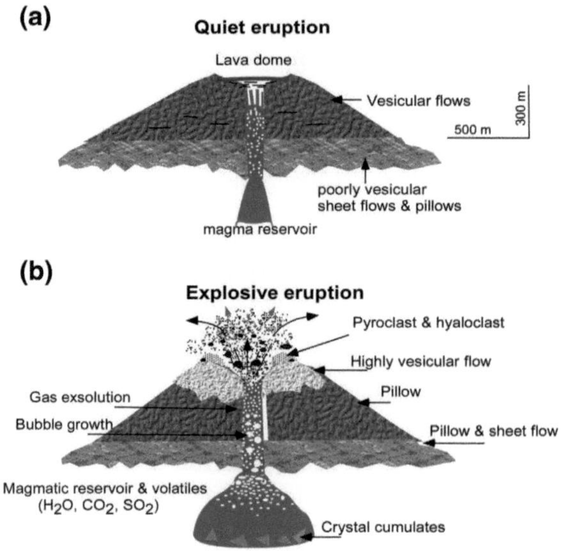

Fig. 5.13 Model of a submarine a volcanic edifice at different eruptive stages: **a** A quiet stage of eruption gives rise to a lava dome with gentle slopes. The magma column is accompanied by a small amount of gases and the lava erupted consists essentially of basalt. **b** An explosive stage is characterized by the presence of a large amount of gases and volatiles that have segregated at the top of the magma column. The volatiles consist of H_2O, CO_2, sulfuric acid (H_2S), hydrofluoric acid (HF) and other trapped volatiles, which enhance bubbles of gas producing explosive volcanism. During explosive events fragmented rocks from the edifices are extruded on the surface forming pyroclastic deposits. Explosion will take place when the pressure of the gases has exceeded that of the surrounding material

Volatiles and Gas in Magma

Magma is a viscous mix of molten rocks that rises towards the Earth's surface. Magma also contains highly volatile elements such as H_2O, C, and S, which are related to the primordial composition of the Mantle (Hart and Zindler 1989). If magma degases during volcanic eruption, this will give rise to emissions of H_2, H_2O, CO_2, SO_2, H_2S, S_2, HCl, CO, HF, He, Ar, N_2, CH_4, NH_3, in decreasing abundance. The concentration of these volatiles will vary according to the type of eruption. Satellite measurements have shown that hydrogen sulfide (H_2S) released by volcanoes is estimated to be as much as 1–25 Tg (megatons) per year along with sulfur dioxide (SO_2).

Partial melting of mantle peridotite gives rise to the material that supplies the magma reservoirs. Inside a reservoir, the molten material will subsequently undergo a process of crystal-liquid fractionation and therefore gas complexes will be formed (Fig. 5.13). These gaseous phases are either re-dissolved in the melt or they could escape during volcanic eruptions. The major constituents of volcanic gases are H, C, O, S, N, Cl, F, Br and rare gases such as He, Ne, and Ar. Hydrogen is largely present as H_2O along with other hydrogen bearing species such as methane (CH_4) ammonia (NH_3). Carbon (C) is mainly released as carbon dioxide (CO_2), methane (CH_4) and carbon monoxide (CO). Other gases such helium, argon and xenon are found in trace amounts and are highly volatile thus they seldom form solid compounds with other elements since they are easily lost during degassing.

Helium is believed to be a primordial element inherited from the condensation of the solar nebula and stored within the Earth's mantle during the formation of our planet. However, it is also probable that much of the Earth's helium is produced from the radioactive decay of uranium and thorium within the Earth (Broecker 1985). The Helium anomalies detected in the water column in the South Pacific near 15°N were attributed to hydrothermal and/or volcanic emanation along the EPR that was traced to being more than 100 km away from its source (Lupton and Craig 1981). All these gaseous phases are found in the hydrosphere and atmosphere and can be used as tracers for detecting volcanic and hydrothermal activities under the sea.

Experimental work has shown that explosive eruptions are theoretically impossible at depths greater than that of the seawater's critical point, that is at 315 bars, which is equivalent to a depth of 3000 m for a temperature of 405 °C (Francis 1995; Bischoff and Pitzer 1985, 1989). Rittmann (1944) was the first to point out that above seawater's "critical point", gaseous explosions are unlikely to take place and the lava will merely flow quietly out onto the sea floor. The critical point is where all three phases for water (i.e. gas, liquid and vapor) can coexist together. Indeed, at a temperature and pressure above the critical point, water and vapor are indistinguishable. Below this critical point, vapor phases show a volume expansion at a given temperature. However, increasing the salinity of the vapor phase will favor an increase of the depth range at which any vapor and liquid phases could coexist. If boiling occurs, density differences between vapor and liquid could lead to a spatial separation of the two phases and give rise to explosion. When seawater is trapped in a porous conduit and sealed during alteration, it could be heated and form water vapor, which will also give rise to explosive activity. Furthermore, volatile-rich magmas may form significant pockets of gas, thus, if they are concentrated in the confined environment of a magmatic conduit, they could expand and burst. In addition, crystallization taking place in a magmatic reservoir will concentrate the volatiles in the residual melt, and this may cause large amounts of exhaled bubbles to migrate to the top and the margins of a magmatic conduit, thus increasing the magma's fluidity and/or gas pressure.

The degree of rock vesicularity plays an important role in indicating the content of volatiles in magma. Field observations show that rocks composed of magma that is enriched in volatiles, such as the silica-rich and alkali-rich lava from hotspot provinces, are highly vesiculated (>30 % vesicles) even if the magma was extruded at depths greater than 3000 m (Binard et al. 1992; Hekinian et al. 1991). These vesicles, which are small cavities caused by residual gas bubbles remaining in a lava before it solidified into a rock, are filled with gas that originated in the lithospheric environment prior to the lava's eruption. The volatile contents in oceanic rocks usually form minor constituents when compared to the other elements. For example, volatiles such as HCl, CH_4, CO_2, CO, SO_2 H_2S and H_2O, constitute less than 1 % of the bulk rock and usually only 0.50 % of sea floor basalts.

The amount of dissolved water and carbon measured by Pineau et al. (2004) on Mid-Ocean Ridge Basalts (MORBs) collected from the Mid-Atlantic Ridge segment at 34°50′N during the *Oceanaut* cruise ranged from 1125 to 5253 ppm (part per million or 0.11–0.52 % of the bulk rock) up to 20–119 ppm (0.002–0.011 %). The basalts most enriched in their incompatible elements (E-MORB = enriched MORB) have the highest contents of water and carbon. Even with such a low concentration of volatiles, repetitive magmatic intrusions trapping the volatiles within the magmatic column are likely to increase their concentration and pressure with respect the surrounding formation.

It was shown (Hekinian et al. 2000) that there is a correlation between an erupted rock's vesicularity and its composition (such as P and K) as well as a correlation between the amount of total water and the CO_2 content. This implies that both gaseous carbon and water contents will depend on the initial concentration of these elements in the mantle and this will also be related to the rate of partial melting in the mantle. Pineau et al. (2004) studied the CO_2 content of lava erupted on the MAR and have shown that carbon saturation is reached at increasing depths and pressure. They found, on the basis of laboratory measurements, that carbon is saturated in magma at pressures from 2.6 to 12.8 kbar, which corresponds to depths of 9–44 km in the mantle. Thus the degree of vesicularity is mainly conditioned by the magma's original carbon enrichment and its carbon exsolution during magma ascent (Bottinga and Javoy 1989). When volatiles are trapped within a melt, eventually vesicle expansion will lead to explosive activity and the formation of pyroclastites, vitric tuffs and hyaloclastites (Fisher and Schmincke 1984).

Pinneau et al. (1998) and Hekinian et al. (2000) have shown that the partition between water and CO_2 is such that during magmatic production water tends to dissolve in the melt and will remain in the magma, while CO_2 will rise more rapidly and accumulate at the top of the magmatic column. In the magma column, CO_2 will form bubbles and the rapid ascent of volatile enriched melts in a magmatic column will produce an increase in vesicularity (Fig. 5.13). Thus, CO_2 is the main gaseous phase existing in the vesicles just before their quenching (rapid solidification). An increase in vesicularity will accelerate the decoupling or separation of the magma-bubbles system. Consequently, the gas pressure exerted on the magma column (CO_2 in the bubbles) will rise until it overcomes the load pressure overlying a volcanic construction and thereby enhance an explosive event.

One of the best documented studies on the distribution of volatile contents in sea floor lava was made on a suite of samples from an eruption on the East Pacific Rise at 9°50′N which occurred in 2005–2006. The samples were collected by submersible at about 200 m intervals along the pathway of a single eruptive event and were examined by Soule et al. (2012) in order to determine their degassing process. Measurements were made concerning the vesicularity, crystallinity, and the volatile and helium contents of the collected samples. Observations on the CO_2 loss from the melt and a subsequent bubble growth suggest some constraints for the dynamics of eruption. Based on the amount of CO_2 exsolution between proximal and distal samples, Soule and his colleagues estimated the duration of the eruption to be about 30 h (Soule et al. 2012). For a flow thickness of about 1.5 m, a mean eruption rate of $\sim 25\ m^3$ s/km was proposed (Soule et al. 2007). This eruption rate is similar to that of the Hawaiian lava flows. These results showed that the most recent EPR eruption must have occurred during a series of short-duration pulses rather than during a continuous eruptive event. A lava flow velocity of 0.12 m/s was estimated for the early eruption; however, that flow velocity rapidly declined to 0.02 m/s for the entire period of the eruptive event.

Popping Rocks

The first time I witnessed the release of gas that had been trapped in volcanically erupted material was in 1972 during the *Midlante* cruise with the N. O. JEAN CHARCOT. The MIDLANTE cruise was a preparatory survey cruise for project FAMOUS in the North Mid-Atlantic ridge near 36°49′N and 33°15′W. After a dredge haul was collected on the axis of the rift valley at 2480 m depth, Jean Francheteau and I were looking at the rocks that had been dumped on the back-deck. To our surprise, we heard noises that sounded like pops, just like the sound of popcorn being cooked. We realized we were dealing with samples that were jumping and popping due to the pressure of degassing. We called them *"popping rocks"*. Suddenly, an oval-shaped rounded glassy rock fragment shot up into the air a couple meters away from us. Then several other chunks did the same thing. We figured out that the rocks were degassing because of the difference in atmospheric pressure when they were brought to the surface from the sea floor's depths. We immediately stored these samples inside a rough-woven potato sack along with a tape-recorder inside the bag to record how long it would take the samples to degas. The bag was stored in the science laboratory on board the ship. Unfortunately, we were not equipped with any sealed gas containers to store the samples and collect the gas that was being released, but we made sure to bring such material for future missions. The rocks exploded and popped for 3 days giving off a sharp smell of rotten-eggs as they split open (Photo 5.14). This preliminary discovery was reported in an article published in *Nature* (Hekinian et al. 1973) .

When the popping rocks dredged from the 1972 cruise were taken back to a specialized CNRS (*Centre national de Recherche Scientifique*) laboratory in Paris (*Institut de Physique du Globe*), M. Chaigneau, a specialist in gas analyses, discovered that the volatile content of these rocks consists mainly of carbon dioxide (39–45 %) and carbon monoxide (13–17 %), hydrogen dioxide (27–35 %) and sulfide oxides (2–11 %) (Chaigneau et al. 1980). Later, Marty et al. (1983) studied

Fig. 5.14 Photograph of glassy basalt that are called "popping rocks" or "jumping stones". Exfoliation takes place on the glassy surface of a recently erupted flow crust during degassing and gives rise to glassy shards such as these

the noble gas contents such as helium (He) and argon (Ar) and they also discovered that traces of these noble gases are associated with the vesicles containing mantle-derived CO_2. Other "*popping rocks*" of basaltic composition with a vesicle content up to about 17 % in volume were recovered at about 3,500 m depth on the Mid-Atlantic Ridge rift valley near 14°N. The gaseous phases of these samples consisted essentially of H_2O (23–26 %) and CO_2 (29–77 %) and were studied by Javoy and Pineau (1991).

Hyaloclasts and Pyroclasts: Their Origin

Explosive and/or forceful eruption of lava on the ocean floor is evidenced by the presence of both hyaloclastites and pyroclast debris found close to their source. The increase in CO_2 forming bubble pockets on top of the magma column is the primary cause for the formation of vesicles in a mush-like (partially solid/partially liquid) magmatic environment. In this type of environment, the gaseous phases, H_2O (in the melt) and CO_2 (forming bubbles that can create vesicles in solid rock) are separate from the melt. An explosion is initiated at the top of the magmatic column after any gas vesicles or gas pockets have broken up.

During an explosive magma extrusion on the sea floor, the larger droplets of hot melt containing gas bubbles are quickly chilled and fragmented at contact with seawater. The vesicle walls will then break and form small glassy shards, which, after deposit and a subsequent mixing with sediment and/or rock debris, will become cemented together on the sea floor due to the action of hydrothermalism. A hardened

Fe–Mn crust could subsequently cover the cemented shards. At the same time, the coarser fractions (found inside the extruded lava) are made up of fragmented rocks stripped from the magma chamber's walls or conduits. These larger pieces will solidify more slowly than the finer shards. Because of size differences between the small millimeter-size of the shards and centimeter size of the larger fragments, they are sorted in the water column during eruption and will settle and solidify at different rates to form hyaloclastites (fine-grained glass shards and rocks) and pyroclasts (coarser debris). The pyroclasts, which have most of their vesicles preserved, probably got their vesicle-forming gas bubbles from a deeper, higher-pressure zone in the upwelling magma (Pineau et al. 1998). The highly vesicular pyroclasts found in lava are easily fragmented during volcanic explosion and extrusion.

The observed lithological sequence shows pyroclasts lying under hyaloclasts (on the Pitcairn hotspot volcanoes, at the MAR at 34°54′N, and in other locations) due to the different rate of sedimentation in the water column related to the size, morphology (shape) and density of the particles projected into seawater. Such types of explosive events are directly linked to the magmatic volatile content and are similar to the "lava fountaining phenomenon" observed in subaerial Hawaiian volcanoes during a high rate of magmatic upwelling in a narrow conduit where exsolved volatiles are concentrated in large bubbles (Jaupart 1996).

Deep-sea explosive events can also take place in a conduit within a fractured and collapsed volcanic cone filled with fragmented debris. Seawater circulating in such an environment could be trapped after hydrothermal precipitates and/or lava sealed the voids of the fractured edifice. The arrival of a new magma batch within the edifice will enhance the formation of a vapor phase caused by heating the trapped seawater leading to a subsequent increase in pressure, which could trigger an explosive event. This can explain the presence of mixed hydrothermal precipitates and explosive volcanic rock debris, which have been found on the sea floor after volcanic activity.

Sub-Aerial Versus Submarine Explosion

Sub-aerial (above the surface of seawater) eruption gives rise to explosive events that are more dramatic than volcanism taking place in an underwater environment because they are frequently devastating for nature and for human beings. Many sub-aerial volcanoes have their roots below the ocean floor while others, created on continental crust, have nevertheless undergone similar processes to what takes place inside submarine volcanoes. Numerous examples of the "fire-fountaining" eruptions associated with Hawaiian volcanoes have occurred during sub-aerial eruptions (Macdonald 1982) as well as in undersea environments.

The observed "fire-fountaining" is due to explosive eruptions ejecting volcanic ash (small size debris) and rock fragments that vigorously flow down slope in sub-aerial terrain as well as in submarine environments. This style of eruption is also called "strombolian" (the name is given after the Stromboli Volcano on Lipari

Island, Italy). It is a term used to describe an eruption of fluid, basaltic lava. An explosive eruption in which fragmented magma and gas is released from a volcano at high velocity is called a Plinian style of event. The etymology comes from Pliny the Younger who first described this type of eruption taking place on Mount Vesuvius in 79 A.D. Small-scale sub-aerial explosive volcanic eruptions inject more than a million cubic meters of ash into Earth's atmosphere every month (Simkin and Siebert 2000).

Back in the sixties, scientists (such as McBirney 1963) thought that submarine explosive volcanism could not take place at depths greater than a few hundred meters. It was postulated that depths of no more than 500 m for depleted melt and 1000 m for alkali enriched melts, would be the limits for this type of volcanism. Thus, it was believed that the fragmented debris (pyroclasts and hyaloclasts) resulting from explosive eruptive events would be absent because the pressure of the overlying water column would be sufficient to suppress juvenile gas exsolution. In fact, during previous recovery of deep-seated ash and pyroclasts, the presence of these products under the sea were often attributed to sub-aerial eruptions whose explosive clasts and ash could have been subsequently transported either in the atmosphere or by ocean currents.

Although fine ash is likely to be transported in the atmosphere as has been observed during modern sub-aerial eruptions, it is evidenced that in situ deep-sea explosion can also give rise to this type of deposit. Most of the records of oceanic explosive events come from Archean ancient pyroclastic deposits found in continental areas. Others are from back-arc basins surrounded by island volcanism such as the Okinawa Trough, the Lau Basin and the Tonga-Kermadec Ridge submarine volcanoes. The pyroclastic deposits found at 1440–2250 m depths were interpreted as being formed in the submarine environment (Gill et al. 1990; Wright and Gamble 1999). When coring submarine sediment layers around the islands of Sumatra and Java, it was found that deposits of pyroclastic flows more than 20 m thick poured out during the eruption of Krakatau. This deposit was carried on the sea's surface and was deposited over a radius of about 70 km forming a semicircle to the northeast of the volcano, in the same direction as the tsunami wave of 1983 after the explosion of the Colo volcano on Una-Una Island.

Submarine Volcanic Explosion

We now know that underwater volcanism could include both explosive and quiet types of events. Because seawater's critical boiling point (315 bars and 374 °C) corresponds to about 3000 m depth, it is impossible that explosive events could take place at deeper levels, no matter how many volatiles are concentrated in the magma. Even for the highly vesicular lava flows encountered on the ocean floor, it would be impossible for the vesicles to nucleate, expand and grow at depths greater than 3000 m.

However, there is compelling evidence that explosive volcanism can and does exist up to 3000 m depth. It is of common knowledge that vesicular lavas with gas exsolution forming submarine flows are prevalent. The fact that highly vesicular lavas and fragmented scoriaceous blocks (pyroclasts) are also found on the ocean floor at relatively great depths (>1500–3000 m) has changed scientists' point of view concerning the occurrence of under water explosive activity.

The main difference between submarine and sub-aerial explosive volcanism is that when we approach atmospheric pressure, water is the main volatile to degas from magma, whereas in a deep-sea environment (>1000 m depths) about 85 volume percent of the volatile which degases is CO_2. This is because H_2O dissolves more easily in molten magma than CO_2. The formation of CO_2 bubbles (nucleation) can start to take place at depths of about 60 km under pressure of 20 kbar (Blank and Brooker 1994). This was extrapolated by measuring the amount of bubbles trapped in magma (bubble density per total volume of gas) (Burton et al. 2007).

The "quiet type" of volcanic eruptions are the commonest form of volcanism taking place on the sea floor giving rise to basaltic pillow lava and sheet flows with minor gas exsolution and very little vesicle formation (<5 %). These types of samples are found along spreading ridge segments. However, the increase in the degree of vesicularity (>15 %) and the occurrence of fragmented debris and ash products at great depth (>1500 m depth) suggests that explosive types of volcanism must also have taken place. Intra-plate volcanism giving rise to highly vesicular lava containing up to 15–30 % vesicles is commonly found in the Pitcairn, Austral and Society hotspot areas, as reported by Binard et al. (1991).

Proof that submarine explosive events took place at depths greater than 1000 m was found on several sea floor structures located at a significant distance from emerged landmasses. The presence of hyaloclastites on off-axial volcanic edifices in the northern Pacific were reported by several authors (Batiza et al. 1984; Bonatti and Harrison 1988; Fornari et al. 1984; Smith and Batiza 1989; Hekinian et al. 1989). During two IFREMER cruises on slow spreading ridge segments of the Mid-Atlantic ridge (*DIVA 1 cruise in* 1994 and *OCEANAUT cruise in* 1997), evidence of explosive activity was collected by submersible. Accidental rock debris of pyroclasts and hyaloclastites were observed and sampled by the submersible *Nautile* at more than 1000 m below the summit of volcanic edifices in the rift valley of the Mid-Atlantic Ridge.

Submarine explosive deposits of pyroclastic debris are often associated with hyaloclastites. The first pyroclast-hyaloclastite deposit was observed on-site in 1997 on the MAR (Mid-Atlantic Ridge) at 38°20′N (site called "Menez Gwen", meaning "White Mountain" in Breton) and at 37°18′N (Lucky Strike site) southwest of the Azores on the central volcano of the rift valley's inner wall. These two sites had layered volcaniclastic deposits at least 400 m thick (Ondreas et al. 1997; Eissen et al. 2004). The nature of the clasts found in these deposits consists of highly vesicular altered shards of scoriaceous glass associated with other rock fragments. In the *Menez Gwen* ridge segment (38°20′N), the pyroclastic deposits extended over a large area of centrally located volcanoes in the rift valley of the

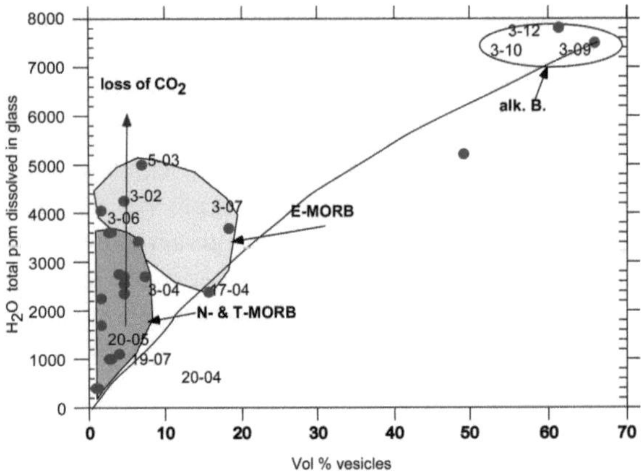

Fig. 5.15 Volume percent versus water (H_2O total ppm) distribution of basaltic rocks collected during the Oceanaut cruise from the MAR (Mid Atlantic Ridge) segment at 34°50′N after Pineau et al. (2004). *N-MORB* Normal MORB (Mid Ocean Ridge Basalt), *E-MORB* Enriched MORB and *AlkB* alkali basalt. The enrichment in volatiles (i.e. water) increases with the degree of rock vesicularity

MAR axis at 550–1750 m depths. The summit of these edifices had been cut by a graben about 2 km wide and 350 m deep (Ondreas et al. 1997) showing altered, coarse-grained pyroclastic flows covering basaltic pillow lava.

The second pyroclastic deposit, also discovered by the submersible *Nautile* during the *Oceanaut cruise*, forms the crater of a volcanic cone that had collapsed during an explosive eruption that took place at about 1400–1900 m depth near 34°50′N on the MAR (Hekinian et al. 2000; Bideau et al. 1998) (see Chap. 7). These deposits were collected from the Median Ridge located on the axis of the MAR at 34°50′–35°N that was formed by several volcanic cones which had coalesced together to form a volcanic ridge. Two of the edifices that were visited contained poorly cemented fragmented rocks of pyroclasts and layered glass shards of hyaloclastites forming the upper part the volcanic structures.

Laboratory analyses on the composition of these pyroclasts showed that they consist essentially of alkali basalt and E-MORBs and they are among the most enriched in incompatible elements and volatiles. They contain high K/Ti (>0.6–0.8) and total alkali ($Na_2O + K_2O > 3$ %) contents and show a higher content of gas vesicles (>50 %) than the less vesiculated samples (Normal and Transitional MORBs) found near the base of the Median Ridge (Fig. 5.15). When compared to the area near 35°50'N, the pyroclastic flows reported by Eissen et al. (2004) from 38°20'N on the MAR also contain more alkali basalts and E-MORBs than depleted MORBs.

On a microscopic scale (during petrographic and microprobe analyses), the glassy droplets found in the gas cavities had the same composition as the surrounding bulk glassy matrix, which formed the lava. This indicates that the magma producing the pyroclasts was a basaltic melt whose volatile contents had increased during its ascent towards the sea floor surface (Fig. 5.13). Prior to eruption, the basaltic source material for these explosive, volcanic rocks from the MAR, had crystallized less than 15 % solids within the parental liquid.

Hence, pyroclasts and hyaloclastites resulting from submarine explosion could occur on both fast and slow spreading ridge systems and seamounts. Finding these pyroclasts and hyaloclastites at some distance away from emerged landmasses indicates that underwater explosion can occur at less than 3,000 m depth, and this type of explosive volcanism occurs more frequently than had been thought.

Growth of a Volcano

The construction of an underwater volcano is related to the eruption of molten lava as well as to the propagation of sub-crustal melt forming a magma reservoir and its conduits such as dykes and sills. From the reservoir, magma will flow through both vertical and lateral fissures to form the skeleton of a volcanic edifice (Figs. 5.15 and 5.16). If the magma solidifies within the edifice during its intrusion, it will form a series of vertical dykes representing the conduits of ascending melt. These magmatic conduits supply the lava flow that will be extruded on the volcano's surface. Each eruption is the result of magmatic injection taking place at different levels within the volcanic edifice. As a volcano grows, its mass will increase, therefore it will rise taller and as a result, less magma will erupt from the summit while more lava will erupt from lateral vents causing the creation of adventive or parasite cones. In order for the extrusive flow to continue piling up, it also has to overcome the surrounding lithostatic pressure.

During crystal-liquid fractionation, a melt in the magma reservoir will evolve and become more enriched in volatiles. The volatiles and the lighter elements will tend to rise towards the roof and/or the cooler borders of the magma chamber, while the heavier components will sink. Also, the replenishment of a magma reservoir is an important consideration after a sequence of eruptive events since if there is an arrival of new melt further eruptions will eventually take place.

Cycles of volcanism, and the repetitive nature of volcanic cycles, will contribute to build a volcanic edifice. Volcanic stratigraphy of successive lava flows is easier to observe on intraplate volcanoes of hotspot origin and on slow spreading ridges where magmatic activity builds tall volcanic cone structures. (See Chap. 9). On fast spreading ridge segments, volcanic cycles and successive magma upwelling takes place during fissural eruption along the strike, so the resulting structure seems to be flatter, and forms somewhat continuous dome-shaped features. The cyclical nature of volcanic events is observed on sub-aerial volcanoes where there is evidence of short, sequential volcanic activities (over periods of days) as well as longer sequences (over periods of thousands of years).

The successive accumulation of lava poured out on top of previous flows will build a volcanic edifice that could eventually reach above sea level. However, it is obvious that the growth of a submarine volcano will depend on numerous factors.

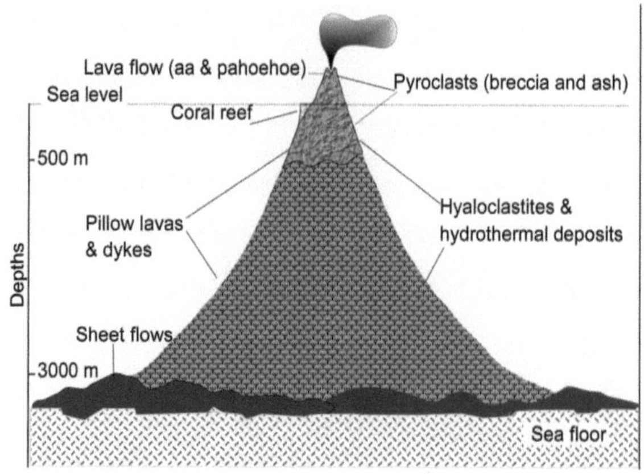

Fig. 5.16 The formation of a volcanic edifice sketched from observations made on individual edifices constructed in intraplate abyssal plain regions. The different styles of volcanic activities give rise to various types of eruptive flows. The initial stage of eruption on the deep sea floor gives rise to fluidal lava or sheet flows containing very few volatiles. These are lobated and flat flows. At shallow depths (<700 m depth) explosive eruptions give rise to pyroclasts and ashes

These factors are the supply rate, the rate of magma ascent, size of the conduit, the fluid density and the time scale of the eruptive cycles. The total lava flow delivered along the belt of submarine mountain ranges formed at the spreading ridge segments around the World is on the order of about 4 km^3/year. It is assumed that during passive upwelling, the average rate of mantle flow is about 13 mm *a year* along the MAR. Such upwelling rates vary according to the tectonics of spreading and the supply of mantle plume-driven decompression melting, which could give rise to hotspot volcanoes. The magma ascent velocity for the Mt. St. Helen's eruption was about 2–3 m/s in May 1980 and dropped to 0.08 m/s 5 months later.

The seismic monitoring and field observation of volcanic edifices has permitted us to propose an estimation concerning their growth rate. Earthquakes are commonly associated with magma ascent giving rise to volcanic eruptions. Thus, the construction of an isolated submarine volcanic edifice starts on the surface of the sea floor and it will take between 10,000 and 1 million years to built a sizable (>1000 m high) structure. It is calculated that building a submarine edifice of 3500 m tall such as the Bounty Island in the Pitcairn hotspot region, took about 58,000 years with an eruption rate of about 0.54 km^3 per 100 years. The Kilauea volcano on the Big Island of Hawaii is one of the most active volcanoes on Earth. It is about 1000 m above sea level and about 6000 m above the sea floor, yet it was built in less than 1 million years since we know that the present day lava eruptions are about 10–18 km^3 per 100 years (Clague et al. 2000). From geodetic and seismological observations, it has been suggested that magma rises through conduits located underneath volcanic edifices and propagates through fissures in rift zones. The supply of magma to the rift zones is sometimes accompanied by shallow (1–4 km depth) swarms of earthquakes corresponding to magma intrusion forming shallow dikes. Thus, the Kilauea volcano's magma supply is estimated to be 0.18 km^3/year (Cayol et al. 2000).

Other sub-aerial volcanoes forming the island of Sao Miguel (in the Azores) have a smaller production rate than Hawaii (0.02–0.03 km^3 per 100 years, Moore 1991) so it probably took longer to form an island in this area. During 1977, the Karla volcano in Iceland was monitored at the same time as an earthquake while the volcano deflated and magma poured along the flanks forming rift zones. It was calculated that the tremor hypocenters migrated at a rate of 0.5 m/s along the rift zone (Rutherford and Gardner 2000).

One of the tallest volcanoes in the world is located on Big Island of Hawaii. This now inactive volcano is called Mauna-Kea and is about 8000 m high from its base on the sea floor. It is taller than mount Mt. Everest in the Himalayas and its summit culminates at 4100 m above sea level. Because of its privileged location standing above the clouds, Mauna-Kea has become the site of an international astrophysical observatory. Seven nations have built observatories on top of this volcano. On the flank of the volcano at an altitude of 2000 m, there is a relay-zone rest station for the staff to get accustomed to the high altitude while they spend the night before going up to work in the observatories on the top.

Historical Eruptions

Even if there are fewer active volcanoes on emerged landmasses than under water, they are the most spectacular because of their direct impact on human society and our economy. During the past 500 years, volcanic eruptions have been responsible for more than 200,000 deaths. It is of primordial interest for human beings to have knowledge of the sea floor environment in order to better understand the physical processes involved in producing natural hazards. For example monitoring deep trenches in subduction zones and recording tectonic activity in active major faults that can trigger earthquakes, such as the earthquake causing the tsunami of Fukushima in March, 2011, might help to inform people before other natural catastrophes occur. These tectonic stress-induced activities may be interrelated and could affect structures that are thousands of kilometers away from each other.

Society expects Earth scientists to provide adequate and accurate warnings for natural hazards. In order to estimate the risks of eruptions and/or earthquake hazards, it is important for geologists to have a good knowledge about the history, the composition and the structure of the area at risk. There are national and international associations, which are involved in surveying and monitoring natural hazards, but this is not yet operational everywhere in World. The lack or absence of a continuous monitoring system in naturally high-risk areas could be disastrous to humans. Blong (1984) estimated that on an average about 640 people per year died from natural hazards in the twentieth century. Unfortunately, it is only after a major disaster that government findings are made available for studying natural hazards. For example, we now know that the Pacific ocean is one of the best seismically monitored regions of the World with stations on the Hawaiian islands, in French Polynesia, and on the west coast of the USA, while the other oceans such as the south Atlantic, the Arctic and the Indian oceans, are relatively deprived of seismic monitoring and warning systems.

Today about 10 % of the Earth's population lives near the 1,500 active subaerial volcanic edifices. This is due to the nature of the fertile soil, which is excellent for growing crops. In 1983, prior to the eruption of the Colo Volcano on Una-Una Island of the Sulawesi province North of Java (Indonesia), an Earthquake swarm followed by a volcanic eruption occurred on July 14–18, 1983. Because Indonesian geologists had gained experience from past eruptions of a similar type of volcano, they recommended that their government evacuate 7,000 people from the Island, which they did, and thereby saved their lives. A devastating tsunami and pyroclastic flow destroyed the island completely (Katili and Sudradjat 1984). In 1980, when the Mount St. Helens volcano erupted in Washington State, it was known that this volcano had been active about every 100–140 years since 1400 A.D. (Fisher et al. 1998).

The composition of the magma that feeds an edifice will also determine the type of eruption. The most dangerous for humans are the so-called pyroclastic (explosive) types of eruption that carry a cloud of broken fragments and gas. These flows move faster than wind-driven hurricanes and can reach temperatures as hot as 800 °C. Lava will flow as quickly as 100 km an hour and over distances of more than 100 km away from the source. Such types of volcanoes can damage cars, bury towns and burn entire forests. It was this type of pyroclastic eruption from Mont Pelée which destroyed the city of St. Pierre in Martinique in 1902, causing 30,000 deaths in only a few minutes. New methods of detection, such as satellite images and local observatories on or near volcanoes and which have been used since the end of the twentieth century, have enabled increased understanding concerning catastrophic volcanic eruptions. (Fisher et al. 1998).

Looking at the examples of the Mt. St. Helen's eruption in 1980, Mt. Unzen in Japan in 1991, and Mt. Pinatubo in the Philippines in 1991, we know that these volcanoes' activities were detected and the evolution of their volcanic clouds was closely monitored. In other areas, volcanoes are still killers. For example, the eruption of Nevado de Ruiz in Columbia in 1985 was more disastrous than what happened after the explosion of Mont Pelée 80 years earlier. A glacier capping the volcano at an altitude of 5400 m high collapsed and the subsequent mixture of muddy water and volcanic debris swept down the slopes of the volcano causing a devastating mud flow that killed 25,000 people 5000 m below the summit of the edifice.

The Pinatubo volcano, after being inactive for more than 500 years, suddenly came alive on June 15, 1991. The day before, Philippine officials ordered the population of more than eighty thousand people to evacuate the surroundings (Sigurdsson et al. 2000). The explosion of the Pinatubo volcano deposited voluminous pyroclastic flows extending up to 16 km from the vent and distributed over an area of more than 400 km^2. Only three hundred people were killed by this eruption, thanks to early warnings by scientists. I was in the south Pacific and saw the sky was completely red due to the sun's effects on the huge volcanic cloud.

In June 1991, the memorable eruption of Mt. Unzen in Japan took the lives of the internationally famous volcanologists Maurice and Katia Krafft and their colleague Harry Glicken along with forty Japanese journalists. The volcano's summit was fragmented during the explosion, which released a cloud of shattered rocks and produced ashes and gas. As the coarse and denser fragments followed a gorge curved along the flank of a previous eruption, the smaller fragments mixed with toxic gas and ashes rose in a deadly cloud more than 10 m high. The two volcanologists and accompanying journalists were standing about 10 m above the gorge and were caught in the toxic cloud that suddenly enveloped the observers along its path.

Other historical eruptions that were disastrous for the people living nearby are the Santorini eruption in the Aegean Sea that probably was the cause of the disappearance of the Minoan civilization in 2000 B.C., the eruption of Vesuvius in 79 B.C., and the Laki volcanic ridge in Iceland, which erupted in 1783 causing about ten thousand deaths. In Indonesia, the Tambora volcano in 1815 and the Krakatau volcano in 1883 killed 92,000 and 36,417 persons respectively (Fisher et al. 1998). The eruption of Tambora released about 50 km^3 of silicic lava (trachyandesite) reaching 43 km in altitude (Francis 1995).

The explosion of Krakatau in 1883 was heard on Rodriguez Island in the Indian Ocean at more than 4653 km away from Java and quakes were felt on Singapore at 840 km to the west. The dust cloud released by this volcano circled around the World. In 1928, the "child" of Krakatau, Anak Krakatau, emerged from the sea and has since grown to a height of 315 m.

The 1982 eruption of El Chichon volcano in Mexico was tracked by remote-sensing satellites and showed a continuous eruption cloud circling the World for twenty days in April 1982 (Rampino and Self 1984). More recently, on May 23, 2011, the eruption of the Grimsvotn volcano in Iceland sent thousands of tons of volcanic ash into the sky to an altitude of more than 12 miles. The cloud forced the closure of Icelandic airspace and spread fears of a repeated eruption while global airplane traffic was seriously interrupted for several weeks.

Volcanic clouds emitted during eruption have considerable effects on the composition of the atmosphere. For example, ozone, which is made of 3 oxygen atoms (O_3), forms a layer in the atmosphere and this is important for living organisms because the ozone layer can absorb ultraviolet radiation from the sun. However, ozone reacts with natural and man-made chemicals such chlorofluorocarbon (Halmer 2005). The sulfur bearing oxides, sulfates and chlorate emissions released during volcanic eruptions form complexes with the oxygen of the atmosphere and are also responsible for the ozone layer's destruction. Following the monitoring of the eruption of Pinatubo in the Philippines from 1991 up to 2005, a considerable increase in the chlorine content in the stratosphere has been noted. (Halmer 2005) This could have been another cause of the ozone's destruction during that period. Whether the observed increase of chlorine in the atmosphere is due to the two major volcanic explosions of 1991, or is due to man-made chemicals, has not yet been clearly resolved.

Unfortunately, volcanic eruptions are still very hard to predict. With all we know, we still cannot inform others about how magma can or will flow into underground conduits nor how long it will take before generating volcanic activities.

Underwater Eruptions: First Observation

Reports on the eruptions of submarine volcanoes are rare. However, a few historical cases have been reported. In 1925, a Japanese research vessel sank above a submarine volcano in the northwestern Pacific. In 1968, a sailing boat passing near a submarine volcano in the south Pacific in the vicinity of Tubuai in the Austral island chain felt under water earthquakes due to the eruption of a shallow seamount, named the Macdonald seamount (See Chap. 9). In 1963, a submarine volcano rose to the surface at about 50 km southeast of Iceland forming the island of Surtsey. Another active submarine edifice, the Kicken Jenny, in the Lesser Antilles rose to a height of about 230 m above sea level in 1939 and it has been active ever since (Lopes 2005).

In November 2008, the "Vents Program" discovered two on-going eruptions in the area of the Tonga-Kermadec ridge in the southwestern Pacific, one at West Mata Volcano near Tonga Island, and the other on the northeastern Lau basin spreading-ridge area. The following year, in May 2009, deep-seated submarine volcanic eruptions were observed from a ROV during the "*Ridge cruise and Margin*" project funded by the National Science Foundation and the National Oceanographic and Atmospheric Administration (NOAA). This expedition was supposed to explore the sites of these two eruptions aboard the R/V Thompson using the ROV JASON 2 at 1174 and 1208 m depths.

A report written by David Clague was put on the Web:

> The two Jason's ROV dives at West Mata volcano discovered two active volcanic vents. The first and deeper vent, named the Hades vent, on the southwest rift was erupting both effusively and explosively at the same time on both days. Small bursts were occurring at one end of an erupting fissure perhaps 5 m long while pillow lavas were being extruded from the other end. By the next night the activity had become more vigorous, sometimes blowing glowing bubbles as much as a meter across. The second shallower vent, named Prometheus, was located very near the summit of the volcano and about 100 m away from the first vent. The eruption here was entirely explosive with low-level, but nearly continuous fire fountains throwing ejecta into the water during both dives. Both vents were obscured much of the time by billowing sulfurous gas emissions, but bright orange lava was seen in both vents. The orange glowing lava was visible for minutes at a time.
>
> Joe Resing, chief scientist of the expedition wrote:
>
> Here, in the seven days of this cruise we have seen active extrusions of lava on the seafloor. We have seen explosions with flashes of light. We have seen molten rock forming new earth.

The scientific party on board the R/V Thompson using the ROV JASON and the video camera observed two eruptions that lasted only few minutes on the West Mata volcano near Tonga Island. This was great news for the scientific community. I had always been on the look out to see visual evidence of volcanic activities on the sea floor during all my diving cruises. I must have come close to this, especially during our dive in the south Pacific Rise spreading center near 17°26′S. Finally, in 2009, explosive eruptive events taking place at more than 1000 m depth on a submarine volcano were able to be observed by human eyes.

Tsunami

Tsunami is a word from the Japanese language meaning "harbor waves". A tsunami wave travels at a speed of more than 800 km/h when crossing the great depths of the ocean. Sometimes a *tsunami* is also erroneously called *a "tidal wave"* however tsunamis have nothing to do with tides. A tsunami is generated by underwater ground motion due to earthquakes or massive landslides. The speed of a tsunami waves reaches about 10–100 km/h in coastal shallow water regions. But seismic waves travel 20 times faster than tsunamis.

The upward fault thrusting of the oceanic crust generates an upward displacement of the water column producing shock waves that will keep building up until the conclusion of an earthquake. Because of structural upheaval generated on the sea floor, shock waves formed in the water column will propagate through the ocean from coast to coast and, when reaching shallow depths near land, tsunami waves will rise, just like the sea floor rises, and then they will spread out in all directions.

When tectonic activities like upheaval or volcanic activities cause a shaking of the Earth, this will be recorded by surface propagating seismic waves called "T" (tremor) or surface waves. These waves can deliver warnings about the existence of events capable of generating tsunamis. Thus, a quick determination of the seismic moment can help to assess the risk of tsunami. Tremor (T) waves are obtained based on the amplitude in displacement of the surface waves over a large period of time (50–300 s). The causes of tsunami generating waves are due to the displacement of multiple water masses. There are volcanic tsunamis due to underwater volcanic eruptions generating landslides or pyroclastic transported debris and there are tectonic tsunamis that are triggered by earthquakes during faulting and/or fracturing of the oceanic crust-lithosphere.

Tsunami waves up to 40 m above sea level were observed as reported by Simkin and Fiske (1983) in a volume commemorating the 100th anniversary of the Krakatau eruption. They reported in their volume that a steam boat "*the Berouw*" was thrown from its harbor mooring and was found 2½ km inland between coconut trees.

An earthquake along the south coast of Chile in May 1960 caused volcanism and earthquakes during the subduction of a lithospheric plate underneath the

Andes Cordillera mountains near the town of Valdivia. This generated a tsunami and waves of less than 1 m on the sea floor, but when the bottom waves reached shallow water depths the wave's crest was more than 20 m high and killed more than 2,000 people (Cromie 1962).

On December 26, 2004, more than 100 years after the Krakatau volcanic eruption in Indonesia, the Sumatra-Java trench was again the site of a large earthquake with a magnitude of 9.3 on the Richter scale. This earthquake generated a tsunami wave that was felt on the other side of the Indian Ocean striking the coast of Somalia in eastern Africa and killing a total of about 300,000. The epicenter was located on the northern portion of the Sumatra–Java trench at about 100 km away from the coast of Sumatra near the island of Simelue where the relatively denser Indo-Australian plate is moving beneath the lighter Burma-Malaysian plate.

On June 3, 2011, an earthquake started at 58 min after midnight and lasted about 7 min extending northwestward along the Sunda trench for about 1200 km almost reaching the coast of Myanmar (Burma) (Park et al. 2005). The faulted area was about 100 km wide.

An earthquake with a magnitude of 7.2 was felt in the morning of November 29, 1975, in Hawaii and was monitored by the Hawaiian Volcano Observatory located on the Kilauea Volcano (Tilling et al. 1975). The epicenter was located on the flank of the Kilauea caldera on the sea floor at about 25 km from the town of Halapa on the southeast coast of the island. This gave rise to sea waves as high as 12 m. Underwater landslides probably provoked the collapse of the submarine flank of the volcano. This triggered and pushed a large wave into motion due to the uplift of the sea floor at the base of the volcano. New open fissures 50–100 m wide occurred at the same time on the island itself. Since the tsunami expended much of its energy on a sparsely populated area, only two people were killed. About half an hour after the main earthquake shock, Kilauea begun to erupt along a 500 m long fissure formed in the caldera of the volcano. This was a quiet eruption with lava fountaining up to a maximum of 50 m high (Head and Wilson 1989).

Of course the most famous tsunami of recent date took place on March 11, 2011, when a giant wave hit the north coast of Honshu Island in Japan and caused an accidental nuclear catastrophe when an electric power plant in Fukushima was infiltrated by seawater. The source of the movement of the ocean floor was an area in the Japan trench which had been previously explored by Xavier Le Pichon (Kaiko project, 1984–1985). Le Pichon and his colleagues knew this region was tectonically active, so the event of an earthquake was not surprising. The magnitude of the Tohoku earthquake, which attained 9.0 on the Richter scale, was, however, the highest magnitude ever recorded in Japan, and the results were disastrous, as we all know.

Unfortunately, no early warning was possible since the wave hit the shore of Japan only 51 min after the tectonic activity occurred, so the number of deaths after the "Fukushima tsunami" could not have been prevented. Now, however, man needs to take stock of the location of the world's nuclear power plants and decide what to do about them should they be found in tectonically active areas or along the path of potential future tsunamis.

References

Avedik F, Geli L (1987) Single channel seismic data from the East Pacific Rise axis between 11°504′N and 12°54′N. Geology 15:857–860

Ballard RD, Holcomb RT, van Andel TH (1979) The Galapagos Rift at 86°W: sheet flows, collapsed pits, and lava lakes of the rift valley. J Geophys Res 84:5407–5422

Batiza R (1980) Origin and petrology of young oceanic central volcanoes: are most tholeiitic rather than alkalic? Geology 8:477–483

Batiza R (1981) Quantitative data on the abundance and distribution of seamounts in the Pacific Ocean. McDonnell Center for Space Sciences Pub. MCP-498, p 23

Batiza R, Fornari DJ, Vanko DA, Lonsdale P (1984) Craters, calderas, and hyaloclastites on young Pacific seamounts. J Geophys Res 89:8371–8390

Bideau D, Hekinian R, Sichler B, Gracia E, Bollinger C, Constantin M, Guivel C (1998) Contrasting volcanic-tectonic processes during the past 2 Ma on the Mid-Atlantic Ridge: submersible mapping, petrological and magnetic results at latitude 34°52′N and 33°55′N. Mar Geophys Res 20:425–458

Binard N, Hekinian R, Cheminee J-L, Searle C (1991) Morphological and structural studies of the society and Austral hot spot regions in the South Pacific. Tectonophysics 186:293–312

Binard N, Hekinian R, Cheminée J-L, and Stoffers P (1992) Styles of eruptive activity on intraplate volcanoes. J Geophys Res 97(13):999–14015

Bischoff JL, Pitzer KS (1989) Liquid-vapor relations for the system NaCl-H2O: summary of the P-T-x surface from 300 °C to 500 °C. Am Jour Sci 289:217–248

Bischoff JL, Pitzer KS (1985) Phase relation and seafloor hydrothermal systems. Earth Planet Sci lett 75:327–338

Blank JG, Brooker RA (1994) Experimental studies of carbon dioxide in silicate melt: solubility, speciation and stable carbon isotope behavior. Rev Miner 30:157–186

Blong R (1984) Volcanic hazards. Academic Press, New York, p 424

Bonatti E, Harrison CG (1988) Eruption styles of basalt in oceanic spreading ridges and seamounts: effects of magma temperature and viscosity. J Geophys Res 93:2967–2980

Bottinga Y, Javoy M (1989) MORB degassing evolutions of CO_2. Earth Planet Sci Lett 95:215–225

Broecker WS (1985) How to build a habitable planet. ELDIGIO Press, New York

Burton MR, Mader HM, Polacci MM (2007) The role of gas percolation in quiescent degassing of persistently active basaltic volcanoes. Earth Planet Sci Lett 264:46–60

Cayol V, Dieterich J. Okamura A, Miklius A (2000) High magma storage rates before the 1983 eruption of kilauea, Hawaii. Science, 288:2343–2346

Chaigneau M, Hekinian R, Cheminée J-L (1980) Magmatic gases extracted and analysed from ocean floor volcanics. Bull Volcanl 43–1:241–253

Clague DA, Davis AS, Bischoff JL, Dixon JE, Geyer R (2000) Lava bubble formed by submarine hydrovolcanic explosion on Lo'ihi Seamount and Kilauea Volcano. Bull Volcanol 61(7):437–449

Cromie WJ (1962) Earthquakes. Science services, vol 1. Nelson Doubleday Inc. Garden city, p 64

Cyamex Scientific Team (1978) Découverte par submersible de sulfures polymétalliques massifs sur la dorsale du Pacifique oriental par 21°N (projet Rita). Compte Rendu Acad Sci Serie D, 287:1365–1368

Detrick RSP, Buhl P, Vern E, Mutter J, Orcutt J, Madsen J, Brocher T (1987) Multi-channel imaging of a crustal magma chamber along the East Pacific Rise. Nature 326:35–41

Eissen P, Fouquet Y, Hardy D, and Ondréas H (2004) Recent MORB volcaniclastic explosive deposits formed between 500 and 1750 m.b.s.l. on the axis of the Mid-Atlantic Ridge, south of the Azores. Am Geophys Union Geophys Monogr 140. doi: 10.1020/140GM09

Fisher RV and Schmincke H-U, (1984) Pyroclastic rocks. Springer, Berlin, p 472

Fisher VR, Heiken K, Hulen JB (1998) Volcanoes; Crucible of change. Princeton University Press, Princeton, p 317

Fornari DJ, William BF, Ryan WBF, Fox PJ (1984) Sea floor lava fields on the East Pacific Rise geology 13:413–416
Francheteau J, Juteau T, Rangin C (1979) Basaltic pillars in collapsed lava pools on the deep sea floor. Nature 281:209–211
Francis P (1995) Volcanoes: a planetary perspective. Oxford University Press, New York, p 443
Gill J, Torssander P, Lapierre H, Taylor R, Kaiho K, Koyama M, Kusarabe M, Aitchinson J, Cisowski K, Dadey K, Fujiioka A, Klaus K, Lovell M, Marsaglia K, Pezard P, Taylor B, Tazaki K (1990) Explosive deep water basalt in the Sumisu backarc rift. Science 248:1214–1217
Halmer MM (2005) Have volcanoes already passed their zenith influencing the ozone layer? Terra Nova 17:500–5005
Hart SR, Zindler A, (1989) Constraints on the nature and development of chemical heterogeneities in the mantle. In: Peltier WR (ed) Mantle convection. Gordon Breach Publishers, New York, pp 261–387
Head JW, Wilson L (1989) Basaltic pyroclastic eruptions: influence of gas release patterns and volume fluxes on fountain structure, and the formation of cinder cones, spatter cones, rootless flows, lava ponds and lava flows. J Volcanol Geotherm Res 37:261–271
Hekinian R, Binard N (2008) Le Feu des Abysses. Editions QUAE, c/o Inra, RD 10, 78026. Versailles, France, p 175
Hekinian R, Chaigneau M, Cheminée J-L (1973) Popping rocks and lava tubes from the Mid-Atlantic Rift Valley at 36°N. Nature, 245 n° 5425, pp 371–373
Hekinian R, Thompson G, Bideau D (1989) Axial and off-axial heterogeneity of basaltic rocks from the East Pacific Rise at 12°38′–12°51′N and 11°26′–11°30′N. J Geophys Res 94:17437–17463
Hekinian R, Bideau D, Stoffers P, Chiminée J-L, Muhe R, Binard N (1991) Submarine intraplate volcanism in the South Pacific: geological setting and petrology. J Geophys Res 96:2109–2138
Hekinian R, Pineau F, Shilobreeva S, Bideau D, Gracia E, Javoy M (2000) Deep sea explosive activity on the Mid-Atlantic Ridge near 34°50′N: magma composition, vesicularity and volatile content. J Volc Geotherm Res 98:49–77
Jaupart C (1996) Physical models of volcanic eruptions. Chem Geol 128:217–227
Javoy M, Pineau F (1991) The volatile record of popping rocks from the Mid-Atlantic Ridge at 14°N: chemical and isotopic composition of gas trapped in vesicles. Earth Planet Sci Lett 107:598–611
Katili JA, Sudradjat A (1984) The devastating 1983 eruption of Colo Volcano, Una-Una Island, Central Sulawesi, Indonesia. Geol Jahrb Reihe A 75:27–47
Lagabrielle Y, Auzende J-M, Eisen J-Ph, Janin M-C, Cotten J (1994) Geology and geochemistry of a 800 m section through young upper oceanic crust in the North Fiji Basin (Southwest Pacific). Mar Geol 116:113–132
Lopes RMC (2005) The volcano adventure guide. Cambridge University Press, illustrated edition, ISBN 0-521-55453-5, 11 p
Lupton JE, Craig H (1981) A major Helium-3 source at 15°S on the East Pacific Rise. Science 214:13–18
Macdonald KC (1982) Mid-ocean ridges: Fine scale tectonic, volcanic and hydrothermal processes within the plate boundary zone. Am Rev Earth Planet Sci 10:155–190
McBirney AR (1963) Factors governing the nature of submarine volcanism. Bull Volcanol 26:455–469
McLain JS, Orcutt JA, Barnett K (1985) The East Pacific rise in cross-section: a seismic model. J Geophys Res 90:8627–8639
Moore RB (1991) Geology of three late quaternary stratovolcanoes on Sao Miguel, Azores. US Geol Surv Bull 1900:1–26
Murase T, McBirney AR (1963) Viscosity and related property of volcanic rocks at 800–1400 °C. Hokkaido Univ Fac Sc J ser 7(1):487–584

Murase T, McBirney AR (1973) Properties of some common igneous rock and their melts at high temperatures. Geol Soc Am Bull 84:3563–3592

Nayudu YR (1962) The submarine eruption of basalts and the problem of palagonitization. In: International symposium on volcanology, Japan, p. 49 (abstract)

Nicolas A, Freydier Cl, Godard M, Vauchez A, (1993) Magma chambers at oceanic ridges: how large? Geology 21:53–56

Ondreas H, Fouquet Y, Voisset M, Radford-Knoery J (1997) Detailed study of three contiguous segments of the Mid-Atlantic Ridge, south of the Azores (37°–38°30′N), using acoustic imaging coupled with submersible observations. Mar Geophys Res 19:231–255

Park J, Anderson K, Aster R Butler R, Lay T, Simpson D (2005) Global seismographic network records the great Sumatra-Adaman Earthquake. Eos Trans Am Geophys Union 86:57–61

Perfit MR, Chadwick WW Jr (1998) Magamtism at Mid-Ocean Ridges: constraints from volcanological and geochemical investigations. In: Faulting and magmatism at Mid-Ocean Ridges, Buck WR, Delaney PT, Karson JA, Lagabrielle Y (eds) Geophysical Monograph, 106:59–115

Pineau F, Shilobreeva S, Hekinian R, Bideau D, Javoy M (2004) Deep-sea explosive activity on the Mid-Atlantic Ridge near 34°50′N: a stable isotope (C, H, O) study. Chem Geol 211(1–2):159–175

Pinneau F, Shilobreeva S, Hekinian R, Bideau D, Javoy M (1998) Explosive activity on the sea floor purely magmata c or magma-seawater interaction origin? Geochemical and isotopic arguments. Goldschmidt Conference. Toulouse, France, pp 1185–1186

Rampino MR, Self S (1984) The atmospheric effects of El Chichon. Sci Amer 250(1):48–57

Rittmann A (1944) Vulcani, activi e genesi, Napoli, Editrice Politecnica, p 305

Rosendahl BR, Raitt RW, Dorman LM, Bibee LD, Hussong DM, Sutton GH (1976) Evolution of oceanic crust 1. A physical odel of the East Pacific Rise crest derived from seismic refraction data. J Geophys Res 81(29):5294–5304

Rutherford MJ, Gardner JE (2000) Rates of Magma ascent. In: Encyclopedia of Volcanoes, Sigurdsson H, Haougthon BC, McNutt SR, Rymer H, Stix J (eds). Academic Press, California, pp 207–217

Sigurdsson H, Hougthon BF, McNutt SR, Rymer H, Stix J (2000) Encyclopedia of volcanoes. Academic Press, London, p 1417

Simkin T, Fiske RS (1983) A classic geophysical event. Eos Trans Am Geophys Union 64(34):513–514

Simkin T, Siebert L (2000) Earth's volcanoes eruptions: an overview in encyclopedia of volcanoes by Sigurdson H. Elsevier, New York

Sinton JM, Detrick RS (1992) Mid-Ocean Ridge magma chambers. J Geophys Res 97:197–216

Sleep NH, Barth GA (1997) The nature of oceanic lower crust and shallow mantle emplaced at low spreading rates. Tectonophysics 279(1):181–191

Smith T, Batiza R (1989) New field and laboratory evidence for the origin of hyaloclastite flows on seamount summits. Bull Volcanol 51:96–114

Soule SA, Fornari DJ, Perfit MR, Rubin K (2007) New insights into Mid-Ocean Ridge volcanic processes from the 2005–2006 eruption of the East Pacific Rise, 9°46′–9°56′N. Geology 35:1079–1082

Soule SA, Nakata DS, Fornari DJ, Fundis AT, Perfit MR, Kurz MD (2012) CO_2 variability in Mid-Ocean Ridge basalts from syn-emplacement degassing: constraints on eruption dynamics. Earth Planet Sci Lett 327–328:39–48

Tilling RI, Kayanagi RY, Lipman P, Lockwood JP, Moore JG, Swanson DA (1975) Earthquake related catastrophic events, Island of Hawaii, November 29, 1975: a preliminary report. US Geol Surv Circular 740:1–32

Witmarsh RB (1975) Axial intrusion zone beneath the median valley at the Mid-Atlantic Ridge at 37°N detected by explosion seismology. Geophys J Roy Astron Soc 42:189–215

Wright IC, Gamble JA (1999) Southern Kermadec submarine caldera arc volcanoes (SW Pacific): caldera formation by effusive and pyroclastc eruption. Mar Geol 161:207–229

Chapter 6
Hydrothermal Activity and Metalliferous Deposits

Abstract The World's major marine geological structures have been affected by physical and chemical changes during and after their creation at spreading axes and in intraplate regions where islands and seamounts are formed. These changes are largely related to the penetration of the hydrosphere into the rigid lithosphere. A chemical reaction between the hydrosphere (seawater) and the lithosphere (rocks) in the presence of heat gives rise to hydrothermal fluid. The circulation of this fluid is the main cause of the lithosphere's transformation due to the alteration (hydration) of metallic components from rocks and their subsequent precipitation on the sea floor in the form of ore deposits. A theory about the existence of hydrothermal discharge and metal deposits on the ocean floor was based on geophysical constraints related to the Earth's heat budget. The deficit of heat discharge on spreading ridges was thought to be the result of a discharge of hydrothermal fluid on the sea floor.

Hydrothermal activity is a by-product of volcanism. The alteration of rocks due the circulation of hot fluid in the lithosphere gives rise to the precipitation of metallic compounds such as iron, copper and zinc sulfides, and precious metals such as gold and silver as well as other basic compounds, all of which can be utilized by the world's population. Land-based hydrothermal deposits were long known and have been exploited since early periods of human presence. For example, Neanderthal people drew animal pictures with powdery ochreous compounds derived from hydrothermal precipitates found on the uplifted site

Hydrothermal activity on the sea floor has been one of the scientific community's interests since its discovery in 1968–1969 (Degens and Ross 1969). Since the 1970s, we have seen actual evidence of hydrothermal fluid circulation through the solid, porous lithospheric material, and this has been occurring since the early formation of our planet. The three main factors governing a hydrothermal system are the presence of a heat source and water, and also the permeability of nearby rocks. Solid Earth's permeability is an important factor related to the degree of lithospheric fissuring and fracturing as well as to the various types of rock formations. For example, pyroclasts, hyaloclastites and tectonic breccias form incoherent, randomly piled-up deposits, which enhance an

R. Hekinian, *Sea Floor Exploration*, Springer Oceanography,
DOI: 10.1007/978-3-319-03203-0_6,

area's permeability thereby facilitating the circulation of seawater and hydrothermal fluids around and/or between the solid rocks.

Missing Heat

Before the discovery of the ocean's submarine geysers and other formations, now called "hydrothermal sites", scientists suspected the existence of hydrothermalism at oceanic spreading centers. Mid-Oceanic ridges where fresh lava is extruded are also sites of heat production and heat loss as the lithospheric plates move away from their point of origin at the ridge axes. We can expect to observe that the crust cools and contracts as it slips away. When the crust was about 60–80 million years old after its creation, it had already lost much of its original heat and thermal convection in the lithosphere was much lower. Subsequent phenomena of metamorphism, sediment filling and alteration sealed much of the original rocks' porosity and closed some of the fractures of the older oceanic crust during the lithosphere's aging process.

In the 1960s, when scientists calculated the amount of magma formed at ridge axes and determined how long it would take for the crust to cool, the amount of heat produced at any given time was found to be lower than what was expected. Measurements of the ocean crust's thermal regime was conducted on the sea floor using special thermal probes that had the shape of a large needle and which were pushed into the sediment by a coring tube. Probing the crust's thermal temperature was conducted at various distances perpendicular to the accreting ridge axes. These temperature probes revealed that the ridges had a heat flow deficit, which was not what was expected according to the theories based upon conductive cooling models (Parker and Oldenburg 1973; Sclater and Francheteau 1970).

When the crust is heated from below, it will cool due to thermal conductivity through the overlying sediment. During this process of cooling, no material is transported. Therefore, it would be expected that the crust would be colder at a distance from the ridge axis. In fact, the anomaly noted was that the amount of heat lost due to its conduction in sediment was found to decrease towards a ridge axis instead of increase, as would be expected. Thus the heat deficit observed between the theoretical values of a predicted model on conductive cooling and the reality of the temperatures measured on site led scientists to conclude that significant quantities of heat must have been lost by a non-conductive mechanism beneath the ridge crests, and this was interpreted as being due to the circulation of seawater as a heat exchange agent.

It was therefore speculated that seawater circulation was a factor which directly affected the cooling of crust formed on the Mid-Oceanic Ridges (Williams and Von Herzen 1974). The theory postulated that, beneath the ridge axis, cold sea water percolating down through the various porosities of the oceanic crust would descend until it reached the vicinity of hot magmatic upwelling zones, at which point, when the seawater had been reheated after having cooled its immediate environment, it would rise again to the surface. This phenomenon of water's descent and subsequent ascent would take place because of density differences between the denser cold seawater and the lighter, warmer hydrothermal fluid.

Discovery of Hydrothermal Deposits

Hydrothermal metallic deposits forming iron sulfides and oxides containing significant base metals were first recovered from the Atlantis II Deep in the Red Sea in 1968–1969 (Degens and Ross 1969). However, the first hydrothermal site to be directly observed was found during the FAMOUS project in 1974. Scientists observed and sampled chunks of dark-brown colored hardened crust from a mound located at the intersection between the transform fault wall and the Mid Atlantic ridge axis near 36°56′40″ N-33°03′40″W at 2700 m deep. These chunks were picked up by the remote controlled arm of the submersible *Cyana* and placed in the sampling basket on the front. When the submersible returned to the sea's surface, divers brought the sampled material on board the ship. Using just my eyes and my fingers, I realized that this friable, brown colored material was most likely an iron-manganese compound, probably of hydrothermal origin. I later identified the samples as being a hydrothermal Fe–Mn oxyhydroxide crust after examining the material under a microscope and doing some chemical analyses.

Never before seen deep-water organisms colonizing hot springs, better known as "hydrothermal vents", were first discovered during exploration with a deep-towed camera (called ANGUS) from WHOI (Woods Hole Oceanographic Institution) in the rift valley of the Cocos-Nazca spreading ridge near 85°–87°W and 0°47′N in the Pacific Ocean, near the Galapagos Islands. Subsequently, in February–March 1977, during the *submersible Alvin*'s dives in the same area, observers saw a luxuriant "oasis" of life around hydrothermal vents (Corliss et al. 1978, 1979). A year later, the French-US team discovered more animal life on the first polymetallic sulfide deposits found along a segment of a fast spreading ridge of the East Pacific Rise near 21°N, during the second half of the RITA (anachronism for RIvera-TAmayo) project.

The first part of the RITA project took place in 1978 with the French diving saucer *Cyana* and its support ship *Le NADIR* and was called "Cyamex" (CYana-MEXico). This leg was organized and conducted by a team from CNEXO directed by Jean Francheteau (Francheteau et al. 1980). Hydrothermal geysers called "black smokers" were observed and described as fountains or strong jets of mineral-laden water spewing from hydrothermal chimneys several meters tall. In 1979, the RITA project continued with dives using the WHOI submersible *Alvin* and its support ship *Lulu.* At this time, scientists determined that the "smokers" were belching out a homogeneous solution of high temperature (350 °C) fluid since a temperature probe could measure the black water's heat at the mouth of the chimney.

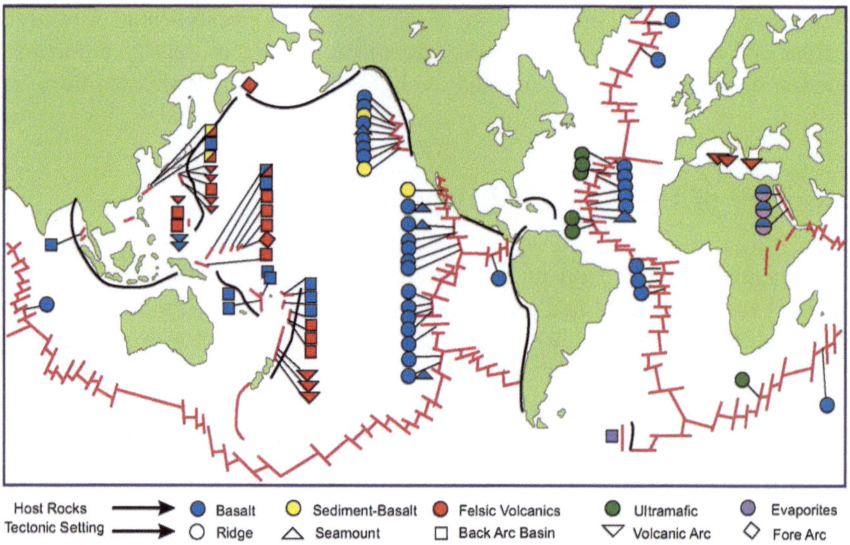

Fig. 6.1 Sea floor distribution of hydrothermal deposits and associated rock types in the World's Oceans is shown (Courtesy of S. Scott 2011)

In 1977, in parallel with these discoveries, Peter Lonsdale (University of California at Scripps Institute of Oceanography in San Diego) found evidence of active hydrothermal springs in the area of high heat flow in the Guaymas basin in the Gulf of California (see *News and Views 289, 9, 1981*). This portion of the spreading ridge axis is buried under several hundred meters of sediment carried off the continent by the Colorado River. Subsequent drilling operations and Alvin dives and drilling operations later confirmed the presence of hydrothermal activity in this area.

Since these early discoveries of hydrothermal deposits, scientists have continued to find hot submarine geysers where living organisms congregate and with various types of deposits of polymetallic sulfides (Fig. 6.1). Hydrothermal deposits have been found along all the World's spreading ridge segments; however, hydrothermal sites are also present in other sea floor provinces such as intraplate volcanic regions and back arc basins and are even found associated with the volcanic ridges of subduction zones (Scott 1985; Scott and Binns 1992).

Hydrothermal Fluid Circulation and Sub-Crustal Alteration

The formation of hydrothermal fluid commences when seawater penetrates into the lithosphere and changes its composition during heating at contact with the hot rocks near a magma reservoir or another source of heat. Cold seawater, consisting essentially of H_2O, Na, Sr, K, Mg, Ca, Ba, SO_4, HCO_3, F and Cl (See Chap. 2), penetrates the rocks beneath the sea floor through fissures and pores as it descends into the crust and lithosphere to where it encounters a source of heat which causes it to react with the surrounding, warmer rocky environment.

To give a simple explanation, fluid circulation in a hydrothermal system can be divided into four zones (Fig. 6.2): (1) The "recharge zone" is where cold (2–13 °C) seawater enters into the oceanic crust and a low temperature (100–450 °C) chemical reaction starts to take place with the surrounding formations. The warm water loses some of its constituents such as Ca^{+2}, and SO_4^{-2} and precipitates anhydrite ($CaSO_4$). Similarly, the Na, K, Sr, Br and F of seawater will enter into the silicate lattice of the rock to replace other cations. (2) The "acidification zone" is where seawater changes its composition and becomes a more corrosive fluid. At this stage, another chemical reaction takes place with the loss from seaweater of Mg^{+2} that is incorporated into the oceanic crust. This occurs when the basaltic rocks and their major mineral constituents such as plagioclase and olivine are transformed into clay. (3) At the "reaction zone" the temperature of the descending fluid rises after it has reached the vicinity of a magma reservoir. The depth could be anywhere from 500 m up to 2 km within the crust. Thermodynamic studies (Von Damm and Bishoff 1987) on vent fluids have estimated that the temperature in the reaction zone is on the order of 390–400 °C. Under these conditions, higher temperature mineral assemblages such as amphiboles, talc and other hydrated silicates will be produced from the alteration and leaching of the basalt-dolerite complexes which form the crust. Minerals such pyroxene and olivine will give rise to ferromagnesian silicates and plagioclase will give rise to epidote types of hydrated minerals. (4) In the "discharge zone" the hydrothermal hot fluid carrying metals that have been leached from the underlying rocks will return to the sea floor and will then react with ambient sea water and give rise to sulfide deposits. Metals can also be contributed from the magma reservoir itself (Yang and Scott 1996).

Other reactions during H^+ exchange with Ca and Ba from the silicate minerals which combined with the SO_4^{2-} of sea water will give rise to anhydrite ($CaSO_4$) and barite [(Ba^{2+} (fluid) + SO_4^{2+} (sea water) = $BaSO_4$ (solid)]. Anhydrite, barite and hydrated silica are compounds that are commonly found coating sulfide chimneys.

In summary, seawater itself is metal deficient and its sulfur content is found in the form of sulfates. Later, after descending into the lithosphere where seawater comes into contact with hot rocks in the reaction zone, the seawater becomes chemically altered and gives rise to H^+ and free Mg^{2+}, Ca and SO_4^{2-}. During such a reaction, the hydroxyls (OH^-) and Mg^{2+} ions freed from seawater will go into the rock to form hydrated silicates (Fig. 6.2). At this stage, seawater will have changed its composition to become a "hydrothermal fluid" enriched in HS (sulfuric acid) with a pH of 3–5. It is now Mg^{2+} free and has become very corrosive with the ability to further leach metals from the rocks found in the reaction zone. The fluid reaches its maximum temperature (450 °C) near the magma reservoir where heavy metals are leached from the rocks. After leaving the area near the heat source, the temperature of the fluid will drop to near 330–350 °C

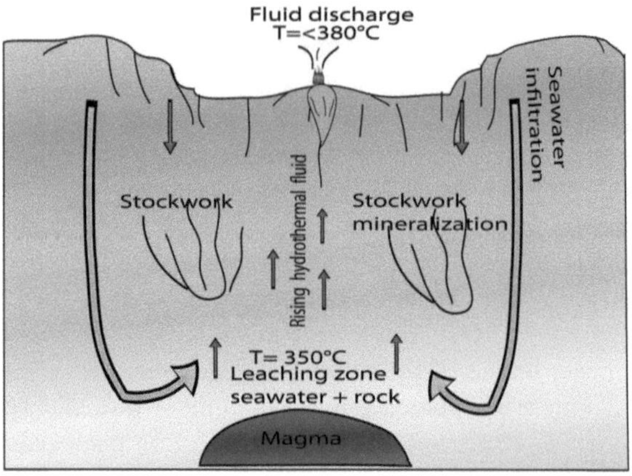

Fig. 6.2 Global view of a hydrothermal system at a spreading ridge axis indicates that the circulation of seawater under the axis takes place through fissures and fault zones, which enhance hydrothermal circulation. Chemical reaction between seawater and the rock formations in the presence of a heat source provided by a magma reservoir is responsible for metals leaching from the rock to form hydrothermal fluid

(Bishoff 1980). The metals most suitable for leaching are the transition metals such as Cu, Zn, Fe, Mn Ag, and Au, which form metallic compounds along with sulfur and chlorine.

The hydrothermal fluid spewing out on the sea floor is heavily charged with Mn, Si $(OH)_4$, H_2S, CH_4, Li, CO_2, and metals such as Fe, Zn and Cu, in amounts that are more than a hundred times greater than what is found in ambient sea water (Charlou and Donval 1993). The fluid rising immediately above active chimneys (hydrothermal vents) is found in a concentrated jet, only a few centimeters in diameter, similar in flow intensity to the fluid exiting a water-hose. As the vent fluid is discharged and rises above the chimney, it is quickly diluted with seawater creating a hydrothermal plume. This hydrothermal plume rises up to a few hundreds of meters above the sea floor. Indeed, plumes were often observed and tracked by human eyes up to about 50 m above strongly active "black smoker" chimneys. Obviously, the fluid will rise higher until it is completely diluted to a density the same as the ambient seawater and can then be transported by undersea currents.

Hydrothermal fluids at contact with ambient seawater near the vents precipitate particles of ore-forming elements at the site of venting. These precipitations form sulfide edifices. Other elements form Fe and Mn oxyhydroxides (which are oxide compounds created with the oxygen extracted from sea water) that co-precipitate with transitional trace elements such as Cr, As, Co and Ni. The hydrothermal fluid comprised of Fe, Mn and transition metals is dissipated in the seawater and then transported in the water column. This transported material will eventually contribute to the formation of polymetallic sediment and/or nodules, which can be found thousands of kilometers away from their sources.

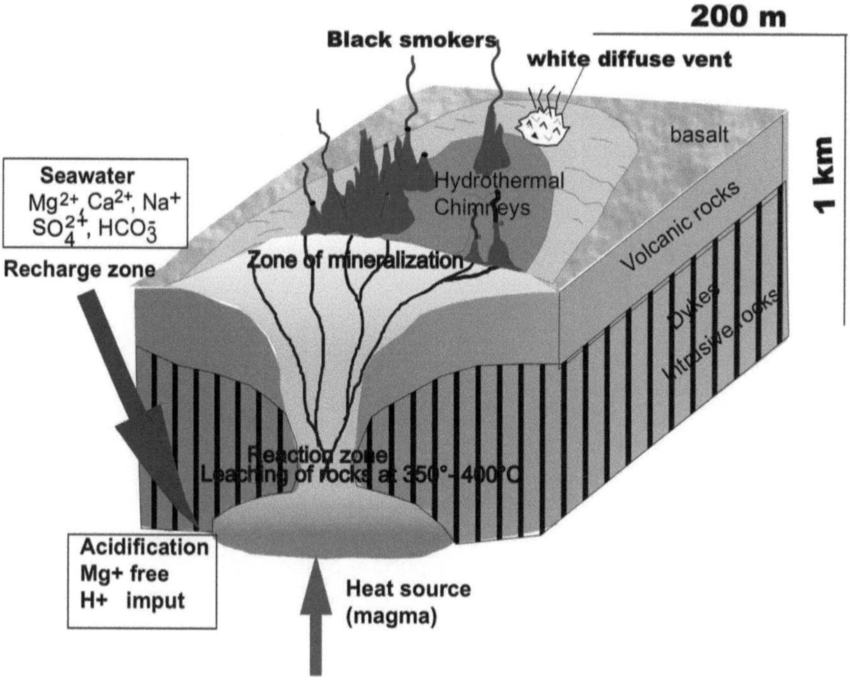

Fig. 6.3 Block diagram shows hydrothermal circulation and precipitation of metalliferous deposits (Hekinian and Binard 2008). The penetration of seawater into the lithosphere underneath spreading ridge axes through fissures and faults gives rise to a chemical reaction between water and the rock formations near a heat source such as a magma reservoir. The acidification takes place by exchanging basic ions (magnesium) and producing corrosive acids responsible for leaching the rocks. The leached elements form compounds in a hydrothermal solution charged with metals and Si. The hydrothermal fluid ascends towards the seafloor where it precipitates its contents. Some of the precipitates form in situ hydrothermal edifices while another portion of the fluid is transported in the water column

For an on-site observer, the color of the hydrothermal exiting fluids varies according to its composition, which reflects various types of element content in metals and/or other elements. The black exiting fluids consist essentially of dark-colored metallic sulfides composed of Zn, Fe, and Cu and their exiting temperature is usually higher than 270 °C reaching up to 360 °C (Fig. 6.3). The white, milky fluids consist essentially of Si, Ba, and Ca. The transparent hydrothermal fluids are obviously more difficult to detect, since they are deprived of metallic compounds and sulfur. Nevertheless, these clear-colored fluids can still exit at high temperatures (up to 300 °C) and have a much lower salinity than seawater since they have undergone a phase separation in the leaching zone. During their phase separation, a vapor phase consisting of relatively fresh water will be separated from a salty phase that gives rise to salt brine. Transition metals such as Zn, Fe, and Cu will form chlorides and sulfide-compounds in the residual salt brine.

Types of Hydrothermal Deposits

Deposits of hydrothermal origin exposed on the ocean floor have been classified according to their structure, their textural appearance and their composition. They include: massive sulfide edifices, hydrothermal mounds, porous deposits, native sulfur, Fe–Mn deposits, ochreous deposits, Fe-oxyhydroxides, siliceous deposits (colloidal form), and silicates (clay-like deposits). Stockwork and veins of mineralization formed within the lithosphere (which are also evidence of hydrothermal activity) will be treated as a separate classification.

Massive Sulfides

Consisting essentially of Zn, Cu and Fe rich metalliferous deposits, massive sulfides appear blocky, as it they were constructed of concrete blocks of various shapes and sizes. The variously shaped blocks are generally composed of a core enriched in chalcopyrite while the outer portion of the blocks is essentially made up of pyrite. The Cu-rich massive sulfides and the Fe-rich massive sulfides differ from each other not only in their relative proportion of iron and copper mineral contents, but also in their degree of porosity because the Fe-rich massive sulfides appear to have a porosity of 10–15 % while the porosity of the Cu-rich massive sulfides is less than 10 %. Nevertheless, both types of massive sulfides are definitely less porous than any other sulfide deposit, and their pores are filled with visible crystalline (idiomorphic) pyrite and/or chalcopyrite as well as with opaline material.

Snow Balls and Beehives (Porous Sulfide Deposits)

"Snow balls" were created by white-colored low-temperature exiting fluids (Speiss et al. 1980; Haymon and Kastner 1981) and are composed of more than 50 % pore spaces. These extremely porous deposits form relatively small (<2 m in diameter) structures, which are colonized by tubular worms (*pogonophoara* and *alvinella*) due to the biological activity associated with hydrothermalism. Fossilized worm remains are also often found in the vicinity of "dead vents" where hydrothermal activity is no longer occurring.

The mineral constituents of snowballs and bee-hives are mainly zinc sulfides (sphalerite) and a smaller amount of pyrite. The worm tubes observed on these structures contain concentric, extremely thin layers (lamellae) of shapeless (collomorph texture) zinc sulfides, Fe-sulfides, and amorphous silica. These types of deposits show a low temperature sequence of mineral crystallization except for the occurrence of some isolated grains of Cu-rich sulfide phases, which suggest that at some point the hydrothermal fluid could have reached the critical temperature of >300 °C, enabling the formation of chalcopyrite.

The beehive type deposits are comparable to the snow balls, however they are gray colored and have a conical shape. They differ from hydrothermal chimneys by the absence of a central conduit, and instead they are made up of of highly porous material enabling hydrothermal fluids to be diffused throughout all the pores exposed on their conical surfaces. The presence of the small pores on these two types of edifices implies a more cloudy, fluid mixing between the exiting hydrothermal fluid and seawater than that encountered in the normal conduits found on hydrothermal chimneys.

Hydrothermal Mounds

Mounds are formed by the degradation of previously formed chimneys that have been subsequently re-cemented together. They are found in association with a single large or several smaller circular chimneys of sulfides erected on the porous sulfide mounds. A continued discharge of low temperature hydrothermal fluid seals the chimney orifices and results in a more massive and larger appearance for the mound deposits.

Columnar Edifices: Hydrothermal Chimneys

Large vertical tubes varying in size between 30 and 50 cm up to 3 m in diameter and from 1 to 45 m tall have been reported on several rifted segments of the EPR axis. When active, they are often surmounted by smaller columns (<30 cm tall) from which hot hydrothermal fluid is discharged. The main body of the columnar edifices is formed during the growth of sulfide chimneys around a centrally located discharge zone.

Barite Chimneys

Barite chimneys are white, as opposed to the reddish-brown sulfide chimneys. Barium occurs as barite minerals in rocks and precipitates after leaching during the mixing of hydrothermal fluid with sea water following the equation:

Ba^{2+} (fluid) + SO_4^{2+} (sea water) = $BaSO_4$ (solid)

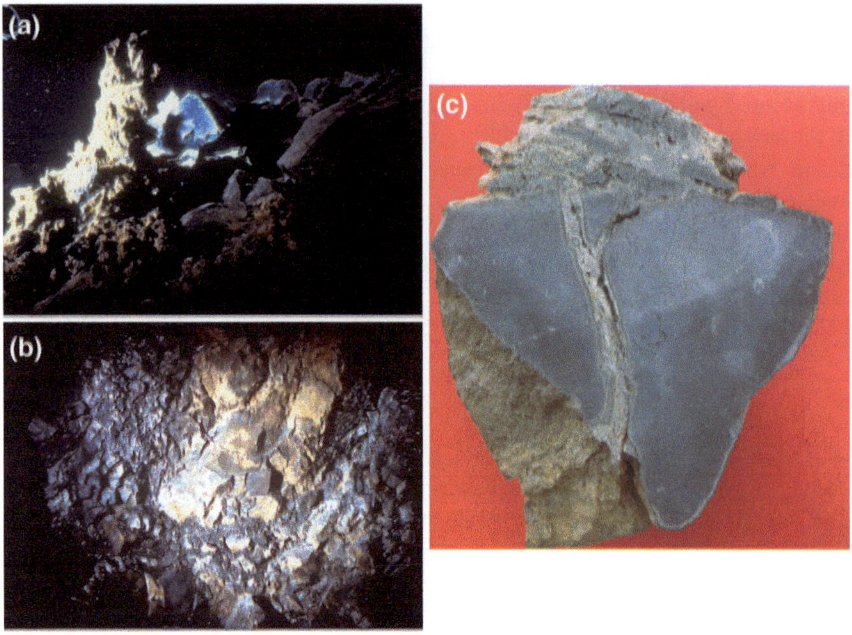

Fig. 6.4 Hydrothermal formations (photos copyright IFREMER *Geocyarise cruise leg 1, 1984, Cyana dive Cy 84-05*) **a** Small (>2 m in height) hydrothermal chimneys made up of zinc-copper sulfides standing on top of a talus pile at the foot of the western graben of the EPR at 18°S at 2627 m depth. **b** Stockwork and fragmented volcanic blocks exposed on the western wall of the East Pacific Rise axial graben at 18°30′S, 2622 m depth showing **c** veins (5–10 cm) filled with sulfide metals

Hydrothermal Breccia

Hydrothermal breccia are caused by the debris formed when hydrothermal edifices become old so they crumble into a pile of fragments. However, hydrothermal breccia could also be formed during the percolation of fluid charged with metalliferous matter through talus piles in the active tectonic zone associated with spreading ridges or normal faulted terrains which are areas of recent volcanic and tectonic activity.

An example of hydrothermal breccia was discovered in the south-east Pacific Ridge segment near 18°30′S at the foot of a vertical wall forming the axial graben where sand, gravel, and boulder sized (0.1–20 cm in diameter) altered rock fragments were cemented by sulfides and Fe-oxyhydroxide precipitates. In addition, small (<1 m tall) zinc-iron-sulfide edifices with active small chimneys were found standing on talus piles at a depth of 2627 m (Fig. 6.4a). Candle-like stalactites formed at the foot of these edifices indicate the fluidal nature of the sulfide precipitates. The cement forming this type of breccia is comparable to that of stockwork material and consists of pyrrhotite, pyrite, marcasite, wurtzite and opal associated with goethite and limonite.

Another type of breccia consists of the debris of stockwork mineralization resulting from the alteration of hydrothermal outcrops during tectonic activity. Talus piles and landslides were created along spreading-ridge graben walls and fissures on the ridge axes. This type of breccia is made of veined-rock fragments where remnants of angular chilled margins and mineral grains are intermixed with Fe-sulfides. These specific types of breccia are discussed in this chapter's section dealing with stockwork.

Ochreous Deposits or Gossan

Variegated purple-red and reddish-yellow Fe-oxyhydroxide-rich material is found associated with hydrothermal chimneys, as a powdery sediment fill, and is also observed forming mounds.

It is often difficult to tell the difference between *gossan deposits*, resulting from the oxidation of pre-existing sulfide edifices, and *ochreous deposit*s whose origin is due to slow venting (exhalative discharge) giving rise to primary Fe-oxyhydroxide deposits. This is mainly true when dead chimneys are composed entirely of Fe-oxyhydroxides and when no trace of sulfides is detected. In the case when this type of edifice covers an extended field (>200 m in diameter), such as are encountered on off-axis seamounts and on intraplate edifices, the *ochreous deposits* are probably due to a primary exhalative (slow venting) origin due to the circulation of Fe-rich fluids. In contrast, many of the ridge axis *gossan deposits* were probably once ancient sulfide edifices formed from sulfide-rich fluid emanations, which have become extensively altered, but will still contain trace amounts of sulfides.

The ochreous and gossan deposits often alternate with other more quickly extruded hydrothermal deposits and form small fields (<20 m in diameter) of powdery mounds which are located near other metalliferous sulfide edifices. Gossan and ochreous deposits are found as frequently as sulfide chimneys but they differ from the sulfides because of their friable nature and due to their intimate association with Fe–Mn oxides. In fact, Fe–Mn oxides are usually present with this type of deposit as a coating on top of red and/or yellow colored oxyhydroxide compounds.

Although both gossan and ochreous deposits contain variable amounts of amorphous Fe-oxyhydroxide powdery material, they also consist of nontronite (smectite/clay) and Fe–Mn crusts such as todorokite and birnessite. Hydromica, illite, and kaolinite also occur as minor constituents of these deposits (Hoffert et al. 1978). Ochreous and gossan deposits contain goethite spherules and amorphous Fe-oxides, but they differ from other Fe-oxyhydroxide material because they contain atacamite (due to the oxidation of a Cu-rich deposit), traces of pyrite and other sulfides (such as covellite and chalcopyrite).

Fe-Oxyhydroxide Deposits

Fe-oxide and hydroxide rich deposits were first recovered on-site from the Mid Atlantic Ridge in 1973 during the FAMOUS Project. They consist of about 40–80 % $Fe_2O_3.H_2O$ forming primary translucent spherules of goethite minerals. These are primary precipitates and not an alteration state of pre-existing sulfides. They differ from other Fe-rich deposits such as the gossans and clay-rich (nontronite) ochreous deposits by their low transitional metal contents ($Cu + Zn + Co < 0.05$ %), their low SiO2 (<20 %), and their higher degree of iron oxidation ($Fe3+/Fe2+ = 400 - 1500$). These deposits are deprived of sulfur (<0.2 %) and do not show any trace of secondary sulfide alteration. Also, the depletion in Mn (0.5 %) content and the lack of manganese crust on newly formed Fe-rich hydrothermal deposits indicate that the decoupling between Fe and Mn precipitation is related to the source composition and to its degree of solubility in water. The relatively high content of Fe with respect to Mn in basalt will enhance the formation of Fe-rich precipitates rather than manganese. The manganese will precipitate as hydrogenous products from the water column.

The primary source of Fe is from the minerals of titanomagnetite and ilmenite in the leached source rocks with the formation of rutile. Rutile is encountered in association with chlorite and serpentine in hydrothermally metamorphosed dolerite and/or gabbro.

The formation of Fe-rich oxyhydroxide deposits could be due either to hydrothermalism or to a simple alteration of the surface of a hot lava flow at contact with seawater. Hydrothermalism giving rise to Fe-rich oxyhydroxides takes place at relatively shallow depths within the oceanic crust in comparison to the source of the sulfide rich deposits. Fe-staining on the broken surface of pillow lavas indicates that a locally low-temperature reaction (at the site of alteration) at the boundary layers between seawater and the cooling of a lava flow during its emplacement on the ocean floor is a common phenomenon. Also, seawater trapped in cavities above sills or dykes may be heated, similar to boiling water in a closed pot on a stove, and the hot seawater will react with the country rock. When fluid circulation in the oceanic crust remains at relatively shallow depths (<1 km), the leaching of deep-seated material will be prevented, and the water/rock ratio in the reactive zone will be higher. Experimental results of Hajash (1975) demonstrate both the presence of copper-bearing sulfides and very high Fe and Mn concentrations in fluids that form oxides where the water has become more oxygenated.

The common occurrence of Fe-oxyhydroxide on seamounts of intraplate origin may also be due to the porous nature of the edifice, which acts some what like a sponge, absorbing seawater whose solution will then have an increased pH and oxygen fugacity after being mixed with rising hydrothermal fluid. This may also explain the lack of polymetallic sulfide deposits on many tall (>1500 m in height) intraplate seamounts where sill and dyke intrusions occur at shallow depths within the volcanic edifices.

Silica-Rich Deposits

Very fragile and porous (50 % porosity) silica-rich deposits with a spongy appearance are formed by strips of silica filaments, which are associated with barite and Fe-oxyhydroxides. Silica-rich deposits have the appearance of colloidal (amorphous, shapeless) material varying in color from milky white to light brown. Other silica-rich deposits found in association with massive sulfides and sulfide mounds show a similar texture, and contain more than 90 % SiO_2. Opal is the major silica-containing phase in the deposits, which probably have more than one origin. For example, the silica-rich deposit found on the Southeastern Seamount (East Pacific Rise, 13°N) contains remnants of pyrite, chalcopyrite, covellite, digenite, chalcocite and idaite and was thought to have originated from the replacement of sulfide minerals (Hekinian and Fouquet 1985), while other silica-rich deposits composed of opaline products with rod-like filaments were attributed to the growth of bacteria-like material (Juniper and Fouquet 1988).

Opal is also encountered as a replacement product of wurtzite associated with the black smokers at 21°N on the EPR. Thus, another origin for the silica-rich products is thought to be primary, low-temperature precipitation of hydrothermal fluids directly on top of or within the interstices of volcanics or forming inter-layered rims in the sulfide chimneys themselves. The chemistry of this type of deposit indicates an enrichment of SiO_2 (81 %), of Fe_2O_3 (5.7 %), traces of Na_2O (1–2 %), MgO (0.3 %) and H_2O (ignition lost 10–11 %). It is also deprived of most transitional metals, large ion lithophile (LIL) elements such as strontium (Sr < 60 ppm), and LREE (Light rare earth elements).

Hydrated silicates have been found to be forming in recent hydrothermal fields on the ridge axis of the EPR at 21°N and 13°N. Low-temperature, shimmering water escaping from small cracks and fissures encountered on pillow lava and sheet flow terrain precipitates milky white and yellowish red coatings on the rock surfaces. Usually the side of the rock exposed to the hydrothermal fluid will show the altered crusts, which range in thickness from less than 1 mm up to 50 mm. The white precipitates have an irregular contact with the surface of the glassy basalt margin. The contact is marked by concentric lamellae a few microns thick made up of various hydrated components such as kaolinite (clay), mixed-layer smectite, serpentine, and boehmite. Other samples show an abundance of mixed-layer chlorite-smectite associated with small amounts of kaolinite (Haymon and Kastner 1986). The exit temperature of the hydrothermal fluids is responsible for giving rise to these alteration crusts and was measured to be less than 60 °C. The elements composing the crusts are amorphous Al- or Mg- silicates and chemical analyses have shown that the MgO is 13–22 %, SiO_2 is 27–43 % and Al_2O_3 is 18–25 % by weight. The minor element constituents are Ca (1–4 %), Na_2O (< 0.5–2 %) and Fe_2O_3 (1–5 %).

Fe-bearing nontronite is one of the most common hydrothermal precipitates found on the sea floor. It is a light-green, yellowish-green soft and/or semi-consolidated aggregate sometimes associated with amorphous Fe-oxyhydroxides but also associated with the silica rich deposits. Chemically, the Fe-nontronite clay is also silica-rich since it essentially consists of SiO_2 (49–54 %) and Fe_2O_3 (28–31 %) and it is deprived of Al_2O_3 (<1 %). This clay is also very low in transition metal content (<200 ppm) so metals such as Zn, Co, and Ca are sparse.

The formation of the Fe-nontronite has been attributed to two processes: (1) direct low temperature discharge from hydrothermal fluids circulating within the oceanic crust and (2) from the interaction with siliceous and carbonaceous ooze acting as a permeable material through which low-temperature fluids (<60 °C) are circulating. Temperature determination from oxygen isotopic studies indicates that the formation of the Fe-nontronite occurred at 25–47 °C (McMurty et al. 1983).

The most plausible hypothesis for the creation of nontronite is related to a hydrothermal precipitation of low-temperature circulating fluids. Nontronite precipitates directly on the seafloor; however, in the case of the Galapagos Mounds area, the nontronite is probably precipitated within the sedimentary layer. The precipitation is the result of the oxidation gradient enhanced by the upward migration of oxygen-poor hydrothermal fluids and the downward diffusion of oxygen from seawater (Dymond et al. 1980). This model of direct precipitation for nontronite is in accordance with the presence of similar products associated with the sulfide chimneys near 13°N and 21°N on the EPR and on intraplate volcanoes.

Stockwork Mineralization

The pattern of fluid circulation in the lithosphere is usually determined though a study of low temperature (<400 °C) fluids and the subsequent hydrated mineral assemblages that have precipitated and/or replaced the more stable original magmatic minerals. These transformations are marked by changes in mineral composition and the precipitation of new minerals.

Hydrothermal fluids charged with metallic particles could also precipitate metal-bearing compounds along their pathway to the surface. These types of precipitates are called "stockwork" mineralization (Fig. 6.4b, c). Stockwork is a term used by mining companies and it usually defines a mineral deposit that has an economic value formed by a closely spaced network of veins and veinlets cutting through the surrounding country rock. Often, these mineralized veins crossing various types of flows (dykes and lava flows) are exposed by faults and fissures on spreading ridge walls and in transform faults walls as well as in hotspot volcanoes. The mineral association found in the stockwork will depend on the existing physical and chemical conditions of the environment where the veins were formed. A low temperature (<150 °C) and high pH (>6) fluid is likely to precipitate oxides and hydroxide phases. More acidic (pH < 5) and higher temperature conditions (>250 °C) are favorable for the precipitation of sulfides. The subsurface of the ocean floor at the sites of discharge for hydrothermal fluid is marked by an extensive alteration of the surrounding rock formations (basalt, dolerite, gabbro and others). The sulfide-filled veins can be seen forming an intricate interconnected network terminating on top of fault scarps underneath sulfide edifices.

A volcanic or hydrothermal outcrop could later become brecciated due to the extensive hydrothermal fluid circulation that has altered the rocks through fissures and cracks, and created veins filled by hydrated Fe–Si clay and metallic sulfides. Such alteration can loosen the coherent volcanic rocks, which are more easily broken into fragments that could then be re-cemented during further circulation of hot fluids and mineral precipitation. Examples of stockwork from the EPR axial graben were observed by submersible near 12°50′N as well as near an ultra-fast spreading segment of the EPR at 18°30′S (Fig. 6.4a–c).

Formation of Hydrothermal Deposits

As explained earlier in this chapter, the formation of hydrothermal deposits occurs through four steps when seawater enters into the lithosphere through pores and fissures in the volcanic rocks, then comes into contact with heat, so its composition is changed and it becomes more corrosive so it can dissolve metals and other elements from the volcanic rocks before being discharged. The hydrothermal fluid forms precipitates of metal bearing sulfides on the sea floor and in the country rock on its way to the surface. When the discharge flow rate of the hydrothermal geysers is high, chimneys could be formed, and they will then serve as conduits for subsequent hydrothermal discharge.

The formation of a hydrothermal chimney will occur when a large volume of fluid with a rapid discharge rate (1–5 m/s) is spewed out onto the sea floor. The contact between the exiting hot chemically altered fluid and the cold ambient seawater produces another chemical reaction between the edge of the fluid flow and the seawater, constructing a thin skin of sulfates and sulfides. More details on the observed growth of a hydrothermal chimney in an area called the "Chain site" are presented in Chap. 7.

The early stage of mineral precipitation varies with the composition and the temperature of the fluid. A rapid discharge of hydrothermal fluid will initiate the formation of a fragile skin of whitish gray colored anhydrate (calcium sulfate), which will form a protective carapace at the interface between the hot fluid and the cold seawater. Below 150 °C, anhydrate starts to dissolve in seawater but, before this happens, there may be a subsequent deposition of amorphous silica, which lowers the structure's permeability and prevents the exchange between the fluid and seawater (Tivey and Delaney 1986). When the protective coating of the chimney walls is built, thereby preventing any extensive mixing with the ambient seawater, several other sulfide phases start to form. The most inner part of the chimney is also the most insulated from seawater and will be formed by the precipitation of Cu-sulfide at high temperatures (>300 °C) while Zn and Fe will precipitate at lower temperatures (100–270 °C). The texture of the mineral grains will also vary inward from dendritic, spikey shaped minerals to more coarse-grained deposits. In the

Zn-rich type of chimney, which is more porous than the Cu-rich chimney, the inner part will become partially clogged. In the Cu-rich chimneys, the inner part consists essentially of coarse-grained chalcopyrite. In the Cu–Zn chimneys, the inner walls will be the sites of Zn-sulfide precipitation before the crystallization of chalcopyrite. The actual construction of a hydrothermal chimney was monitored and is discussed in Chap. 7 in a section about the Chain Site on the East Pacific Rise.

Some chimneys are more than 30 m tall. Commonly the average height of a hydrothermal edifice is about 3–7 m as seen at 20°N, 10–13°N, and 17–21°S on the East Pacific Rise. Sometimes it is difficult to assign a definite geometry to these edifices, and several edifices could be erected on top of piles of fragmented and re-cemented sulfides. Other edifices may have the aspect of a tree with several branching chimneys on top of a bulbous shaped porous sulfide deposit. At their base, sulfide chimneys can be found sitting on an irregular surface and may even cover sheet flows and fissures. However, chimneys do not extend along the entire length of a fissure, and instead are always localized.

Phases Forming Hydrothermal Deposits

Hydrothermal activity will probably take place over several steps within a restricted area of the sea floor surface and this will contribute to the construction of sizable metalliferous deposits. Different events in the formation of an extensive deposit have been observed in various locations on the sea floor. The formation of a metalliferous deposit can be divided into five stages (Fig. 6.5a–e) as follows:

(1) The first phase involves the discharge of low (<150 °C) temperature fluid through pillow lavas and/or sheet flows discharging iron-silicates and some sulfides (Fig. 6.5a, b). This type of low temperature discharge is accompanied by a gathering of the benthic animal community such as tube worms, clams, etc.
(2) High temperature (>350 °C) exiting fluids are found during the next stage of activity, resulting in the construction of chimneys to channel this fluid and provoking the precipitation of iron-copper sulfides and anhydrite ($CaSO_4$) (Fig. 6.5c).
(3) A coalescence of several chimneys leads to the formation of a porous edifice or mound, through which more seawater can enter the lithosphere, and where hydrothermal fluids can more easily percolate, with the precipitation of iron-copper-zinc sulfides. At the same time, the circulation of hot fluids will enhance the alteration of the country rock and the precipitation of metalliferous compounds forming stockwork.
(4) At the declining stage of hot hydrothermal activity, cooler fluid (>200 °C) will be discharged forming mineral compounds that will eventually fill the pores or fissures of the hydrothermal mounds (Fig. 6.5d). This event will give rise to a zonal distribution where copper rich precipitates are more plentiful at the

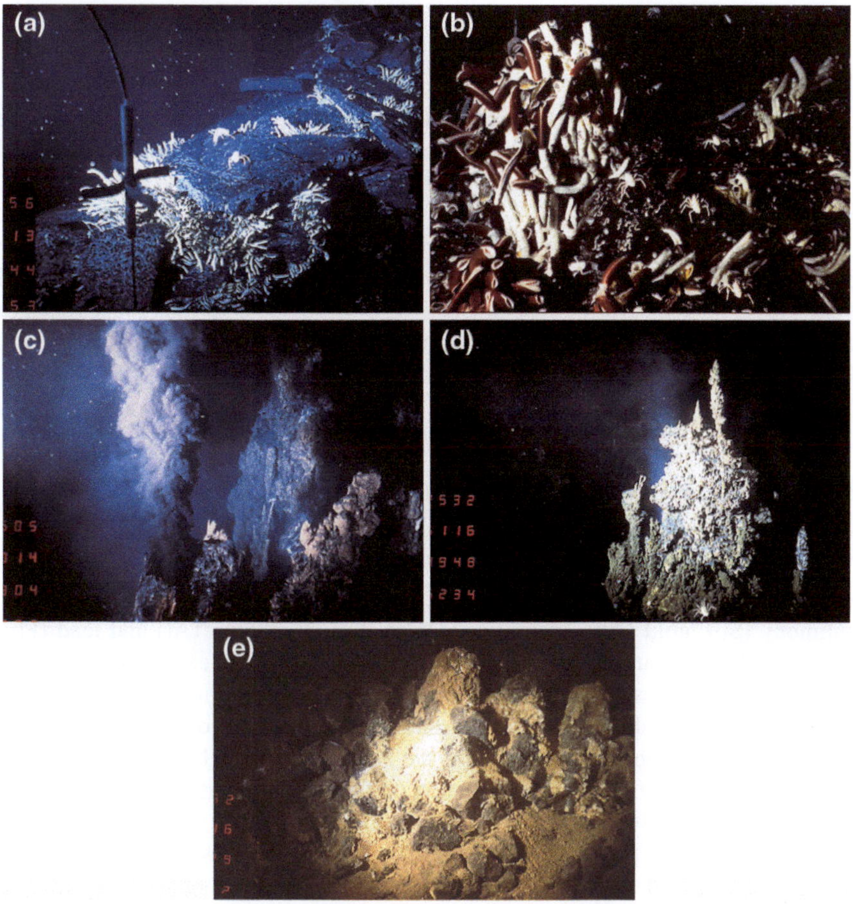

Fig. 6.5 Five photos (copyright IFREMER *Geocyarise cruise 1984, Cyana dives Cy84-07, Cy84-24*, and *Geocyatherm cruise 1982, Cyana dive Cy82-35.*) showing the construction and evolution of a hydrothermal field. **a** Early formation of an active hydrothermal field with the exit of clear warm water throughout a fissured lava flow. A small biological community starts to develop, as seen during dive Cy82-07 at 2525 m depth on the EPR near 11°27′N. **b** Luxuriant biological community of tubeworms associated with the early construction of a hydrothermal sulfide edifice during the precipitation of high temperature (250 °C) to moderate temperature (150 °C) deposits. Example shown was seen during dive Cy82-35 at 2623 m depth on the EPR near 12°50′N. **c** Stage of intense hydrothermalism with the exit of high temperature (>200 °C) fluid and the construction of polymetallic sulfide edifices seen during dive Cy82-07 at 2530 m depth on the EPR near 11°27′N and **d** Late, waning stage of hydrothermal activity with the decrease of hydrothermal fluid flow as observed during dive Cy84-24 on a multiple chimney located on the EPR's eastern wall at about 3 km from the graben's axis near 12°43′S at 2572 m depth. **e** Dislocated and altered dead hydrothermal chimney seen during dive Cy84-24 at 2582 m depth on the eastern wall of the EPR axis near 12°43′N

center while zinc, iron sulfide and colloidal iron sulfide are more plentiful towards the exterior of the mound.

(5) When fracturing of the sea floor takes place during tectonic events, the formation of normal faulted scarps will also occur. The fault scarps can expose the veins of mineralization (stockwork) and a talus of slumped material will collect at the foot of the faulted scarps (Fig. 6.5e)

Thus, a metalliferous deposit is an oval-shaped structure containing more than 3,000 tons of material that originates from the evolution of distinct hydrothermal events (Fig. 6.5a–e). During its construction, it will assume various shapes such as columnar edifices, massive sulfide mounds, breccia, slumped talus material, etc. (See paragraphs on types of hydrothermal deposits).

Exploitable Mineral Resources of the Sea Floor

The major metallic elements found and exploited in metalliferous deposits on land are iron (Fe), cobalt (Co), copper (Cu), zinc (Zn), palladium (Pd), silver (Ag), antimony (Sb), gold (Au), platinum (Pt), tin (Sn), aluminum (Al) and lead (Pb). Many of these occur in seafloor deposits but are not yet exploited. Deep-sea deposits other than sulfides are manganese nodules and manganese crusts. The best-known exploitable undersea deposits of nodules are located between the Clarion and Clipperton fracture zones in the Pacific Ocean. Only about 10 % of the total area covered by nodule accumulation on the sea floor, with an average 2.4 % content of copper + zinc + cobalt, is comparable to that of the entire terrestrial sulfide ore deposits (Exon et al. 1992). The manganese crusts are of particular interest because of their high content in the platinum group elements. They are also easier to mine because they are found in shallower water depths than the manganese nodules (>4000 m depth).

Hydrothermal deposits are important for industry and their exploitation goes back to the dawn of civilization (Scott 1987). Sulfide deposits on the Island of Cyprus have been exploited for copper for more than 2500 years and were mined during the Roman Empire. In fact, the Island of Cyprus was uplifted from the sea floor during the Tertiary period more than 120 million years ago and has the characteristics of a modern oceanic ridge fragment. The metalliferous deposits found on Cyprus were formed over a period of about 12,000 years and contain 4–7 million tons of ore. Cyprus is a good area for geologists to explore exposed, land-based hydrothermal deposits so scientists who want to better understand the sea floor environment often visit the island.

The exploitation of ocean floor hydrothermal deposits is becoming a reality. However, the exploration of the sea floor should continue in order to provide more information on the deposits. This will help to better understand the chemical and thermal processes that are involved during fluid circulation and mineral precipitation. Also, knowing the setting and composition of the environment associated with hydrothermalism may help in the discovery of new land-based deposits.

Another important aspect of hydrothermal activity is related to the bacteria and other biological communities that are produced or which develop near these sites. Until their discovery in the 1970s, scientists were unaware of the existence of such animal communities in the depths of the sea, close to extremely hot geysers of acidic, mineralized seawater. Continued research will increase our understanding of the interaction between mineral precipitation and the animal communities that have formed around the hydrothermal sources.

References

Charlou J, Donval J (1993) Hydrothermal methane venting between 12°N and 26°N along the Mid-Atlantic Ridge. J Geophys Res 98(B6):9825–9642
Corliss JB, Lyle M, Dymond J (1978) The chemistry of hydrothermal mounds near the Galapagos Rift. Earth Planet Sci Lett 40:12–24
Corliss JB, Dymond J, Gordon LE, Edmond JM, Von Herzen RP, Ballard RD, Green K, Williams D, Bainbridge AE, Crane K, van Andel Tj H (1979) Submarine thermal springs on the Galapagos Rift. Science 203:1073–1083
Binns RA, Scott SD (1993) Actively-forming polymetallic sulfide deposits associated with felsic volcanic rocks in the eastern Manus Basin. Papua New Guinea: Econ Geol 88:2226–2236
Bischoff JL (1980) Geothermal system at 21°N, East Pacific Rise: physical limits on geothermal fluid and role of adiabatic expansion. Science 207(4438):1465–1469
Degens E, Ross DA (eds) (1969) Hot brines and heavy metal deposits in the Red Sea; a geochemical and geophysical account. Springer, New York, 600 p
Dymond J, Cobler R, Muratli CM, Chou C, Conrad R (1980) In initial reports, Deep Sea drillin project. In: Rosendahl BR, Hekinian R (eds) vol 54. US Govt. Printing Office, Washington, pp 377–386
Exon NJ, Bogdanov NA, Francheteau J, Garrett C, Hsü J, Mienett W, Ricken SD, Scott SD, Stein RH, Thiede J, von Stackelberg (1992) Group report: what is the resource potential of deep sea ocean? In: Su KJ, Thiede J (eds) Use and Miss-use of the sea floor. Wiley, p 7–27
Francheteau J, Needham D, Juteau T. Rangin C (1980) Naissance d'un Océan, CYAMEX, Publshed by CNEXO (Centre National pour l'Exploitation des Océans) IFREMER 92138, Issy Les Moulineaux France, 85 p
Hajash A (1975) Hydrothermal processes along mid-ocean ridges: an experimental investigation. Unpublished Ph D dissertation Texas A&M University College Station Texas, p 61
Haymon RM, Kastner M (1981) Hot spring deposits on the East Pacific Rise at 21°N: preliminary description of mineralogy and genesis. Earth Planet Sci Lett 53:363–381
Haymon RM, Kastner M (1986) Caminite: a new magnesium-hydroxide-sulfate-hydrated mineral found in submarine hydrothermal deposit, East Pacific Rise, 21°N. Am Mineral 71:819–825
Hekinian R, Fouquet Y (1985) Volcanism and metallogenesis of axial and off-axial structures on the East Pacific Risenear 13°N. Econom Geol 80:221–249
Hekinian R, Binard N (2008) Le Feu des Abysses. Editions Quae: Versailles, France, p 175
Hoffert M, Perseil A, Hekinian R, Choukroune P, Needham HD, Francheteau J and Le Pichon X (1978) In initial reports, deep sea drilling project. In: Rosendahl BR, Hekinian R (eds), US Govt Printing Office, Washington, vol 54, pp. 339–376
Honnorez J (1981) The aging of the oceanic crust at low temperature: In Emiliani C (ed), The Sea. The oceanic lithosphere. Wiley, New York, vol 7, pp 525–587

Juniper and Fouquet Y (1988) Filamentous iron-silica deposits from modern and ancient hydrothermal sites. Can Mineral 26:859–870
Lonsdale P (1977) Clustering of suspension-feeding macrobenthos near abyssal hydrothermal vents at oceanic spreading centers. Deep-Sea Res 24(9):857–863
McMurty GM, De Carlo EH, Kim KH, Kronk LW (1983) Trans Am Geophys Union 64:1018 (Abstr)
Parker RL, Oldenburg DW (1973) Thermal model of ocean ridges. Nature 242(1973):137–139
Sclater JG, Francheteau J (1970) The implication of terrestrial heat flow observations on current tectonic and geochemical models of the crust and upper mantle of the Earth. Royal Astron Soc Geophys J 2:509–542
Scott SD (1985) Seafloor polymetallic sulfide deposits: modern and ancient. Mar Min 5:191–212
Scott SD (1987) Seafloor polymetallic sulfides: scientific curiosity or mines of the future? In: Teleki PO et al. (ed) Marine minerals resources and assessment strategies, Proc NATO Advence Res Workshop Series C 194 Reidel Boston Mass, pp 277–300
Scott SD, Binns RA (1992) An actively-forming, felsic volcanic-hosted polymetallic sulfide deposit in southeast Manus back-arc basin of Papua New Guinea (abstr). EOS Trans Am Geophys Union 73:836
Spiess FN, Macdonald KC, Atwater T, Ballard R, Carranza A, Cordoba D, Cox C, Diaz-Garcia VM, Francheteau J, Guerrero J, Hawkins J, Haymon R. Hessler R, Juteau T, Kastner M, Larson R, Luyendyk B, Macdougall J.D., Miller S., Normark W, Orcutt J and Rangin C (1980) East pacific rise: hot springs and geophysical experiments. Science 207:1421–1433
Tivey MK, Delaney JR (1986) Growth of large sulfide structures on the endeavour segment of the Juan de Fuca Ridge. Earth Planet Sci Lett 77:303–317
Von Damm KL, Bischoff JL (1987) Chemistry of hydrothermal solutions from the southern Juan de Fuca Ridge. J Geophys Res 92:11334–11346
Williams DL, Von Herzen RP (1974) Heat loss from the earth: new estimate. Geology 2(no 7):327–328
Yang K, Scott SD (1996) Possible contribution of a metal-rich magmatic fluid to a sea-floor hydrothermal system. Nature 383:420–423

Chapter 7
Oceanic Spreading Ridges and Sea Floor Creation

Abstract The Spreading Ridge System is one of the most striking features exposed on the seafloor. It is a belt of volcanoes that surrounds the planet both under the oceans and in subaerial terrain. The Spreading Ridge System is where plate divergence occurs when ascending hot material rising from the mantle creates the sea floor. The spreading ridges extend around the World for a total length of 70,000 km and a width of about 1,000–2,000 km. This system is cut by tectonic and magmatic discontinuities such as fracture zones and disrupted spreading ridge segments. When compared to a human body, the spreading ridge system is like a continuous "backbone" and the "rib-bones" are the fracture zones emanating from both sides of the ridge axis.

At the base of the lithospheric plate, at 70–100 km depth within the crust, diverging forces push the plates apart from each side of the ridge axis in order to make space for new molten crust to be extruded. As the crust created on the ridge axis is moved away from its point of origin, it cools and becomes less ductile, which causes it to break due to the tectonic constraints related to spreading. The speed of crustal creation at a ridge axis is not the same everywhere, as indicated by the different spreading rates that have been measured (Fig. 7.1). The different rates of spreading have influenced the ridge landscape, as well as the rocks' morphology and structure. Thus, on a slow spreading ridge (total rate <3 cm/yr) such as the Mid-Atlantic Ridge (MAR), the axis of spreading along its strike has created a large depression, graben or rift, sometimes more than 1 km deep and about 15–30 km wide. On the other hand, faster rates of spreading such as in the Pacific Ocean cause the axial graben to be much smaller and the ridge itself is more volcanically active than the MAR axis. Thus, the East Pacific Rise will be less affected by tectonic breaks because of the larger volume of lava erupted. Also the East Pacific Rise (EPR) is more affected by off-axial volcanism (See Chap. 9), which constructs more individual volcanoes than on the MAR. The MAR segments behave as more rigid structures and show better defined tectonic breaks than the EPR. Also the East Pacific Rise does not follow any continental contour line because it is detached from continental masses whereas the slower spreading

R. Hekinian, *Sea Floor Exploration*, Springer Oceanography,
DOI: 10.1007/978-3-319-03203-0_7,

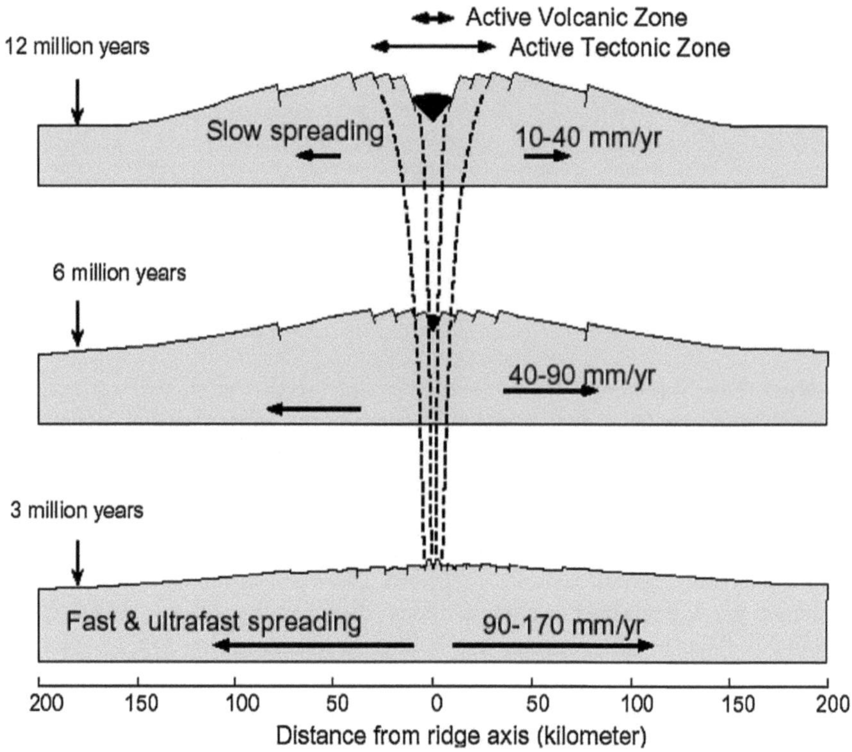

Fig. 7.1 Near-axial structural profiles across accreting ridge systems with different spreading rates showing the variable sizes of their axial rift valleys (Choukroune et al. 1984)

ridges such as the MAR and the Mid-Indian Ocean Ridge follow the orientation of the bordering continental masses.

Other differences between slow and fast spreading ridges are due to the depth and degree of partial melting of their mantle source (Fig. 7.1). Mantle temperature variations have an important effect on the extent of melting beneath oceanic ridges. As mantle rises adiabatically, decompression melting will take place beneath oceanic ridges, but this melting will be affected by plate separation rates. Fast plate separation at the EPR is responsible for a large extent of melting (>22 %).

Ridge Segmentations

Ridge segmentations are the result of tectonic and magmatic processes that are responsible for forming discontinuities along the strike of the oceanic ridges. The existence of these ridge segmentations was inferred from bathymetric contour-line variations and from subsurface structural and compositional changes observed during geophysical investigations along the strike of the oceanic ridges. The evidence of these ridge discontinuities was found along the East Pacific Rise with the multichannel bathymetry system that enabled mapping of a large portion of the sea floor in a single swath, in the early 80s. However, the fact that the strike of the ridge was segmented was first suggested by the pioneering work of Bruce Heezen and Mary Tharp who drew the first physiographic map of the western North Atlantic in 1954, and later for the World's ocean floor in 1967 (Heezen and Tharp 1968). Looking at this physiographic map of the oceans, it becomes obvious that the entire spreading ridge system around the World was marked by ridge segment discontinuities as well as being cross cut and displaced by transform faults, which have offset the accreting plate boundary region for distances of hundreds of kilometers (from about 50 up to 500 km).

The causes of ridge segmentation are multiple:

(1) The early history of continental opening formed major breaks and fracture zones in the lithosphere, which later caused the large-scale discontinuities during the separation of Africa and the Americas in the Equatorial Atlantic where the largest ridge segment offsets are observed.
(2) Instability in the asthenosphere due to differences in rheology (i.e. the force of deformation of a structure, plus the degree of plasticity and the quantity and speed of magma flowing through the crust and upper mantle).
(3) Depending on the spreading rate and different plate tectonic motions, the ridge segmentations will affect the mode and rate of magma delivery. For example, ridges with slow spreading rates and starved magmatism will have a thicker, more brittle lithosphere, which will slow magma delivery even more. Fast spreading ridges, with their faster rate of magma delivery will show smoother breaks between the segments.

Since 1981, scientific ships equipped with multichannel echo sounders have enabled us to obtain detailed bathymetry of the seafloor so we have been able to identify further types of segmentation of the spreading ridge system. The observed topographic variability along a spreading ridge axis, the morphology and composition of the rocks, as well the geophysical observations (gravimetry, seismicity and magnetism) and the distribution of hydrothermal activity have enhanced our understanding of ridge segmentation. The volcanic and tectonic activities along each portion of an individual spreading ridge segment will vary along their strike. Generally speaking, the ends of the segments are volcanically less active than the middle portions.

The EPR fast spreading ridge systems are made up of several segments that have undergone different periods of volcanic, tectonic and hydrothermal activities (Fig. 7.2). The overall observations between transform faults and large "non-transform offset" boundaries have permitted us to define several orders of segmentation on the fast spreading ridge systems of the Pacific Ocean.

The 1st order segmentation reflects large-scale magmatic upwelling in the asthenosphere (Macdonald et al. 1988), which usually occurs between two

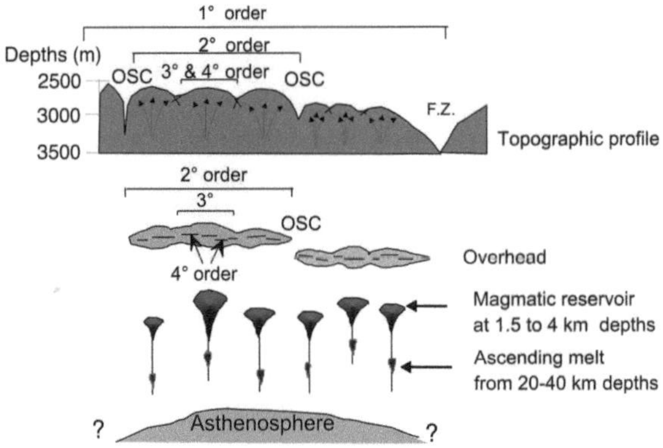

Fig. 7.2 Spreading ridge segmentations show axial discontinuities caused by fracturing and magmatic upwelling. A succession of magmatic reservoirs, hot at the center and cooler on the margin, are supplied from the partial melting of the asthenospheric mantle migrating towards the surface. Ridge segmentations between transform faults evolve and change progressively with time. Several orders of ridge segmentation are assigned in relation to their length, relief and time frame stability: 1° order Segment – length: 30–100 km. Relief: about 400 m. Stability: 10^7 yr 2° order Segment – length: 10–50 km. Relief: <100 m. Stability: 10^4 yr 3° order Segment – length: 10–30 km. Relief: <50 m. Stability: 10^3 yr 4° order Segment – length: <5 km. Relief: <30 m. Stability: <10^2 yr *OSC* Overlapping Spreading Center

transform faults. These 1st order segments range between more than 20 km up to more than several hundred km in length. This type of segmentation has the longest temporal stability (10^7 years) before any tectonic disruption, and the largest depth difference (up to 500 m) at the ridge-transform intersections (RTI). The 1st order discontinuities have offset the ridge axis along distances of about 200 and 800 km.

Superimposed on the first-order segments are smaller discontinuities (which are 2nd, 3rd and 4th orders). These segments divide the ridge into smaller and shorter-lived offsets. The 1st and 2nd order discontinuities are seismically marked by long wavelengths. There are long-lived magma injections along these inflated, volcanically active segments, which have large and steady state magma reservoirs with replenishing and mixing.

The 2nd order discontinuities have a segment length of 30 to 300 km. They have a well-developed Overlapping Spreading Center (OSC) along the ridge axis where their two ends curve toward each other and distances of 1–20 km separate them from each other. These segments have an average relief of about 100 m at their segment ends and a stability of about 10^5yr.

The 3rd order segmentation is represented by 10–30 km long discontinuities and shows a ridge topography that pinches in at the segment ends due to the decrease in magmatism. 3rd order segmentation is characterized by a vertical relief of less than 50 m in height at the segment ends and has a stability of less than 10^4 yr.

These 2nd and 3rd intermediate types of discontinuities (about 10–50 km in length, 1–10 km in width) have offset the strike of the EPR ridge axis by more than

1 km and are more likely to reflect melt segregation at the base of the crust. Whether they represent only an offset of axial topography or ridge segments that are overlapping each other, these intermediate discontinuities, have been called DEVALS (Deviation from Axial Linearity) or "overlapping spreading centers" (OSC). The presence of an OSC is the result of two ridge segments, which propagate towards each other due to the extrusion of lava, but whose two flows fail to meet (Macdonald et al. 1988). Hence, along the segmented ridge crest, two evolutionary paths occur with intermittent and alternate volcanic pulses. Eventually, the dominant spreading center with its larger magma delivery will propagate over a longer distance.

The 4th order segmentation is represented by discontinuities due to en-echelon fissures and collapsed lava ponds running along the strike of the spreading centers. These smaller discontinuities of less than 1 km in length have offset the ridge axis by less than 100 m. They have a vertical relief of less than 25 m in height and a segment length of less than 5 km. The segment's ends are associated with talus piles. According to submarine observations (i.e. at NEPR 13°N and SEPR 18°30′S, Hekinian et al. 1985), the large fissures forming an en-echelon relay zone are separated by older constructional highs (<10 m high). Also these 4th order segmentations are ephemeral and short-lived since they are essentially controlled by dyke injections and hydrothermal circulation (Haymon 1996). 4th order segmentations are characterized by an amagmatic period during lithospheric cooling and cracking which allows seawater circulation and leaching to take place. The duration of the hydrothermal cycle was inferred to be about 100 years for the ridge segment at 9–10°N on the EPR (Wright et al. 1995). Smaller 3rd and 4th order discontinuities are characterized by localized magma propagation from a main magma reservoir during dyke and sill intrusions along the ridge strike.

The segments seen on slow spreading ridge systems, such as the Mid-Atlantic Ridge, are marked by sharper tectonic breaks or discontinuities than on fast spreading ridges, which show more "curved" and settled structural appearances. Segment discontinuities on slow spreading ridge systems compared to those from the fast spreading East Pacific Rise differ in their 2nd order discontinuities showing the offset of their ridge axis rift valley floor. They are deprived of Overlapping Spreading Centers (OSCs) and instead they have developed nodal basins at their segment ends. These structures are longer-lasting (up to millions of years) than those from a fast spreading ridge axis. Also, the 3rd order discontinuities in slow spreading ridge areas are confined to the rift valley and are defined by volcanic constructions, mounds and individual edifices, which are often offset from each other. Fast spreading ridges (total spreading >4 cm/yr) show a more uniform and linear ridge axial topography without a deep rift valley, but with a relatively shallow graben (<300 m depth). The magma budget is more sustained and more extended along the ridge axis than on slow spreading ridges. Lateral, off-axis magmatism giving rise to off-axial seamounts is also more prevalent for fast spreading centers than for slower spreading ridges.

Historically, the first small scale spreading ridge discontinuity (2nd and 3rd order) was found by the German oceanographic vessel FS *Sonne* in January 1982 during a geological survey of the East Pacific Rise near 13°N under the leadership of Harold Bäcker from Preussag (Germany). It was at the same time, during the Cyatherm cruise with the RV *LE SUROIT*, that we met the German scientists on site. I invited Harold Bäcker on board the R.V. *LE SUROIT* to participate in a diving operation with *Cyana*. After his dive, we were invited on board the *Sonne* where we gathered for drinks on the evening of February 9, 1982. It was then that Harold, Jean Francheteau and I had a look at the first multichannel (Hydrosweep) map that the *Sonne* had made of the area at 13°50′N with its curving bathymetric contour line. At the time, we did not understand what this "curving structure" meant. Could it have been an artifact of the multichannel acoustic signal? We did not give the problem any further thought, but this was a mistake. We should have tried to better understand that the structure we saw would later be identified as an OSC. In fact, in the same year during July, the US Oceanographic Vessel *Thomas Washington* sailed from San Diego to explore another area of the East Pacific Rise using its newly acquired multichannel system called "SeaBeam". A few years later, in an article published by Ken Macdonald and others from the University of California in Santa Barbara, we noticed that the features described by the authors were familiar and we realized we had missed this while we were on board the RV *SONNE* in 1982 (Macdonald et al. 1988).

The Mid-Atlantic Ridge

The Mid-Atlantic Ridge (MAR) consists of slow spreading (<3 cm/year total rate) ridge segments and has the particularity of having a variable and more irregular axial topography than that of faster spreading ridges. The MAR segments show marked discontinuities such as well-defined transform and non-transform offsets. The segments have variable and heterogeneous lava morphology. They are the sites of tall seamounts as well as island formations.

Going from the Arctic Sea to the Bouvet triple junction (Bouvet island) in the south Atlantic where it branches off towards the Indian Ocean, the MAR is about 14,000 km in length. The topographic differences are related to the changes in volcanic activity along the ridge segments. The presence of several islands piercing the sea surface, such as Jan Mayen, Iceland, the Azores, Ascension and Tristan de Cunha indicates localized volcanism of hotspot origin.

Two distinct structural provinces are recognized in the Atlantic:

(1) The northern MAR province is related to the long-lasting, highly magmatic and volcanically active ridge segments located north of the Azores including the Reykjanes Ridge near Iceland and extending down to the Oceanographer fracture zone at 35°N. In the south Atlantic, the MAR continues below the Romanche fracture zone at 15°S down to the triple junction of Bouvet and Gough islands near 40°16′S–10°00′W. These southern regions of the MAR are also characterized by an elevated topography of their ridge segments with

an average depth of less than 3,000 m. Similarly to the area in the North, the southern part of the MAR is also associated with disseminated islands having hotspot origins.

(2) The other MAR provinces, between the Oceanographer fracture zone near 34°50′N down to the Romanche fracture zone near 5°S, are associated with magma-starved ridge segments. These zones are less affected by magmatic activity and hence have a relatively thin crust. The average topography in the Equatorial Atlantic is deeper than 3000 m and reaches up to 4,300 m. The province with the coldest lithosphere and deepest topography is in the Equatorial Atlantic between 3°N and 3°S. This is a site with significant lithosphere deformation and alteration where strike-slip transform movements coupled with extensional motions occur. It is an area with the highest concentration of closely spaced fracture zones between the African and South American continents. Some of these major offsets include the St. Peter and St. Paul's Rocks, the Chain and the Romanche transform faults (See Chap. 8) (Fig. 7.2).

In summary, the North and the South Atlantic accreting plate boundary regions are sites of more magma productivity due to the upwelling of hot convective cells within the mantle giving rise to ridge centered hotspots. These hot mantle convective areas are separated by a stretch of colder lithosphere located in the magma-starved equatorial Atlantic region. Detailed studies of these different types of spreading ridge segments were made by submersible during several sea-going expeditions conducted between 1973 and 1998.

Two parts of the MAR that were particularly studied in detail are located in an area near the Azores, also known as the "FAMOUS area" near 37°N and in the OCEANAUT area located near 34°50′N on the ridge segment between the Oceanographer and the Hayes fracture zone, which is also called the "Oceanographer area" (Fig. 7.2).

The FAMOUS project: The Mid-Atlantic Ridge near 37°N

The region of the MAR at 35–37°N has been offset by 13 different spreading ridge axial segments with an average length of about 80 km. This part of the MAR ends with a first order discontinuity called the Oceanographer Transform Fault near 35°N (Fig. 7.2). One of the ridge segments, about 140 km long, is also the area where the FAMOUS project took place. This segment is located west of the Azores between 36°58 and 36°34′N at latitude 33°16′W. The segment has a total spreading rate of about 1.2 cm per year. The Franco-American Mid-Ocean Undersea Survey project (FAMOUS) was designed to make detailed observations on a diverging plate boundary where new crust was being created on a ridge segment influenced by the Azores hotspot.

The project was initiated in March 1971, in Bordeaux, France, during an international oceanographic congress (Océanexpo). The first meeting took place between an American from National Oceanographic Administration (NOA), named Bracket Hersey, and a French administrator, Claude Riffaud, who was director of Marine Technology at CNEXO, plus Xavier Le Pichon, director of the marine sciences department at CNEXO, and Commander Gérard de Froberville, who represented the French Navy. These men met at the hotel Aquitania. At that time, both France and the USA had the capability for deep sea diving with manned submersibles, which is why these two nations took the initiative of launching a collaborative deep-diving expedition called "French American Mid-Ocean Undersea Survey" (FAMOUS). The aims of the project were to explore deep-seated volcanism on a ridge segment where the separation of tectonic plates was also the site for the creation of new crust. Up to the 70s, we had managed to obtain a general overview of the ocean and had rediscovered the spreading plate theory. Now we wanted to analyze the phenomenon that causes the plates to spread. Project FAMOUS was designed to take geologists down to the source where new crust was being generated. It was an exciting goal.

In July 1971, a few months after I had taken my job at the oceanographic center of CNEXO in Brest, France, one morning Xavier Le Pichon came into my office to brief me on Project FAMOUS and the cooperation he had initiated with the U.S. to dive in the Rift Valley of the Mid Atlantic Ridge (MAR). The project was planned to take place in the summer of 1973. When Xavier asked me if I wanted to be part of the team, I was very pleased and honored to be asked to participate in the project; yet, at the same time, I was worried about whether or not I would be able to handle what people expected from me. However, despite my doubts about myself, I did not hesitate to accept Xavier's offer to be a part of this great adventure.

The FAMOUS team was composed of scientists from a variety of disciplines. There were Gilbert Bellaiche, a marine and structural geologist, Pierre Choukroune, a tectonician with experience in the Alpine and Pyrenean mountain structures, Jean Louis Cheminée, a volcanologist experienced in island volcanism, Jean Francheteau, a geophysician, David Needham, a marine geologist interested in sea floor structure and sedimentary processes, and myself, a petrologist specialized in oceanic rocks. The head of the French Scientific Team was Xavier Le Pichon, a geophysician and one of the pioneers in the plate tectonic theory. At that time, the only person among the scientists who had any previous experience diving inside submersibles was Gilbert Bellaiche. He was a tall, thin, very friendly person who worked on the coastal and deep-sea geology of the Mediterranean Sea. Based at a CNRS laboratory in Ville-Franche-sur-Mer (French Riviera), he was one of the few persons in the World to have dived the deepest, down to 8410 m depth, with the bathyscaphe *Archimède* in the Japan Trench in 1967.

I knew most of the scientists on the team and this made me feel more comfortable. I was especially at ease with Jean Louis Cheminée whom I had met for the first time in 1960, on top of a mountain called "Monte Zatta" in the Apennines (Italy), when I was completing my fieldwork as part of my diploma called "*Tesa di laurea*" at the University of Pisa in Italy. Jean-Louis was tall and strongly built and he always wore a friendly smile. He was eager to be involved in new adventures and projects, and he had a great deal of experience in the field of volcanology since he was familiar with most of the World's active, subaerial volcanoes.

Jim Heirtzler, a geophysicist from the Woods Hole Oceanographic Institute in Massachusetts, headed the U.S. scientific team. The American team included Jim Moore, a volcanologist from the US Geological Survey (USGS) at Menlo Park, California, plus Bill Bryan, a petrologist from Woods Hole Oceanographic Institute (WHOI), Bob Ballard, a structural geologist, also from WHOI, George Keller, a sedimentologist from Oregon State University, and Jerry van Andel, who was head of the Oceanography Department at Oregon State University.

At the time, the choice of the French team members did not have the unanimous approval of all the influential professors, also called "*les pontes*" (*big shots*), from the French University system and the Centre National de Recherche Scientifique (CNRS). A particular reluctance for the choice of the FAMOUS team came from H. Tazieff who clearly expressed his reserve about the people who had been chosen by Xavier Le Pichon. However, the head of CNEXO, Yves La Prairie, was confident in Le Pichon and his decisions. La Prairie gave his full support and quickly dismissed the critical attitudes of other people in the French scientific community.

Prior to conducting the diving expedition on the MAR, it was going to be necessary to test some new equipment for the submersibles, and it was also decided to provide some first-hand training for the participants in the project. FAMOUS was a "major first" in the academic world since it would be the first time that a coordinated scientific operation was organized involving international institutions with their ships, diving submersibles and their technical and scientific personnel.

Prior training for sea-going expeditions is no longer necessary, however, for the FAMOUS project, training was worthwhile, because it helped us to get better acquainted with our scientific colleagues and with the accompanying the technical teams. It was also very helpful to be able to take part in geological field trips on subaerial volcanoes in order to observe recently formed volcanic landmasses, which could have some similarities to what we might encounter on the seafloor. Later on, during the dives, we noticed that this approach helped both the pilots and the scientists to express their points of view and enabled better communication.

The months of training helped familiarize the diving teams with the volcanic environment, even if the environment at more than 2000 m depth is not the same as what is encountered on subaerial regions. The other phase of the training program was to have first-hand experience in diving and become familiar with the habitat of a diving sphere and its instrumentation. Several dives were to take place close to seaports for logistical purposes. Thus, two cruises were planned in the abyssal hill region off Madeira Island and close to Corsica in the Mediterranean.

At the time, in 1972, the *Archimède* was ready to handle personnel training and equipment testing. It was also important for members of the FAMOUS project to get acquainted with the submersible's personnel and to learn about the constraints of using the submersible's equipment. However the most crucial questions were concerned with how would the scientists react, and how would they work, while being enclosed inside a sphere with a diameter of only 2.20 m, at more than 2000 m depth below the sea's surface. Would the scientists be able to ignore the huge water column above their heads? How would this situation influence their work on the sea floor? The only way to know was to send these men to the bottom of the sea, and see what happened.

I must say that at the beginning it was not easy for me. During my first dive, I was somewhat uncomfortable and even a little scared about going down to the ocean floor inside the submersible. I recall that during my first dive with *Archimède*, it was very difficult to focus my eyes and my attention on the seafloor. I had a knot in my stomach and I was having trouble describing the geological features outside the porthole. I even had a mental blackout for several tens of minutes.

Visit to the Afar Region in Ethiopia

There are only a few subaerial regions where spreading segments emerge on the surface and these areas are found in the East African Rift and in Iceland. The East African Rift extends from the middle of East Africa up to the Gulf of Aden where it becomes a submarine-spreading center in the Red Sea. Thus the Afar region was the first training ground chosen for us to visit, in order to familiarize the divers with what it might be like when diving on the Mid-Atlantic Ridge. The region of the Afar is situated at the junction of two propagating oceanic ridge systems of the East African Rift, one entering the Gulf of Aden and the other one moving further north and going directly into the Red sea. Recent volcanism is confined today to the axial zone on three volcanic chains: the Erta'Ale, Alayta and Tat'Ali. The Erta'Ale chain is about 80 km long and is composed of seven volcanic cones. Among these volcanic cones, the Erta'Ale volcano, formed during the last 1 million years, is the most active. It contains two lava lakes inside its crater, which have been active since 1960 (Barberi and Varet 1970a, b). This is a shield volcano about 613 m high with a basal diameter of about 50 km. Thus, our main goal was to visit the Erta'Ale. The molten lava lakes on this volcano had been visited each year from 1967 to 1973 by volcanologists from the Italian Centro Nazionale di Ricerche (CNR) and from the French Centre National de la Recherche Scientifique (CNRS).

The summit of the Erta'Ale has an elliptical form 1.5 km long and 0.7 km wide. Jean-Louis was familiar with this region because he had worked for several years in the area to study the various eruptions along the chain. A large eruption covering the two lava lakes had occurred in February 1973. It was just a couple of months later, in the spring of 1973, that we went on our expedition and found ourselves driving along rocky trails in Toyota trucks. Our first camp was in the region of Ardou-Coba located at the northern tip of the East African Rift on the

Red Sea in the Gulf of Aden, north of Djibouti. The weather was very humid and hot, so we set-up our military folding bunks outside on a lava field in order to sleep. Our guide made a fire and we were "protected" by two military guards, who were assigned to the group by the local authorities in case we ran into any difficulties. This was an illusion because the so-called rebels were well equipped and at ease in their territory. But that was the rule imposed on us by the local government in Djibouti.

As we pursued our expedition deep into the Afar region along a rubble road, each stop was marked by the encounter of local people standing in front of the convoy. It looked like the people had "suddenly appeared, out from the rocks". Usually there were just a few men and some children who were begging for water and medicine. The rocky landscape is where this local population lives and hides. Jean-Louis, with his previous experience in the region, had anticipated the beggars, so he had organized our supplies to carry surplus water, flour and medicine to distribute during our journey.

Further away, as soon as we arrived in a greener region with water-filled ponds, the panorama changed suddenly, and there were no more people standing along our path. The flat plain seem deserted until we noticed a cloud of dust rising in a distance. Horsemen arrived from no-where. They were local, armed personnel on horseback, who were galloping towards us while we were busy with our camping gear. Jean-Louis appeared to be relaxed and he seemed to know the chief of the "rebels", probably from his previous trips to the region. We were looking at the horsemen and they observed us for a moment, when suddenly the chief dismounted from his horse. This was a good sign. He approached Jean-Louis and shook his hand before welcoming the rest of us. He asked for medicine and some flour to make bread. One beautiful Afar woman, who was the chief's friend, had a large, fresh gash on her chest. Apparently she was wounded and the chief asked us for help. We gave him some antiseptic cream and a few aspirin pills to attenuate her pain. We do not know if she remained alive after they left us.

Our group started to settle down for the night. While we were deploying our gear we saw a single-engine airplane circling around our heads. The plane's passengers were obviously trying to see who we were. After making two turns in the sky, the monoplane landed on a flat area and two men climbed out. One was tall and heavy, wearing dark sunglasses, and he had a big cigar in his mouth. He approached our camp and asked us if we had seen any starving people around. The men told us in English that they were from the Red Cross, based in Switzerland, and that they had a lot of food waiting to be given away to the starving population at a main crossroads, a few tens of kilometers away. I was disturbed by their attitude and told them that if they wanted to find the starving local people, then they had to be on the ground and travel along the rocky paths instead of looking for them from the sky.

While we were getting settled, we saw an antelope running along the rocky desert. When the cook of the expedition saw the animal he told us, with a big smile, that he could make a good meal if someone were willing to hunt it down. Jean-Louis shot the animal with a rifle and the Afar cook's helper ran towards the wounded animal by jumping from rock to rock. As soon as he reached the antelope, he cut its throat and turned the animal to bleed in the direction of Mecca. This is a Moslem custom. In fact, when we finished setting up our camp, the cook was already busy cleaning the kill.

From our campsite we could see the Erta'Ale Volcano that dominates the desert. Our camp was set near an oasis where water and palm trees could keep us away from some of the desert's heat. The French army stationed in Djibouti had loaned us a military helicopter so we could fly over the volcanic crater. When the helicopter came to take us up to volcano, it flew about 200 meters above the crater. We could feel the volcano's heat coming in from the door that was left open for us to better see the active, red-glowing lava lake.

Iceland Field Trip

In the middle of the Atlantic Ocean, the spreading ridge axis has reached the sea surface and created the volcanic island of Iceland. Iceland is geologically and socially one of the most interesting islands that I have visited during my sea-going career. Because of its latitude around 60° North, Icelandic summers are marked by long days since the summer sun sets, or comes close to the horizon, for only a couple hours each day. This gave an opportunity for travelers like us to spend more time in the field and enjoy the nature of the volcanic landscape. Iceland is part of the slow spreading Mid-Atlantic Ridge located above a mantle plume. The country is unique with its empty volcanic landscape, reminding us of the lunar surface. The island is a present day vision of what the Earth must have looked like 4.5 billion years ago when volcanoes, fire and water were the main components shaping our planet. Beautiful people whose ancestors were the Vikings inhabit this island. They are very hospitable and always ready to help.

Diving off Madeira Island

Madeira Island is located in the eastern Atlantic Ocean near the African coast. It is a volcanic cone that rises to more than 4000 m above the seafloor. This site was chosen for our practice dives because we did not have to go far from the port of Funchal before reaching water that was 4000 m deep. This cruise was planned to involve the three different groups of our FAMOUS team: the ship's crew, the submersible's engineering and technical group, and the scientists, all of whom were to be trained together. The main technical objectives were to test the hydraulic system of the *Archimède* and also to verify the water bottle sampling capability of the submarine to ensure success for our scientific goals. In addition, it was a good opportunity for the group's scientific observers to become more familiar with volcanic landscape as it appears under the sea.

The support ship, *Marcel le BIHAN*, and the submersible *Archimède* left Toulon on June 17, 1972, for the Port of Funchal in Madeira. Eight dives took place between the 3rd and the 26th of July 1972. The scientific party included Jean-Louis Cheminée, Jean Francheteau, Gilbert Bellaiche, and myself. The diving group included Gibaudan and Tissieres, who were our electronic engineers, plus Berthelot and Bernasconi, who were our electronic experts and frogmen, and the mechanics were Serrant and Chaix, who later went on to become a pilot of *Cyana* and *Nautile*. The technical engineers were Jean Jarry, Michel Drogou, Dominique Semac and Jean-Louis Michel. The two pilots of *Archimède* were Commanders de Froberville and Harismendy. The area of the dive was chosen according to a previous bathymetric survey that the NO *JEAN CHARCOT* had done in 1966 on the southern coast of Madeira, about 20 km away from the port of Funchal.

The first series of dives were technical and descended to 4,000 m depth on an abyssal plain, in order to test the hydraulic system in a low temperature environment of about 2 °C. The first dive was also aimed at verifying the light probes and their orientation. On July 10 and 13, Jean Louis Cheminée went on two dives followed by Gilbert Bellaiche on July 17. Between dives, the ship and the submersible entered the port except for the first two technical dives when we stayed at sea. This was to test the capability of refueling at sea. Jean Francheteau and I made the last three dives with the pilot Harismendy at 1,700–2,500 meters depth. I remember that during my dive we were on a heavily sedimented bottom at the foot of Madeira Island, at more than 2,000 m depth. For the first time, I saw several large fish called "spada" in Portuguese. Apparently during the day, they dive very deep and at night they come to the surface, which is when the Portuguese fisherman in their lighted boats are able to catch them.

The cruise ended the 29th of July and the ship reached the port of Toulon on August 1, 1972. In fact, the geology was not too fascinating because of the heavily sedimented terrain so it was difficult to see the volcanic outcrops. Nevertheless, this mission definitely helped us to get an idea about the observations we would be able to make during later dives. Certainly, seeing the sea floor so close up, we had to be fast in getting all sorts of information in a very short time. At first, I found it hard to concentrate, but I soon got used to the main task of the scientists, which was to observe, make quick notes, and talk into a cassette tape recorder as much as possible before the target disappeared from in front of the port hole. I learned that the most important thing to be done is to acquire the maximum amount information in order to reconstruct the events and observations of a dive after returning on board the ship. I carefully took notes of "time seen" and "depth" for each geological outcrop that I observed, and I also made sure that the video camera was recording. The next cruise scheduled for our training period was going to take place in October and plans were already being made on board.

Diving off Corsica and Being Caught in an "Avalanche"

Before our FAMOUS cruises took place on the Mid-Atlantic Ridge, another series of training dives were scheduled off the coast of Corsica in October 1972 in order to test more equipment and give more experience to the divers. The *Archimède* was always towed to the diving sites by the French Navy ship *Marcel Le BIHAN.* This ship was originally a German vessel given to France at the end of World War II. It was a "tender" supply ship and had been used to follow and supply hydroplanes in high seas. It was designed with a flat keel, so it rode low, near sea level, and it had a large back deck with plenty of empty space. After being used during the war in Indochina (Viet-Nam), the *Marcel Le BIHAN* was painted white and was assigned to tow the *Archimède.* Its port of call was Toulon.

Copper plates with lettering in German still remained on the toilette doors and also on the bridge to describe the various equipment. On October 3, 1972, the day after than my 37th birthday, the *MARCEL LE BIHAN* left Toulon. After the first three dives, we had to return to Toulon in order to get some special oil (myoline) that was necessary for the submersible. Jean-Louis Cheminée and Jim Heirztler from Woods Hole (head of the FAMOUS project from the U.S. side) made the first three dives. In all, seven dives were made around the Corsican continental slope near the port of Calvi at depths between 1,160 and 2,700 m. These dives were designed to test the transponder navigation network and to improve rock-sampling techniques with the *Archmède*'s mechanical arms. Two dives (C-5 and C-7) were made at the limit of the continental slope and the abyssal plain.

It was during one of my dives, C-5, that two scientists were first allowed to dive together inside the bathyscaphe. This was the first and last time that two scientists dove together, although this is a common procedure for the U.S. submersible *Alvin.* The persons responsible for the French submersible team were against having a second scientist on board because they believed that an engineer and a pilot were more adapted and useful for the job. In addition, it was argued that training a new pilot would be more effective during real dives, which is another reason why only one scientist at a time has been allowed to dive to the ocean floor after dive C-5 off Corsica.

We left the bay of Calvi on Saturday, October 21, 1972. The weather conditions were poor and the sea state was too high for diving, so Captain de Froberville decided not to dive that day, and we had to wait until the next morning. The goal of the dive was to test the maneuverability of the *Archimède* for the job of collecting the three moored bottom transponders. On October 22, two scientists (Jean Francheteau and I) along with the pilot (Gilbert Harismendy) were scheduled to make dive C-5. We put on our yellow, fireproof suits and went down into the zodiac inflatable boat to be taken to the *Archimède* where two engineers were getting the submersible ready for the event. We descended into the airlock and entered the sphere at 9:30 AM. Captain Harismendy was ready and waiting to dive.

At 10:28 AM we started to sink, which is normal when the bathyscaphe dives. We reached the edge of the continental slope at approximately 2,500 m depth and

Gilbert, after communicating to the surface ship that we had touched bottom, started to head forward at full speed with the main propeller. The bottom was abundantly sedimented and undisturbed. The objective was to reach the bottom of the slope at 2450 m depth and then go westward to explore the base at contact with the abyssal plain. We were going full speed (about 40 m per minute) and we were in visual contact with the bottom. It seemed as if we were racing down the continental slope. The gentle slope of about 6° was covered with sediment forming parallel sedimented ridges about 2 m wide and less than 1 m high. These ridges increased size down slope so in order to keep contact with the bottom the pilot had to make a special effort. Suddenly, we felt a motion in the sphere: the keel of the Archimède had touched one of the sedimented ridges. At that moment, we did not know what was happening. It was only after continuing for more than two hundred meters down slope, when the bathyscaph became completely immobilized, that we realized what the problem was. In addition, there was no longer any visibility. We saw only darkness through the porthole as well as on the back television camera located on the tail of the submersible.

Harismendy communicated our position to the surface ship and said in a calm voice that we were momentarily stuck at the bottom and that we could not move. Both Jean and I were silent and we tried to keep busy by writing up our comments of what we had seen through the porthole before we were stopped. Due to the fact that we could not use our main engine and because the television camera on the back was not sending any images, we came to the conclusion that the keel of the bathyscaph was probably buried under sediment. We thought that when we hit the slope, which we barely felt from inside the sphere, we must have triggered an avalanche of sediment, much like turbidity current.

Gilbert did not want to release the ballast of the vessel, because this would eventually make the submersible rise and that would be the end of our dive. So, at first he started to use some of the other propellers in order to try to move the submersible. In the mean time, we began to see the water and the sedimented bottom in front of us because the suspended sediment was clearing off. We took several pictures and continued to keep busy with our notes and by looking at the radar screen and checking the video camera and our cassettes. Gilbert was still busy trying out several ways to move the submersible. Then finally he decided to empty the ballast. He pushed the electromagnetic release button, but nothing happened. We had already been stuck for over an hour on the sea floor, and it looked like we might be there even longer.

The three of us were sitting there, trying to make conversation. We waited a few tens of very long minutes before we felt a slight motion. *Archimède* was rising up, leaving the sea floor, and below the submarine we could now see the circular traces of the iron pellets that had been dropped during Gilbert's several attempts to empty the vessel's ballast on the sea floor. Then, to our relief, the submersible started to rise irreversibly and we reached the surface at 14:28. It was certainly a nice feeling to see the sun's rays filtering through the water column near the surface. On the basis of this dive, we all learned that submersibles should not explore by going down slope but they should always travel up slope.

Four days after this incident, another dive was programmed to go back to the same place as dive C-5 in order to investigate the impact of *Archmède* on the bottom sediment. Dive C-7 was made by Xavier Le Pichon and Jean Jarry with Gilbert de Froberville as pilot. The report of the dive and their observations were written in an article published in Earth and Planetary Science Letters (LePichon et al. 1975).

Since the transponder field had been left at the site, the submersible did not have any problem finding the impact area. They arrived at 2,500 m depth and made a first pass at 30 m above the impact zone in order to make a profile with an echo sounder, which is a mud penetrator, and which helped to measure the thickness of the sediment. They descended further down, and maintained visual contact for their observations and to take pictures. They crossed the sedimentary ridges parallel to the contour line of the slope that the previous dive C-5 had hit. Further down slope they found the zone of impact and the disturbed sediment (Fig. 7.3). These sedimentary ridges were probably 10 m thick and may have been hundreds of meters long. They had a rughose and folded appearance when compared to the surrounding, whiter and undisturbed sediment. The whiter fine-grained sediment was incidentally burrowed by animal dwellers. This sedimentary horizon was partially covered by darker, coarser-grained and reworked sediment. Thus it was interpreted that natural, gravitational slumping producing avalanches of unstable sedimentary layers along the continental slope had been the cause of producing the elongated sedimentary ridges we first observed. When *Archimède* hit the summit of one of these unstable sedimentary ridges, this triggered an avalanche that caught and trapped the submersible on the sea floor.

Diving near Toulon

Prior to our Project FAMOUS dives in the middle of the Atlantic, it was also decided by the submersible team based at La Seynes, near Toulon on the Mediterranean Sea, that we should undertake a few dives in the diving saucer *Cyana,* near their base. My first dive was in the port of Toulon at about 10 m depth in a dirty and muddy environment. This was probably a test to see if I suffered from claustrophobia. The next day, we went off the coast of Toulon to dive near the Island of Porqueroles. Raymond Kentzy, nicknamed "Canoë", who was the senior pilot of the *Cyana* submersible team, piloted this dive. Kentzy used to be a pilot on the diving saucer that Jacques Cousteau had used with the Research Vessel *CALYPSO*. He was an excellent, well-trained pilot, and I trusted him with my life.

Canoë was a relatively short, thickset man with short hair and a constant smile on his face. He also appeared to have considerable strength. On this dive, there were only the two of us on board the small submersible: Canoë and Roger in the *Cyana*, diving on the continental slope at less than 70 m depth. We reached bottom quickly after our launch from the support ship, *Nadir*. As soon as we arrived on the bottom, Kentzy blasted off at full speed above the sediment and suddenly we

Fig. 7.3 During one of its dives on the continental slope off Corsica Island, the bathyscaphe *Archimède* left traces of its ballast on the sea floor (LePichon et al. 1975). The ballast (iron pellets) was dropped during maneuvering in order to free the submersible after being grounded by an avalanche along a sedimented slope. (Copyright IFREMER, *Project FAMOUS 1973, Archimède dive Ar73-7*)

approached a rocky formation where he stopped. He said, "Look carefully and you might see something coming out from underneath the rocks." Canoë started to deploy the mechanical arm and extended the claw just underneath a rock. Suddenly I saw a giant lobster crawl out and start to run. Off we went with the *Cyana*, chasing the lobster. The bottom was a little rough with several rocky outcrops partially buried by a sedimented bottom and it was impressive and scary to zigzag in between the rocks following the lobster. Finally Canoë had to give up the chase, because the lobster lost us and had disappeared in its familiar environment.

Diving on the Mid-Atlantic Ridge in the FAMOUS Area

The area of the Mid-Atlantic Ridge (MAR) spreading center near 36°54′N-33°16′W is located southwest of the Azores between two small, second order ridge discontinuities. In this area more than 25 cruises had been carried out by French, American, Canadian and British institutions before the diving expeditions of 1973–1974 occurred. The base map used for conducting our dives was made by the French navy oceanographic vessel *D'ENTRECASTAUX* of the "Service Hydrographique de la Marine" in May 1973 and May–June 1974. The Navy ship was chosen because, at that time, it was equipped with a narrow beam echo-sounder capable of visualizing structures within a 200 m contour line at 2,500 m depth. In addition, bottom moored transponders were used to increase the bathymetric precision by means of triangulation. Farcy, Plassereau, Vernet and Vincent Renard from IFREMER conducted the mapping project in association with the crew and the officers of the *D'ENTRECASTAUX*.

Fig. 7.4 The French Navy ship MARCEL LE BIHAN served as a support ship for the Submersible Archimède. After each dive during the FAMOUS project, a zodiac went to recover the divers from the submersible

During the Famous diving expedition, the segment of the MAR was explored by three submersibles: the French bathyscaphe *Archimède* and the diving saucer *Cyana* and the U.S. submersible *Alvin*. The bathyscaphe *Archimède* was on the site in 1973, and the submersibles *Alvin* and *Cyana* came to join *Archimède* in 1974.

The N.O. *Marcel Le BIHAN* towed the bathyscaph *Archimède* from its port of call in Toulon to the middle of the Atlantic Ocean. Captain Le Conte was the master on board and all the personnel were French Navy sailors (Fig. 7.4). The "Captain de vaisseau" Gérard Huet de Froberville was the chief pilot of *Archimède*. De Froberville, an officer in the French Navy, had come from Tahiti in 1970 to take on the responsibility of the submersible group in Toulon. He wanted to promote the *Archimède* as a modern tool for the scientific community and make a connecting link between the French navy and the civilians involved in ocean exploration. Gilbert Harismendy and Philippe de Guillebon, also called the red baron because he always wore a red scarf during his dives, were also part of the team. On the N.O. *Le SUROIT*, Captain Guy Paquet and Claude Riffaud, Technical chief of the project, joined the expedition in 1974. Jean Jarry was the engineer, and Guy Sciarrone and R. Kientzy (also called "Canoé"), were the pilots of *Cyana* and were also on board the N.O. *Le SUROIT* as part of the submersible team.

The first phase of project FAMOUS started with the French team on August 1, 1973. In the framework of the Franco–American cooperative program, Robert Ballard, a colleague from Woods Hole Oceanographic Institute (WHOI), was invited to participate on this first diving cruise along with the French scientists: Xavier Le Pichon, Jean Francheteau, Jean-Louis Cheminée, David Needham, Pierre Choukroune, Gilbert Bellaiche, and myself. The first series of dives on the MAR were scheduled to take place inside the Rift valley at about 2,700 m depth near latitude 37°50′N. The area lies along the North American and the African plates, south west of the Azores archipelago on a segment of the rift valley, which was 2,200 m deep, 40 km long and 30 km wide, and was over hanged by faulted walls rising up to 1,500 m high. An active fracture zone striking East–West interrupted this segment of the rift valley, and was called transform fault "A". These dives were critical for improving our understanding of the creation of new oceanic crust.

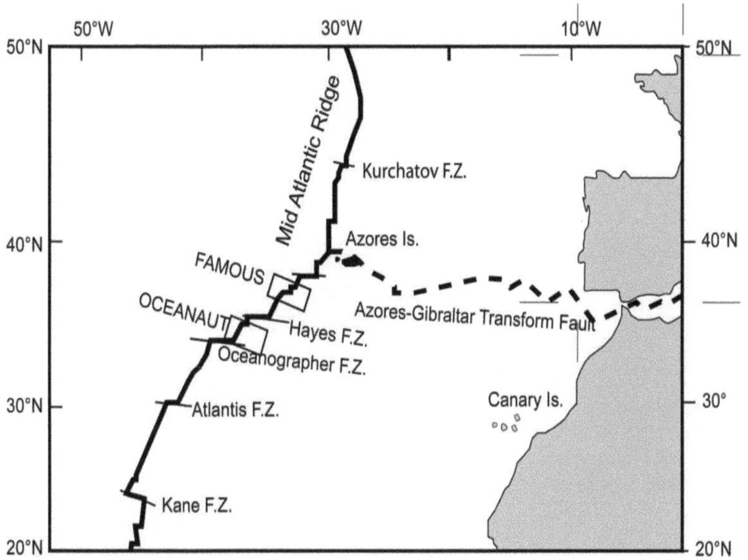

Fig. 7.5 North Atlantic spreading ridge systems and fracture zones. The boxes indicate the dive area of projects FAMOUS (1973–1974) and OCEANAUT (1995)

The first dive was conducted by the *Archimède,* piloted by Gérard de Froberville, with Jean-Louis Michel, co-pilot, and Xavier Le Pichon as the scientific observer (Fig. 7.5). The goal was to locate the most recent volcanic event, which could confirm the separation of the two plates. We were looking for proof that the North American plate was moving to the west and the European plate to the east. Since it was the first time that the eyes of a geologist were able to see the area of crustal spreading, a great deal of excitement and animation took place on board since all the scientists and engineers had gathered around the radio communication cabin to hear the first-hand description of the seafloor where the plates were separating (Fig. 7.5).

A total of six dives were made in the middle part of the rift valley and revealed the existence of recent volcanic activities taking place on the central highs at 2,727 m depth. Mounts Venus and Pluto were seen to be made up of picritic basalt and normal basaltic pillow lava, which were undisturbed by tectonic events (Fig. 7.6). The pillow lava showed fresh, glassy surfaces, meaning that they were relatively young when compared to other collected samples. However, it was necessary to make field observations across the entire rift valley floor, in order to verify the extent of the volcanic activities.

The second phase of project FAMOUS took place in the month of July 1974 and involved two French ships and two French submersibles: the *MARCEL LE BIHAN* with *Archimède* and the RV *SUROIT* with *Cyana*. In 1974, the *Archimède* covered a total of 12 km on the sea floor while the diving saucer *Cyana* explored more than ten kilometers in Transform fault "A" located north of the rift valley

Fig. 7.6 Pillow lava photographed on the eastern flank of Mount Venus during an *Archimède* dive at 2733 m depth on the MAR rift valley at 37°N in the FAMOUS area (Copyright IFREMER, *Project FAMOUS 1973, Archimède dive Ar73-10*)

Fig. 7.7 Iron-silica oxyhydroxide hydrothermal material recovered during a dive (Cy74-76) in transform fault "A" near 37°50′N on the Mid Atlantic Ridge (FAMOUS Project)

during 14 dives. It was during one of the *Cyana*'s dives (Cy74-76) at the Intersection of the rift valley and the transform fault that the first site of hydrothermalism on the Mid Atlantic Ridge was discovered (Fig. 7.7). Although no hydrothermal venting was observed, a chunk of friable material was collected and brought to the surface, and I identified the sample as a being a hydrated Fe–Si hydrothermal deposit.

In addition, the submersible *Alvin* and its support ship the RV *KNORR* were involved during the same period in 1974. The submersible *Alvin* made 13 dives in the Rift Valley a few kilometers south of the French sites. The American scientists involved during the *Alvin* dives were Jerry van Andel, Bill Bryan, Jim Moore, George Keller, Jim Heirtzler, and Bob Ballard. They collected 320 kg of rocks and took 12,000 bottom photos for a total of 62 h while traveling over 19 km on the sea floor.

The objective of the dives was to observe several structural features that had been recorded by the bathymetric studies. From some of the previous dives, we had learned that the most recent volcanic activity had taken place on the volcanic highs in the middle of the Rift Valley such as on Mount Venus and Mount Pluto (information from the Alvin dives in 1974). We knew that this volcanic activity occurred within a narrow zone of less than 1 km in width. It was important to see if recent volcanism had occurred elsewhere, such as at the margin of the rift valley. The bathymetric map showed a sub-circular volcanic edifice located on the western edge of the rift valley and we decided to explore it during the first year of the project.

On the morning of September 4, 1973, dive AR 73-13 took place with Gilbert Harismendy as the pilot, Jean-Louis Michel as the engineer, and I was the scientific observer. As the *Archimède* sank down, Michel was busy communicating with the surface ship using the underwater telephone called "TUX" in order to get a fix with respect to the transponders sitting on the sea floor. The transponder was at an altitude of 523 m above the sea floor and at a depth of 2,000 m, as indicated according to the pressure gauge. The continuous positioning of the bathyscaphe every 300 m as it sank allowed us to relocate ourselves with respect to the transponders moored on the sea floor.

When we arrived near the bottom, about 300 m above the sea floor (or in other words, at 2,294 m depth below the sea surface) we were able to see something on the radar screen. It was a semi-circular scarp about 200 m away from us in the direction of 143° N. It was 10:05, we were at the bottom at 2,636 m depth and as *Archimède* approached the sea floor, it stirred up a lot of powdery sediment, so the water outside the porthole was cloudy. The cloudy water blocked our visibility, so we couldn't see much until the current had cleared away the sediment. We landed on a scarp and on the starboard side I could only see water. The slope was on the port side so Harismendy was able to observe a tubular flow of solidified lava oriented down slope. The large lava tubes were about 2 m in diameter and partially buried by sediment. The pilot reoriented the submersible along the side of the slope in order that I could have better visibility of the sea floor from my porthole. I saw that we were in a field of pillow lavas that were partially buried by sediment (<15 %). Based on the surface morphology and the presence of some sponges growing on the lava flow, it appeared obvious that volcanic activity on this structure had ceased a long time ago.

As we stopped for sampling, the current pushed us away from our target and the pilot started the engine to re-position the submarine, thereby stirring up a lot of sediment and hence losing visibility once again. He had to wait until our visibility was restored before attempting the sampling operation. In the meantime, I took pictures of the site and described the lava flow in my hand-held tape recorder. At 2,637 m depth I noticed a scarp in front of us, and took a picture. The bottom had become rough with giant pillow lava filling up the field of view. Then we crossed a depression and noted that the direction of the flow followed the depression as if the lava had been pouring along a riverbed. After about 30 min, we had traveled 300 m and a strong reflector on the radar screen indicated the presence of a major linear scarp running in the direction of 010–210°, parallel to the ridge axis. The area was highly fractured with collapsed blocks. No recent lava was observed so I was pretty sure that we were no longer on the volcanically active zone. The structure we had visited was an ancient volcanic edifice that had been completely dismantled by tectonic activity. The area appeared chaotic with tall, nearly vertical cliffs. I named the area "Mount Jupiter".

The OCEANAUT Project: Mid Atlantic Ridge at 34°50′N

The objective of the OCEANAUT project was to study magmatic and dynamic processes near 34°50′N. These processes are related to asthenosphere upwelling in a relatively colder environment with less volcanic activity than in the FAMOUS area and at a distance (>1,000 km) from the influence of the Azores hotspot (Figs. 7.5, 7.8). From previous petrological, geophysical and geochemical investigation, it had been shown that the influence of the Azores hotspot was limited or non-existent south of the Oceanographer transform fault. In addition, "cold areas", with thin or absent basaltic crust, are ideal places for the exposure of deep-seated upper mantle material such as peridotite.

This type of area is most appropriate for interpreting original structures and their evolution over time (structural reconstruction) and for learning about the asthenosphere-lithosphere interactions. Studying amagmatic areas and/or areas that are poorly supplied by magma will help to: (a) determine the magmatic budget for the whole crust and upper mantle, (b) be a way for us to model magmatic upwelling and the mode of partial melting and (c) define the structure and composition of magmatic segregation through the upper mantle and lithosphere.

In the early 90s, several detailed geophysical investigations with modern seismic equipment coupled with gravity studies had taken place along the North Atlantic spreading ridge segments. They were intended to determine the existence of magma upwelling zones and define the geometry of magma chambers along several slow spreading ridge segments. Following an earlier seismic experiment with fourteen three-component Ocean Bottom Seismometers (OBS) deployed for 46 days during November and December 1992 (Barclay et al. 1998) and after a micro-seismicity experiment, six seismic refraction and tomography experiments were carried out in October–November of 1996 (Detrick et al. 1995) in the same area as the Oceanaut cruise at 34°50′N on the Mid-Atlantic Ridge. Because of the international interest in this region, the 1991 FARA-Sigma cruise of the French R.V. *L'ATALANTE* (Needham et al. 1992) had occurred in the framework of an international cooperative program. Swath bathymetry, acoustic backscatter (SIMRAD EM12 dual) and gravity data were systematically acquired with a track spacing of 12–16 km (Detrick et al. 1995).

After these geophysical surveys, Daniel Bideau from the department of Marine Geosciences at IFREMER was able to conduct a diving program in August 1995. Daniel was a brilliant scientist who unfortunately died too young at the age of 55 in April 2006. I had followed his career while he was doing his research at IFREMER. He did his PhD dissertation under my supervision and we shared the same office and laboratory where we engaged in many conversations about geology and life. I was impressed by his logical approach in dealing with scientific problems. His main interest was the study of rocks that had undergone transformation (metamorphism) during tectonic and magmatic processes. Daniel was easy going, friendly and had the type of personality, which makes people feel comfortable. This is an important quality for a chief scientist.

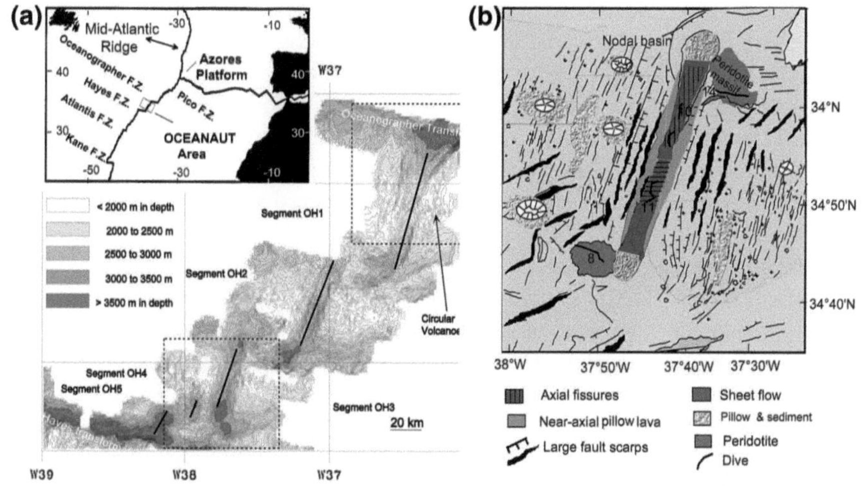

Fig. 7.8 **a** Location map (inset) and bathymetric map of the Mid Atlantic Ridge segments between the Oceanographer and the Hays Fracture zones obtained from the Simrad EM-12 swath bathymetry data (Bideau et al. 1998). The contours are every 200 m. Sites of OH1 (Oceanographer-Hays fracture zone) and OH3. **b** The Mid Atlantic Ridge segment OH3 at 34°50′N, reproduced from bathymetry and dive observations during the "Oceanaut cruise" after Bideau et al. (1998). The base map was obtained during the Sigma cruise (Courtesy of H. D. Needham) with the N.O. L'Atalante equipped with a multichannel SIMRAD

The OCEANAUT (an acronym for OCEAn and NAUTile) *cruise* was the first diving cruise for which Daniel was fully in charge and I must say that he conducted it with great enthusiasm and passion. The *Oceanaut* program was designed to observe the surface expression of ridge segments under different magmatic regimes.

We embarked on the NO *Nadir* on August 23, and disembarked on September 20, 1995, in the port of Punta Delgada on the island of Sao Miguel (Azores). It took several days to transit to the dive area on the Mid-Atlantic Ridge spreading center near 34°50′N. The OCEANAUT cruise was intended to study two short spreading ridge segments, which were previously recognized to be the sites of contrasting geological processes. These two segments were called OH1 and OH3 (short for O = Oceanographer and H = Hays fracture zones respectively). These segments separate the Oceanographer and Hayes Fracture zones.

The segment closest to the Oceanographer transform (called OH1) was the site of a robust magmatic upwelling zone as defined by previous gravity measurements. Indeed, the presence of a strong "bull's-eye" gravity-low suggested that there is a deficit of mass, which was interpreted as being the presence of a magma chamber underneath the spreading ridge. The other segment, closer to the Hayes transform (called OH3), was considered as a quieter magmatic zone with a smaller gravity-low interpreted as being due to a more fractured crust and a lack of volcanism (Fig. 7.8a, b). In order to verify the "tectonic versus magmatic" implications in the construction of these ridge segments, it was important to visit them in situ in order to observe the ground-true geology.

One of the goals of the cruise was to test previous geophysical (Detrick et al. 1995) observations that showed a low gravity spreading ridge segment, which indicates the presence of a magma chamber. In order to text such a hypothesis, it was important to make in situ geological observations. The details on the geology, structural setting and composition of the rocks recovered are found in Bideau et al. 1998, Gracia et al. 1998, Hekinian et al. 2000).

A sea-surface magnetic survey was also undertaken at night, in addition to the submersible dives. Bertrand Sichler, a geophysicist from IFREMER who is particularly handy for manipulating and fixing almost any electronic instrument, was eager to find out how a detailed magnetic survey with small track spacing (<2 km apart) would augment our knowledge about the structure of the sea floor. More than 2,400 km of closely spaced (1.8 km apart) magnetic profiles (40–55 km long) were made and two-thirds were corrected for diurnal variation using a reference magnetometer moored at 1500 m above the seafloor. Bertrand wanted to find out if a particular structure would give a distinct magnetic signature, which eventually could be used as criteria for exploring different types of structures. It was found that the strongest magnetization signals near the axis are linked to the serpentinized peridotites forming a massif at the inner corner of the spreading axis. The deficit of magnetization in the axial area, which has slightly drifted northward with respect to the center of the segments, corresponds to the "bull's eye" feature defined previously from the gravimetric measurements of Detrick et al. (1995). This lack of magnetization is related to a decrease in thickness of the magnetized layer as a result of an uplift of the Curie isotherm (about 550 °C). This suggests that magmatic activity is present at depth, even if segment OH3 is believed to be a magma-starved segment. The more magnetized areas are found in the axial valley, north of OH1 and south of OH3, and are probably related to the proximity of the major transforms.

The *Oceanaut* dives were concentrated in three areas: (1) the axial Rift Valley where most volcanic activity occurs, (2) on the inside corner of the rift valley where tectonic uplift has emplaced deep seated peridotite, and (3) on the wall of the rift valley where individual volcanoes are formed. During about 100 h of direct submersible observations, a total of 250 rock samples, high-quality video data and photographs were collected from the axis, the rift-valley walls and from several off-axis volcanoes.

Axial Rift Valley

The ridge segment is 90 km long and has an hourglass shape which narrows to about 4 km wide at the segment's center at 2,200 m depth, and widens up to 12 km and deepens to 4,100 m depth towards the northern and southern segment

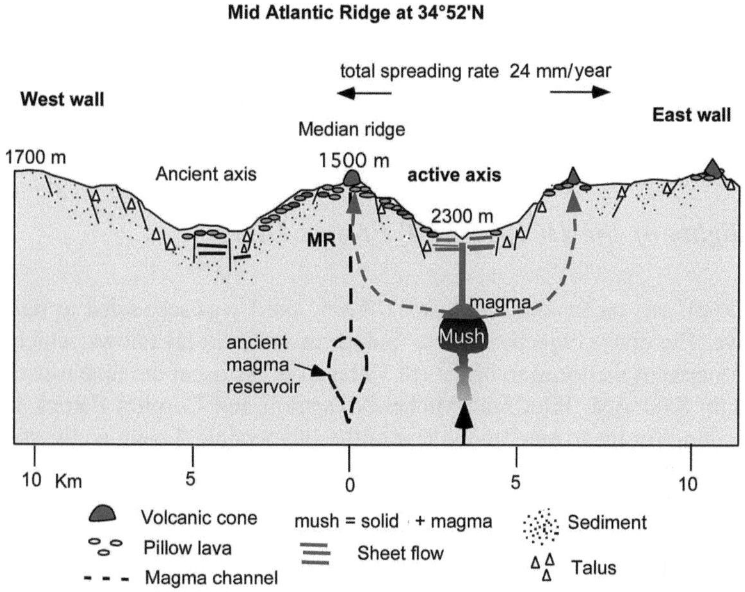

Fig. 7.9 East–West interpretative sketch of cross-section profile of the Mid Atlantic Ridge axis at 34°50′N shows the structural setting of the recent active spreading center and the Median Ridge's construction. Volcanic activity on the Median Ridge was supplied by a short-lived magma chamber and from channels from the axial magma chamber. Recent volcanic cones are formed on the eastern wall as well as on top of the Median Ridge. The fossil western spreading ridge is abundantly covered by sediment

ends. The most recent lava flows occupy a narrow (less than 1 km wide), flat and slightly depressed (a few meters) area in the axial rift valley.

The dives took place in the middle of the ridge segment at 34°50′N–36°30′W. During dives OT01 and OT02, we were able to recognize the differences between fresh lava flow with its glowing, shiny surface as seen in the submersible's projector lights from some older flow, which appears to be duller looking, with some rusty staining and sediment dust (Fig. 7.9). The recent sheet flows have partially buried older (thicker sediment and Fe–Mn crusts) disseminated pillow edifices. Debris of weathered hydrothermal deposits was observed along with shallow collapsed lava ponds with small pillars. Commonly found on fast EPR spreading centers, lava lakes were seen less often on slow spreading ridges where the magmatic budget is lower. This is also another indication that the OH1 ridge segment was recently volcanically active as opposed to the OH3 segment. The shallowest part of OH1 is devoid of faulted terrain and there were localized areas (a few hundred square meters) showing recent volcanism.

The rift valley of the MAR at 34°50′N is characterized by volcanic constructions built during several closely spaced eruptions forming constructional highs with coalescent volcanic cones and called the Median Ridge (MR). Based on morphological variation, distinct eruptive events due to both quiet and explosive volcanism were detected during the *Oceanaut* diving cruise in the axial rift valley. The cones are well preserved in the southern portion of the ridge where a collapsed crater, due to explosive volcanism, was noticed during dives OT02 and -03 (Fig. 7.9).

Highlights of the OCEANAUT Project Dives

Dive OT01 was on Sunday, August 27, 1995, and I was scheduled to be on this first dive. The dive's objective was to find the most recent lava flows, which would help us pinpoint the location of the rift valley axis. As usual the dive was to begin at exactly 8:00 AM. Pilot Jean-Michel Nivaggioli and Co-pilot Patrick Cheilan were waiting for me to enter the sphere of the submersible. I went up the stairs and climbed up the side of the submarine. Once again, I cleaned off my shoes before entering the sphere where Patrick was waiting to give the signal "go". Diving again on a Sunday meant that we were missing a good meal on board the Nadir. It was hoped that on our return, the chief steward, Michel Squiban, also called "Aldo", would have kept some leftovers for our supper.

We arrived at contact with the sea floor at 10:10 and found ourselves in a field of sheet flow, with ropy and "scoria-like" lava. The first dive is usually critical for the rest of the cruise. We did not have any clear ideas about where the most recent volcanic activity would occur. When we reached the sea floor, we tried to relocate ourselves and asked the command post on the surface about our position. Looking through the porthole, I could tell that we were not on an active part of the ridge, but instead we had landed on the first faulted step of the eastern wall where there is abundant rock debris forming talus. However, as soon as we got our position fix from the surface and started to progress towards the west, we hit fresh lava flow. This indicated the presence of a recent flow located on the eastern side of the rift valley floor near the eastern wall. As we proceeded further west, we encountered several mounds up to 20 m high and 100 m wide with older looking (dull appearance) pillow lava in the middle of the fresher sheet lava. Suddenly, there was a sharp contrast in the shiny nature of the lava flow that we were crossing. I asked the pilot to follow the boundary line of the two flows that we had crossed. It was then that we found a recently formed sheet flow terrain of 1 km long by 0.2 km wide that had invaded an older pillow lava terrain. It became clear that the most recent flow had covered the pillow lava, leaving only a few older and taller volcanic constructions. We were heading towards the eastern slope of the Median Ridge, which became more distinct as a major feature on our radar screen. We traveled with continuous visual contact with the sea floor for about 2 km before we arrived at the foot of a fractured slope with abundant talus piles due to major slumping.

The next day's dive (OT02) with Daniel Bideau accompanied by Pilot Pierre Triger and co-pilot Olivier Cipriani also took place in the rift valley. However, instead of crossing the valley from the East to the West, the plan was to start at the level of the first dive OT01 where the freshest lava flows had been seen, and then proceed southward in order to follow the track of the most recent eruption. During the dive, Daniel identified several types of freshly extruded lava. He found "scoria"-looking lobated lava, plus flat sheet flows and ropy lava along a gentle, south oriented slope between 2,243 and 2,252 m depth. At about 2,230–2,242 m depth in a small hilly area, the morphology of the lava flow changed. Indeed at sample site OT02-05 older terrain of dull looking pillows covered with thin manganese crusts and interstitial pockets of sediment between the pillows were observed. Then, as the depth increased to 2,252–2,257 m, another area of fresh sheet flow appeared until arriving at another shallower area (at 2,242–2,200 meters depth). This along strike exploration permitted us to define the extent, 1 km in length, of at least two freshly erupted lava flows covering an older, sediment covered pillow lava terrain. From the previous dive (OT01) with its east west passage across the rift valley, we had defined the width of the freshly extruded lava to be about 1 km wide. When assuming a minimum thickness of 1 m it was inferred that each eruptive event extruded about 0.001 km^3 of lava (see Chap. 5).

The third dive (OT03) took place on the east flank of the Median Ridge (MR) where it was previously inferred that loose material was covering up a large area of the edifice's summit (Fig. 7.9). This was suggested from the bottom imagery obtained from acoustic sonar during a previous cruise with the N.O. *L'ATALANTE* and we wanted to have ground true observations. If this loose material were related to an explosive event then we should be able to find a crater.

Marc Constantin, a geologist from Canada, University of Quebec, who was doing his PhD with me at IFREMER-UBO, was assigned to dive as the scientific observer to explore the volcanic edifice built on top of the Median Ridge. Pilot Jean-Michel Nivaggioli and Co-pilot Patrick Cheilan accompanied Marc. The dive started on the morning of August 29, 1995, at the usual time of 8 AM and the party arrived on the sea floor at 10:42 at 2,330 m depth. After taking the first sample of a fresh ropy lava fragment they started westward, and moved upward along the slope. Marc described the presence of sheet flows and small lava ponds of 20 m in diameter showing the pillars left after the drainage of the molten lava. They traveled about 500 m westward before reaching the foot of the Median Ridge (MR) at 2,330 m depths near the intersection of the rift valley floor, which is marked by a North–South oriented fault. At 12:00, a little more than one hour after landing, they started their ascension of the Median Ridge. The terrain had changed its aspect and was more chaotic with abundant pillow lava fragments and coarse-grained debris forming large, 90 m-large talus piles up-slope. Several step-faulted scarps made up of avalanche debris of pillows as well as pillow-lava tubes were observed up to 1,800 m depth. Then suddenly, to their surprise, the rock debris became more abundant and unstable along the slope. The pilot stared to slow down and get back away from the slope to avoid any unexpected avalanches.

At 14h51 the divers observed the presence of gravel and heterogeneous scoria-like debris mixed with large blocks. This is comparable to what is encountered on subaerial volcanoes after an explosive eruption. Indeed, the material consisted of accidental rock debris also called a "pyroclastic" flow, which was generated

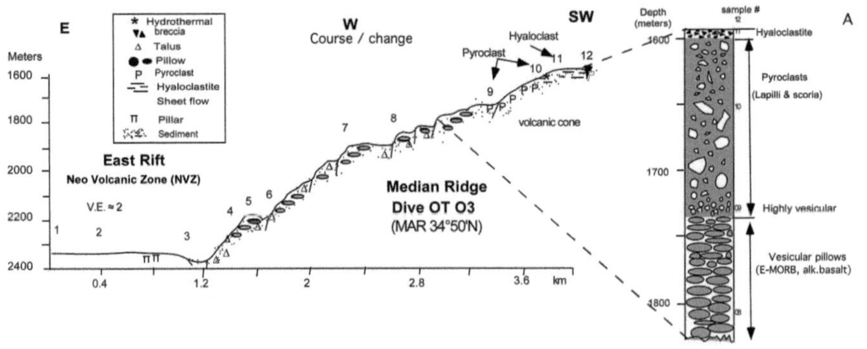

Fig. 7.10 Geological dive OT03 profile (*V.E.* vertical exaggeration) made during the "OCEANAUT" cruise on the Mid-Atlantic Ridge on segment OH01 at 34°50′ N showing volcanic sequences of tubular pillow lava and pyroclastic deposits observed along dive track OT 03 going from the recent rift valley floor up to the volcanic edifice forming a portion of the Median Ridge. C/C indicated ship's course change. The geological stratigraphic column is shown on the *right* of diagram

during volcanic explosion (see Chap. 5). This type of material was mainly observed on a fairly steep slope between 1,742 and 1,634 m depth (Figs. 7.9 and 7.10) therefore it was difficult to sample any sizeable representative material because of its fragile nature (samples were easily crumbled under the mechanical arm of the submersible). Finally one sample (OT03-09) was taken from a vertical wall with a 15 m relief at 1,730 m depths where unconsolidated scoria ejecta were interbedded with gravel size and centimeter large blocks of lapilli-like debris (Fig. 7.11a, b). The fragment consisted of abundantly vesicular (>50 % vesicles) scoria-like lava containing a few visible calcic plagioclase megacrysts. The vesicles showed interconnecting cavities as opposed to the dispersed and isolated individual cavities found in the other less vesiculated flows.

Further up-slope near 1,640 m depth, several sequences of loose pyroclastic and semi-consolidated crusts of iron–manganese cementing volcanic ash were observed. One sample (OT03-11) of an ochre-colored, broken slab of a hyaloclastite was collected at 15:52 at 1,597 m depth. The summit of the edifice at 1,594 m depth consisted of a volcanic crater whose walls were made up of pyroclastic debris. Dust of pelagic sediment partially covered the fragmented debris scattered on the side of the crater. The outcrops of pyroclasts observed during dive OT03 were more than 15 m high and showed crude graded bedding. These pyroclasts formed the rim of an elongated crest surrounding the volcanic crater on the summit of the volcano (Fig. 7.11). The dive ended at 16:13 on top of the volcanic crater at 1,592 m depth. Because of the limited amount of sediment cover and the degree of freshness of the scoria debris, it was inferred that an explosive eruption must have occurred relatively recently, perhaps during the past 1,000 years.

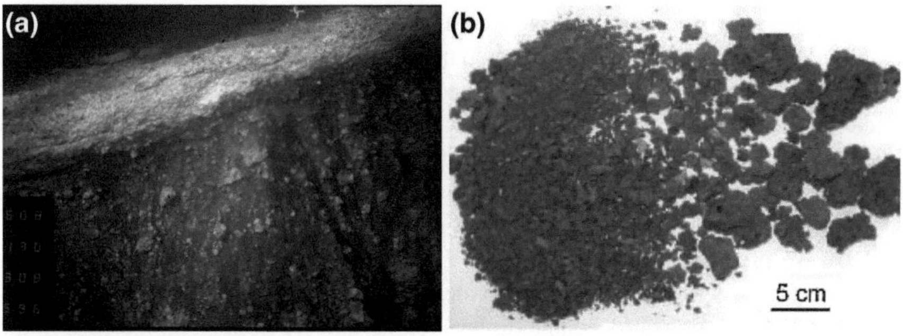

Fig. 7.11 **a** Photograph of an explosive crater's wall at 1730 m depth on the Median Ridge of the Mid Atlantic Ridge valley shows imbedded fragments of glassy pyroclasts. (Copyright IFREMER, Oceanaut cruise 1995, Nautile dive OT03) **b** Sample of fragmented pyroclast debris collected by the Nautile during dive OT03 on the MAR at 34°51′W-36°28′W at 1730 m depth

In the meantime on the bridge, the captain was concerned about a storm warning that had been broadcast. Several tropical storms (Humberto, Luis, Iris and Karen) with a radius of 125 nautical miles had formed near 12–15°N and were forecast from August 28 to September 1st. On September 7, the storm called 'Iris' went up the coast of Brittany (France) with winds up to 140 km/h. Thus, on August 30, our captain decided that we had to move from our diving site on segment OH1, because another tropical storm called "Humberto" was heading towards the Azores at a speed of 29 nautical miles. Thus, we were obliged to move further south to 34°49′N to explore a second ridge segment.

The other ridge segment that was explored during the *Oceanaut* cruise was called OH3 and is located about 100 km further south of OH1. The rift valley of segment OH3 is shorter, about 50 km long, deeper (>3,000 m depth), and it has abundant fissures and faults. From the dives we made, it was shown that the volcanic activity has been limited, as indicated by the presence of small mounds and haystack types of features suggesting localized events. This segment does not have any median ridge construction but is a series of gentle rolling hills in the middle of the rift valley. The most recent volcanism also occurs on the eastern side of the rift valley over a width of about 1 km. The extremities of this ridge segment are located at the intersection of non-transform discontinuities characterized by the presence of two topographic elevations composed of uplifted peridotite.

By 1999 the submersible *Nautile* had already made more than 1500 dives. However, in 1995, I had the opportunity to participate on its 1,000th dive during the *Oceanaut* cruise. Indeed, to my surprise, the chief scientist, Daniel Bideau, wanted me to be on board the submersible for this special event, even if it was not supposed to be my turn to dive. That's when I realized that I was getting older, but at the same time, I was very happy to inaugurate another round of a thousand dives for this superb engine, *Nautile*.

Dive OT-09 was the 1,000th dive of the *Nautile* on September 5, 1995. For this occasion, the engineers of the diving group had prepared a plastic panel with the name of the scientific team to take down with us to be left on the sea floor. The panel with its floats was carried over board by the divers and put into the arms of the submersible. At 10h42 AM we arrived on the sea floor and looked for a good place to install the panel. Jean-Michel Nivaggioli found a flattish surface at 3,033 m depth on top of a lobated flow and we all agreed to leave the panel there. Jean-Michel then drove the submersible in a circle around the plastic panel in order to take pictures before leaving the site.

Then we proceeded on our course in a westerly direction to make sure that we were in the area with the most recent volcanism. After crossing the rift valley axis for about 1 km, the terrain became duller looking and we noticed an increase in the amount of sediment suggesting that we had passed beyond the most volcanically active area of the rift valley. The dive lasted about 5 h at direct contact with the sea floor. As we traveled along the rift valley axis, we encountered several small volcanic constructions made up of pillow lava. They were in the middle of flatter flows such sheet and lobated flows with a more shiny aspect and believed to be more recent. The young flows were associated with fissures less than 2 m wide.

At 15h53, after being on the sea floor for more than 6 h, we were asked by the ship to leave our site and return to the surface. Having covered more than 4 km on the sea floor, with about six stops to collect rock samples and take video footage, we felt that we had accomplished a good day's work since we seemed to have found the most recent active axis along the OH3 ridge segment. This would enable the next dives in the area to proceed to other targets.

When arriving on the surface, and after setting our feet on the deck of the *Nadir,* we were welcomed with glasses of champagne. The entire crew enjoyed a social gathering that the personnel of the mess had prepared to commemorate the *Nautile*'s 1,000th dive.

The next morning on September 6th, another dive (dive OT-10) was scheduled to take place along the axis of the same OH3 ridge segment in order to complete the information obtained during OT09 on which we had found the eruption of sporadic recent flows along the deeper axial valley. This new dive was intended to verify the extent of "volcanic" versus "tectonic" activities. Eulalia Gracia, a very enthusiastic structural geologist from the Science department of the University in Barcelona, Spain, was scheduled to dive together with Pierre Triger and Olivier Cipriani. Eulalia was a thin, short, pretty girl with a nice smile and a very dynamic personality. All the members of the submersible group appreciated her presence, and they were pleased to dive with her. Because of her scheduled dive, she had to leave the 1,000th Dive party early, in order to get a good night's rest.

The *Nautile* reached the sea floor at 10h53 at a depth of 3,008 m on the first step of the east wall. After going down several fault scarps that cut through pillow lava terrain, they reached the rift valley floor where lobated and flat flows occur. At 11h09 (at 3,100 m) they stopped to collect their second sample on a "drapery-like" lava flow, which had broken in situ. Moving further northward, Eulalia observed a series of collapsed lava ponds about 3–5 m depth and less than 100 m long which alternated between more ancient, dull looking pillow lava flows. From wall to wall, the sea floor depth varied between 3,105 m to 3,217 m depth and the rift valley along this crossing consisted of small volcanic constructions less than 50 m high alternating with collapsed lava ponds and fissures. When approaching close to the first western wall's scarp, fissures and faults became more prominent in a pillow lava type of terrain. Based on the observations made during this crossing, it was concluded that the volcanically active portion of the rift valley is very narrow, less than 200 m, over a total width of about 2 km wide. Thus, as was seen on dive OT-09, the OH3 segment of the ridge is in a quieter period of volcanism. It appeared that spreading in this part of the ridge is mainly the result of tectonic activity, which gives rise to fissures and faults with only a small amount of magmatism.

Volcanoes on the Mid Atlantic Ridge Walls

The OCEANAUT cruise was the first time that our group included three women scientists (Claire Bassoulet, Christèle Guivel and Eulalia Gracia) so we named three of the new volcanoes with their first names. Three of the explored volcanoes are located on the eastern ridge flank, up to 25 km away from the present-day ridge axis, and one was found on the western flank, about 20 km from the axis (Fig. 7.8b). The edifices are not presently active and the flanks of the off-axis volcanoes show heavy sediment cover on gentle slopes (>50 % in surface). Small depressions are filled with pelagic sediment and tend to concentrate the debris of sessile fauna, such as dark fragments of gorgonians, organisms encrusted with manganese oxyhydroxides, and siliceous sponge needles. The abundance of debris clearly increases with distance from the axis. Large communities of living sessile organisms are only observed on top of the volcanoes. Their debris appears as dark patches sparsely distributed along the Median Ridge (MR) and around the Claire volcano. All the volcanoes have flat summits made up of pillows and occasional altered sheet flows encrusted with iron and manganese coatings. Based on submersible observations, all four volcanoes consist mainly of highly vesicular pillows (up to 25–50 % vesicles). The youngest volcano, called "Claire" (dive OT-05), is closest to the ridge axis and terminates in small haystacks (Fig. 7.8b, 7.9). The rocks sampled from this volcano have moderately palagonatized, glassy margins and thin Fe–Mn coatings (<0.5 mm). The "Eulalia", "Christèle" and "Fara" volcanoes are found further away from the ridge axis, and have an older appearance with their abundant sediment cover (>30 %).

The summit (>900 m depth) of the oldest-looking "Eulalia" volcano (Dive OT04), located at about 22 km east of the axis, shows altered basaltic rocks with smooth surfaces, without any Fe–Mn oxyhydroxide coating nor visible chilled margins on their top. There are thick (cm) Fe–Mn coatings in the rock interstices, which indicate the effect of seawater alteration or hydrothermalism. The samples contain abundant clay and iron hydroxide with soft, external alteration halos (0.5–2 cm thick) and carbonate material fills their vacuoles.

Similar observations were made on top of the "Christèle" volcano as well as on the most western "Fara" volcano (Dive OT07). These structures are compositionally similar to the surrounding oceanic crust. Hence, we could conclude that they were formed during short-lived, localized magma upwelling.

Inside Corner-High Peridotite

Another task of the *Oceanaut* project was to determine the type of rocks that are associated with the tips of the ridge rift valley segments, and to observe their differences when compared to an adjoining spreading segment. Based on the sea floor images observed with the side scan sonar, it was inferred that the rift valley's inside-corners of the spreading segment-ends showed semi-circular features with steep slopes. The steepness of the slopes seemed to suggest the existence of an uplifted crust. If this were the case, then we could expect to observe deep-seated material, such as peridotite and/or gabbro, exposed on the sea floor.

Two dives (OT-08 and OT-14) were devoted to exploring the inside corner of segment OH3 (Fig. 7.8b). The segment tips are marked by a deeper trough or fracture, also called "non-transform". The tips are not a real transform fault, but rather they seem to be a discontinuity or a break between two spreading ridge segments.

Marc Constantin, as the scientific observer, Pierre Triger and Olivier Cipriani as pilot and co-pilot respectively, were participants on dive OT08 on September 4. As soon as the divers reached the first scarp of the rift valley's inside-facing southwest corner at 3836 m depth, they observed slumped material made up of dark looking peridotite, which was detached from the upper part of the slope. At 12h38 at 3,543 m depth, Marc observed the presence of in situ peridotite together with a dolerite dyke. Serpentinized peridotite was observed between 3836 m up to the summit at 2550 m depth.

Confident with dive OT08′s discovery of exposed peridotite on one corner of the rift valley, we decided to try to find out if the opposite, northern inside-corner end of this rift valley segment would have the same geological setting. Thus, dive OT14 was conducted on the northeast corner of the OH3 segment. The diving party was composed of Pierre Triger and Olivier Cipriani, as pilot and co-pilot, and I was the scientist on board. We touched down on the sea floor at 10h46, and found the same general setting, which comforted our original ideas concerning the nature of this particular type of spreading segment. We made a complete geological profile from the rift valley floor upwards along the slope of this uplifted structure. We reached a slightly sedimented area with pillow lava and lobated flows at 3,568 m depth. The lava flows were not as fresh as those encountered in the middle of the rift segment during dive OT09. The lava flow extended all way to the intersection with the inside corner's high, at a depth of 3,300 meters. After traveling eastward for about 2 km, at 12h54, we arrived in front of a faulted scarp with

Fig. 7.12 Photo taken at 2855 m depth on the slope of the inside corner-high located at the tip of the rift valley's west facing wall near 34°50′ on the MAR showing a contact between a peridotite outcrop (**a**) and a dyke of dolerite (**b**). (Copyright IFREMER, *Oceanaut cruise 1995, Nautile dive OT14*)

abundant slumped angular and prismatic rock debris, which I recognized as being dyke fragments.

I thought that these intrusive dykes must be exposed further up slope. Pierre slowed down and, without stopping the submarine, he very quickly collected our fifth sample, which was a piece of loose debris (sample OT14-05). As the *Nautile* continued its ascension along the slope, we could see green colored gravel-sized fragments mixed with the larger dolerite blocks. This indicated that within the avalanche's debris, fragments of green colored peridotite and chunks of a darker, massive dolerite dyke were mixed together and were most likely going to be exposed higher up along the slope. However, it was not until we reached a depth of 2,875 m at 14h40, that we noticed the real in situ exposed outcrop of foliated peridotite (Fig. 7.12). This peridotite was exposed at about 420 m above the rift valley sea floor. When we reached the summit, flat lying ledges of peridotite were seen at 2,621 m. It was now clear that the massif of deep-seated peridotite had been uplifted, like a diapir, and had cut through the gabbroic and dyke formation to reach the seafloor's surface. Thus we had clear evidence that the inside-corners of the rift valley were definitely created by uplifted massifs of a deep-seated peridotite-dyke association.

The OCEANAUT project headed by Daniel Bideau ended successfully and permitted us to determine the nature and the mode of volcanic eruption of the two short spreading ridge segments formed at the slow accreting plate boundaries of the Mid Atlantic Ridge. We learned that individual ridge segments along slow spreading ridges are formed alternatively during tectonic events of lithosphere opening followed by magma upwelling. Spreading ridge segment OH1 is in an active magmatic stage, while segment OH3 is in a more tectonic stage of faulting and fissuring. Further details on the structure and composition of the two OH1 and OH3 slow-spreading ridge segments have been published by Eulalia Gracia (Gracia et al. 1998), Daniel Bideau (Bideau et al.1998) and Roger Hekinian (Hekinian et al. 2000).

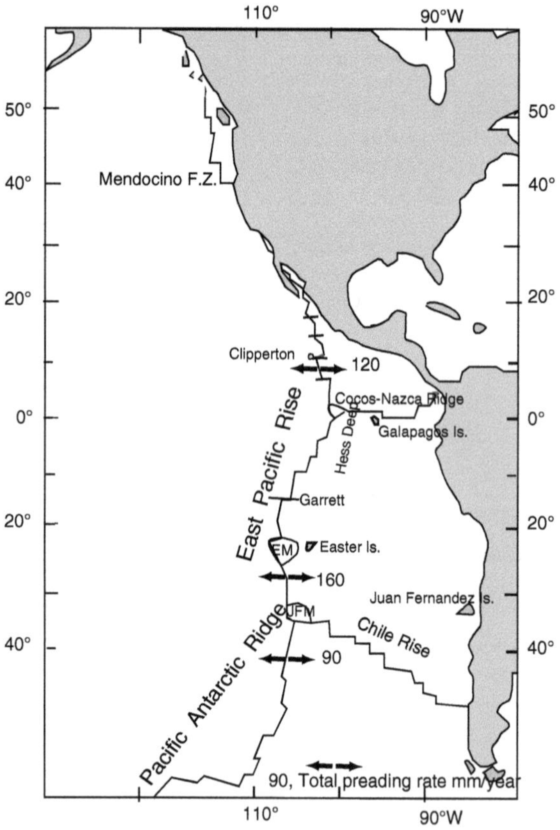

Fig. 7.13 East Pacific ridge segments interrupted by the major discontinuities, fracture zones and small microplates (Galapagos, Easter Microplate *EM* and Juan Fernandez Microplate *JFM*). The various spreading rates are indicated (total rates in mm per year)

The East Pacific Rise

The East Pacific Rise (EPR) extends for more than 9,000 km in length from about latitude 51°N off the coast of Canada to south of the Juan Fernandez Microplate near 36°S (Fig. 7.13). Further south of this latitude, the EPR it is called the Pacific Antarctic Ridge (PAR) and it continues in a southwesterly direction between Antarctica and Australia. The EPR is responsible for creating the Pacific lithospheric plate that plunges underneath the American continental plate to the east, and is subducted at the west underneath island arc formations. Both the east and west margins of the Pacific plate are marked by seismically active zones giving rise to earthquakes and volcanic activities, and they are often referred to as the "Circum-Pacific Fire Belt".

The East Pacific Rise segments are cut by transform faults. One of the most famous is the San Andreas Fault, which starts in the Gulf of Mexico and joins the southern segment of the Juan de Fuca ridge off the coast of northern California. In the Oligocene period, about 30 million years ago, before the ancient Pacific

Farallon plate was subducted under the west coast of the American continent (Lipman et al. 1971), the EPR segments continued off the coast of California and Oregon and joined with the Juan de Fuca and other northern spreading ridge segments. The San Andreas Fault, which is subaerial, roughly follows the trace of the subduction zone at the west of the northern part of California before it joins the undersea Mendocino fracture zone and the Gorda Ridge spreading segment near 40°20′N.

The consequences of the Pacific plate plunging underneath the North American continent gave rise to the Cascade Mountains and contributed to the volcanism of the Cascade Range provinces (DeMets et al. 1990; Flueh et al. 1998) in Arizona, California, Utah, Oregon and up to the State of Washington. As a result of the Pacific plate's plunge, there has been recent Quaternary volcanism in the Cascades Range such as that observed at Mt. St. Helens, Mt. Rainier in Washington (Olympic Mountain area), Crater Lake in Oregon, and Lassen Peak in California.

The EPR consists of ridge segments that have the most variable spreading rates when compared to those from other Indian and Atlantic accreting ridge systems. The spreading rates vary between 5 cm/yr to 18 cm/yr (total rate) all along its length. The variable spreading rate reflects the speed and volume of magma delivery, which also influences the ridge segments' axial morphology and volcanic activity. On the basis of bathymetric results obtained between 21°N (Gulf of California) down to the colder ocean near the coast of Antarctica around 50°S, it is now well established that the accreting plate boundary of the EPR is neither a linear nor a continuous feature. On the contrary, the EPR is formed by short volcanic segments varying in length between a few kilometers (<10 km) up to more than hundreds of kilometers (100–500 km).

East Pacific Rise at 21°N

After the success of project FAMOUS with its international group of scientists and personnel using surface ships and submersibles for a detailed investigation in the North Atlantic, the French and Americans decided to continue a similar collaboration in order to study fast-spreading ridge segments. The idea for this new project was first discussed among the scientists involved on project FAMOUS, while they were on board the *Marcel Le BIHAN* in 1974.

From 1975–1977, scientists from CNEXO-IFREMER, Scripps Institution in San Diego (California), the U.S. Geological Survey, and WHOI (Woods Hole Oceanographic Institution) were involved in a project called "RITA" (acronym given for two transform faults named the Rivera and Tamayo after two Mexican painters) (Fig. 7.13). The leaders of the project were both experts in geophysical prospecting : Jean Francheteau (for the French group, from CNEXO) and Fred Spiess (for the American group, from the Scripps Institution, University of California in San Diego). Previous work of mapping and deep-towed geophysical investigations in the area had been mostly done by U.S. institutions and had primarily involved scientists from Scripps. These studies allowed us to define the ridge axis and have a broad understanding of the topography of the area.

The French team headed by Jean Francheteau included David Needham from IFREMER, Thierry Juteau (University of Strasbourg) and Claude Rangin (University Pierre-Marie Curie, Paris 7). The French team was to be the first on the site in the summer of 1978 with the diving saucer *Cyana* and its support ship *Le Nadir.* The cruise was called "Cyamex" for CYAna—MEXico, since the dives were to take place in the Mexican economic zone. Scientists from a Mexican University were also invited to participate in the dives.

A total of 22 dives were performed of which twelve were located along the ridge axis and ten were perpendicular to the axis on the flank of the ridge. The main goal of the dives was to understand how the lava morphology and the tectonics of a fast spreading ridge differed from those of a slow spreading ridge segment. Were recent eruptive events more abundant at a fast spreading ridge, and was the creation of the crust more extensive? What are the relationships between lava types and the styles of eruption in relation to spreading rates?

After the CYAMEX cruise of 1978, another sea-going expedition conducted by U.S. scientific teams took place in November 1979 with the R.V. *Gilles* from southern Florida (Rosenthal School of Oceanography) and the catamaran *Lulu,* which carried the submersible *Alvin* from the Woods Hole Oceanographic Institution (Fig. 7.14). The R.V. *Gilles* essentially served as a back-up ship for hosting the scientific team that was scheduled to dive with *Alvin.*

I did not participate with the French team, because the chief scientist, Jean Francheteau, had decided to involve a land-trained petrologist, Thierry Juteau. The logic behind his choice was that a land geologist might view the sea floor differently. However, in 1979, I was invited by the U.S. team headed by Harmon Craig to participate on their diving cruise with the submersible *Alvin* at the site of the CYAMEX project. Another French scientist invited to participate on the cruise was Francis Albarede, a geochemist from the *Institute de Physique du Globe* (IPG) in Paris.

John Edmond and Harmon Craig were both geochemists working at MIT and Scripps Institution of Oceanography, respectively. They were the Principal Investigators (PI's) of the project. The chief scientist of the cruise was Harmon Craig. Water chemists, Chris Measure and Karen Von Damm, from MIT (Massachusetts Institute of Technology in Boston, USA) were also involved. Karen did comprehensive work on the hydrothermal fluid samples collected during the expedition, which served as her Ph.D. thesis published in 1983. This was the first contribution made from field experience concerning hydrothermal fluid circulation on the sea floor.

The R.V. *GILLIS* was poorly equipped for mapping since it only had a single-channel depth sounder. Francis Albarede took charge of the dredging stations that we had to carry out in the evening after seven o'clock when the *Alvin* was no longer on the bottom. Also, in 1979, navigation fixes were not as accurate as they are today. In order to locate the sample sites (stations) while we were in the submersible, it was necessary to take fixes with the *Lulu,* which was positioned

Fig. 7.14 The catamaran *Lulu* operated by Woods Hole Oceanographic Institution (USA). Aerial view of the support ship *Lulu* with its submersible *Alvin* hoisted up on deck. (Photo courtesy of WHOI Archives)

within a field of bottom transponders. This was the method we used to navigate *Alvin* on the sea floor, which meant that during our stations to gather samples or take photos, we had to take our bearings by measuring our distance from *Lulu* and then ask *Lulu* to give us her position via radio.

Used for carrying the submersible *Alvin* on a platform (like a baby in a hammock), *Lulu* was a strange looking catamaran that reminded me of a barge (Fig. 7.14) (Ballard 1983). It was built with two converted navy minesweeper pontoons about 30 m in length that were held together by a connecting arch made of steel (Ballard and McConnell 1995). The *Lulu* did not have sufficient autonomy for long journeys at sea and was often towed by a larger vessel. *Alvin* was put on a steel platform that could raise or lower down the submersible between the two pontoons. The port pontoon served as an engine room and galley and the starboard pontoon was used as quarters for accommodating the personnel.

The first time I dove with the WHOI submersible *Alvin* was in 1979 when its support catamaran *Lulu* was still operational. In the evening of November 5, 1979, the night before the dive, I moved to take my quarters on board the R.V. LULU. I was to spend the night on this strange vessel in order to be ready early the next morning for diving. The starboard pontoon's lower level occupied by the diving scientists was stuffy, small and uncomfortable; I felt like I was inside a coffin. However, I still managed to have a good night's sleep, rocked by the motion of the sea. I remember the attractive part of being on board was the smell of a barbecue being prepared in the kitchen where thick Texas-style sirloin steaks were served along with corn and baked potatoes the evening before the dive. This was my favorite meal. If comfort was primitive, the atmosphere was still extremely pleasant, friendly and relaxed with the *Alvin* team and *Lulu*'s crew.

I was scheduled to dive in the morning at 8:00 AM (dive # 981) on the axis of the EPR. The purpose of my dive was to monitor the fluid temperature, sample hydrothermal fluids, and obtain rock samples. The samples were mainly to be taken for Harmon Craig and for Chris Measure. Both men were internationally recognized geochemists. An earlier diving exploration during the CYAMEX cruise in 1978 had been able to define some precise hydrothermal targets. This facilitated our reconnaissance work in finding a hydrothermal field with black smokers. Indeed, about 15 min after the landing on the sea floor, we had already arrived on an active hydrothermal site. This first site showed five active vents, which had formed on a collapsed lava pond with flat slabs and sheet flow formations. We stopped on station and started to deploy our sampling gear. The sample basket with its compartments was in front of the porthole so it was easy to see. The first sample had to be a hydrothermal fluid recovery, which was collected inside a titanium bottle that *Alvin*'s mechanical arm held above an active chimney. This was a difficult task for the pilot, because he had to position the submersible in a way that it would be close enough to the orifice of the chimney where hydrothermal fluid was spewing out at 350 °C, but not too close, since the temperature gradient was strong enough to melt *Alvin*'s plastic portholes. Another site in the same general area was sampled for fluid, which had a lower exiting temperature of 105–110 °C. A third site was visited after having crossed a sheet flow formation associated with several collapsed pits bordering a lava pond. At this site, two tall, skinny active chimneys spewing out warm water were sitting on top of a pillow lava flow at the edge of the lava pond.

For the first time, the team of geochemists found that the acid hydrothermal solutions containing H_2S, SO_4 plus metals and silica in suspension were exiting at a temperature of 273–355 °C on the sea floor. The hydrothermal fluids that were collected contained mineral compounds that had been leached from the oceanic crust at about 0.5–3.5 km depth (*See* Chap. 6).

I had the opportunity to observe the volcanics formed on the seafloor of the Pacific and compare them to what I had witnessed in the Atlantic Ocean. The volcanic landscape looked relatively recent. It did not have the older appearance (dusted with sediment and dull as opposed to shiny), which was commonly viewed on the Atlantic sea floor. The lava flows were more fluidal types, forming draped and sheet flows rather than bulbous pillows. This was an indication that the lava had a faster rate of extrusion in this part of the World's ocean.

The sulfide samples recovered from 21°N on the EPR were studied by Michèle Février, my student who was working towards her PhD degree at UBO (*Université de Bretagne Occidentale*, France). She was a very enthusiastic person who was totally dedicated to her work, which concerned the hydrothermal processes forming ore deposits. In 1981, Michèle participated on an *Alvin* dive with Bob

Ballard and had her first diving experience on the site of hot hydrothermal vents at 21°N on the EPR. Her PhD dissertation presented the first detailed mineralogical and geochemical data compiled from the ocean floor's hydrothermal system (Février 1981). Michelle used mineral paragenesis to detect the sequences of mineral precipitation in the water column and on the seafloor. Thus, it was found that the origin of the sulfur binding the metallic phases and forming copper, zinc and iron compounds came in part from seawater and in part from deep mantle sources. The sulfur formed from seawater is released during a transformation of sulfates due to seawater circulation and its reaction with the crustal material. The work done on the hydrothermal material from the EPR has shown that the black colored fluids spewing out of the chimneys have precipitated mainly high temperature metals such as chalcopyrite (copper rich sulfides) and sphalerite (zinc sulfides), while the lighter, white colored fluids have precipitated silica and calcium enriched phases.

This early work revealed the fact that metal-enriched sulfide precipitates are short lived and are inclined to alter within the seawater-oxygenated environment. Thus, when the hydrothermal source is no longer being supplied, the deposits are transformed into powdery products that can be easily dispersed by bottom currents (see Chap. 6). The evolution of the individual hydrothermal chimneys was inferred on the basis of the mineralogical studies. The size and composition of the crystals will vary from the exterior towards the interior as the temperature of the exiting fluid changes.

Long-Range Survey of the East Pacific Rise

After our exploration of fast (EPR 21°N) and slow (MAR 36°N) spreading ridge segments and the discovery of hydrothermal precipitates, it seemed wise to conduct further exploration on accreting ridge axes. The rational behind the new projects was to try to find out if hydrothermalism and volcanism were common phenomena inherent to accreting spreading ridges. Our questions were concerned with the extent of volcanism and its relationship to hydrothermalism. What was the extent of hydrothermalism? How are volcanic events related to hydrothermal activity and what is the relationship of volcanism to the geological setting of a particular region?

Several large-scale bathymetric and topographic surveys along the crest of fast spreading ridge systems were planned by the international scientific community, especially French, German and U.S. institutions, prior to engaging more detailed investigations with manned and unmanned diving engines on the sea floor.

In 1980, the first French long-range sea-going expedition along the axis of the East Pacific Rise was called the SEARISE cruise. It was dedicated to exploring the EPR axis between 21°N to 21°S, and then pursuing a further crossing over the Pacific basin intraplate region all the way to Tahiti. The cruise was sponsored by IFREMER and headed by Jean Francheteau on board the N.O *JEAN CHARCOT*.

This long-range survey was possible because the ship was equipped with a newly acquired multichannel SeaBeam system, which was able to follow and map a band of ridge axis 3 km wide at a depth of 2500 m during a single pass. This long-range survey covered about 2,500 km of the EPR axis. The cruise was divided into two legs: the first leg left Mazatlan, Mexico on May 10, and arrived a month later in Hao, French Polynesia, on June 8, 1980.

The first Leg (SEARISE 1) surveyed the northern EPR between 21°N to 20°S along the ridge axis. In addition to having a general view of the ridge, the cruise was aimed at surveying the CYAMEX Project diving site (at 21°N) in more detail. Further south, a few stations of water sampling and dredging were performed near the ridge axis at 13°N and along various other segments in order to find signs of hydrothermalism. A particular emphasis was made on sampling and obtaining bathymetric coverage of the Galapagos Triple Junction.

The second Leg (SEARISE 2) started on July 3, 1980. It was to leave from the Island of Hao, which at that time was an advanced military base for the French nuclear testing that was being conducted on the Island of Mururoa, about 1000 km further to the east. Hao is an atoll and since it was controlled by the military, special permission was required to be on the island. Members of the French Foreign Legion and other navy and military personnel were often in transit on the island between Mururoa and Papeete in Tahiti. David Needham, Felix Avedik, Jean-Louis Cheminée, Jean-Paul Foucher (a specialist in heat flow measurement), Jean Francheteau, Bob Ballard and I joined joined the the R.V. *Jean Charcot* in Hao. After having a swim off a sandy beach, we left port early in the morning (at 2 AM) and passed through the narrow channel between the coral reefs which gave access to the open ocean. One hour later, we started setting up our watches. As soon as we reached deep water, we switched on the SeaBeam and placed our magnetometer into the sea.

Some interesting observations were made along the various segments of the EPR during this long-range axial exploration. For example, this study enabled us to better correlate the axial topography with the rock composition and some structural settings such as accreting ridge discontinuities (segments). It was detected that in the vicinities of these discontinuities there is a topographic low while the highs, where the Ridge crest is bulging about 50–200 m above its ends, are most often located in the centers of the segments. This situation is found elsewhere along other spreading ridge segments (Francheteau and Ballard 1983). Another interesting discovery consisted in finding that transform faults displacing segments of the EPR were the sites of a particular type of volcanic activity giving rise to porphyritic basaltic lava enriched in early-formed large (>0.1 mm in diameter) plagioclase crystals (Eissen et al. 1981). This observation suggested the presence of crystal growth in magma reservoirs.

The scientific results of these two Legs were produced after long-term research in various national and international laboratories. One of the results concerned the compositional variability observed in the volcanic samples collected. The existence of geochemical and petrological provinces was confirmed. These provinces had direct implications on the partial melting of different types of mantle located underneath the East Pacific Rise. In addition, this large-scale SEARISE survey along the EPR allowed us to select several targets for future, more detailed studies of volcanic and hydrothermal activities. Among these future cruises, the area that supplied the most results to the scientific community was related to the missions carried out on a segment of the EPR at 12°50′N.

After the 1980 SEARISE project, two other consecutive cruises called "CLIPPERTON" with the N.O *JEAN CHARCOT* (1981) and "Geocyatherm" (1982) with *LE SUROIT* took place. For the first time, a relatively small portion of the East Pacific Rise near 12°50′N and 11°20′N was going to be explored in great detail by surface ships and submersibles.

East Pacific Rise at 12°50′N

The East Pacific Rise (EPR) axis (at 2600 m deep) near 12°50′N is located between two first order ridge discontinuities (the Orozco transform fault near 11°N and the Siqueiros transform near 13°N), as well as between two smaller 2nd order discontinuities, also called Overlapping Spreading Centers (OSC) at 12°43′N and 12°53′N (Fig. 7.15).

The second order discontinuities are marked by a sudden change in direction of the axial ridge segments, and with a bending at their tips. Several sea-going expeditions were conducted on the ridge segment at 12°50′N. One of the most significant cruises that prepared the area for further diving expeditions was the "Clipperton" cruise, followed by the diving cruises called "Cyatherm" (1982) and "Geocyarise" (1984).

The "CLIPPERTON" cruise took place in 1981 and was aimed making at a detailed exploration of a segment of the EPR where hydrothermal activity had been detected during the previous year's SEARISE expedition. The Clipperton cruise began on May 9, 1981, when the NO *Jean Charcot* left the port of Manzanilo (Mexico) and ended on June 6 in Panama. The *Jean Charcot* was the largest sea going vessel of CNEXO at the time. Its name was given in honor of the explorer Commandant Jean-Baptiste Charcot who had explored Antarctica's shore in a ship called "Pourquoi Pas?" in 1908. The *Jean Charcot*, able to carry a total of 48 crew members and scientists, was 74.5 m long, had a weight of 2200 tons and a maximum speed of 15 kn, using a diesel-electric propulsion engine.

For the first time, I was in charge of a scientific expedition on board the R.V. *JEAN CHARCOT*. This made me nervous when I arrived on board. I was going into unknown territory and I was afraid of not being able to do a good job. I did not know how the crew and the scientific party would accept me. The scientific party consisted of people from different disciplines, such as water chemists from the *Centre National de Recherche Scientifique* (CNRS) and scientists from the Universities of Paris 6 and 7 and from the *Centre d'Energie Atomique* (CEA) of Saclay. There was a sedimentologist from Woods Hole Oceanographic Institution and a geophysicist and structural geologist from IFREMER. However, when I

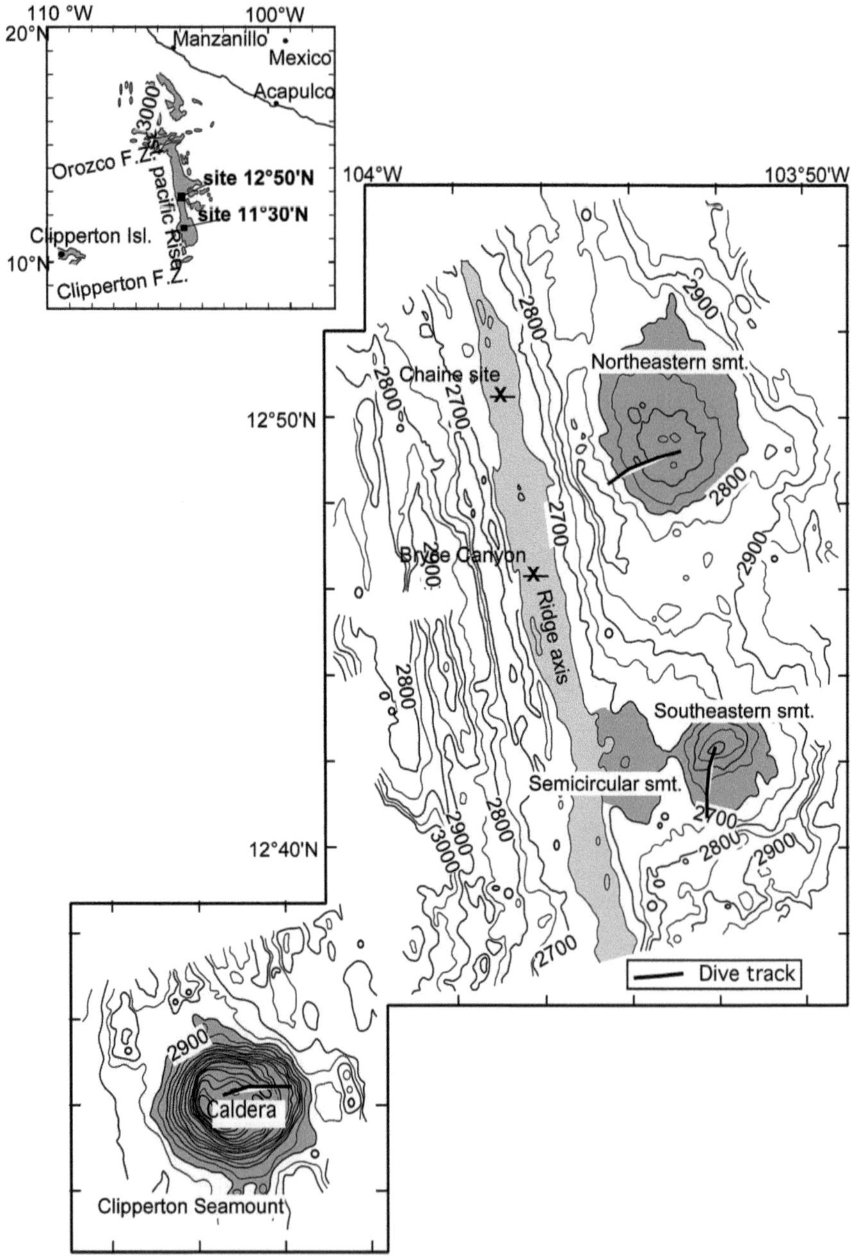

Fig. 7.15 Bathymetric contour map of the East Pacific Rise (EPR) near 12°50′N shows the off-axial seamount and the ridge axis (Hekinian and Binard 2008). The main dive sites (Chain and Bryce Canyon) at about 2500 m depth on the EPR are indicated as well as the dive tracks on the off-axial seamounts. The inset shows the general location of the 12°50′N EPR ridge segment

went to the bridge, I was happy and relieved to see a familiar face. Guy Paquet was in charge of the ship and it removed some of the pressure on me to find myself on board with this experienced captain. I had first sailed with Guy on the same ship, during a cruise called "Walda" off the west coast of Africa in 1972, and we had always gotten along very well.

The vessel had been recently equipped with a multichannel sounding system called the SeaBeam, which was able to accomplish large swath bathymetric coverage. The equipment had been tested in the spring of 1977. This sounding equipment was able to cover a corridor as wide as ¾ of the depth during one single passage of the ship. A bathymetric contour map was instantaneously obtained during the ship's transit. Previously, the use of this type of multi-channel bathymetric equipment (SeaBeam) had been the privilege the U.S. Navy. It had been used for military purposes and all the information that had been gathered was classified and inaccessible to the scientific community. During that period, satellite fixes (such as GPS, Global Positioning System) were also unavailable to civilians, so it was quite difficult to have accurate information for navigating on the sea floor.

Since we did not have access to continuous satellite positioning, we decided to deploy an array of bottom transponders on the ocean floor at 2,600 m depth. A transponder, like a beacon device, generates a signal at a given frequency, which is relayed to a receiver on the ship. The ship receives the occasional satellite fixes and then the navigation team calibrates each transponder by determining its position on a geographic grid. Since satellite fixes were not received continuously, all this recalibration required a lot patience and effort, and it was time consuming. It used to take about six hours to have the field of exploration set and the grid ready to be surveyed. A narrow band about 7 km wide and 35 km long was set for exploring part of the ridge segment between 12°55′N and 12°45′N (Fig. 7.15, 7.16). This was the first map to be made of a fast spreading ridge segment.

After about 5 days when the map was finished, the result was astonishing. We noticed the linearity of the ridge segment, which looked like a highway. The axis was more or less at the same depth (about 2,650 m) with only about 150 m of relief. Such a detailed map was very useful for relocating the samples and any other data obtained from the sea floor. This enabled us to correlate the topography with the composition of the eruptive events and any bottom instrumentation or sampling devices that were laid on the sea floor could be precisely noted on our mapped grid.

This marked the beginning of new era in scientific exploration. We were now able to make a bathymetric map of the sea floor and at the same time superimpose the geological observations obtained with bottom imagery. This facilitated exploration and made any recovered data more accurate.

From the reconstruction of the bathymetry, it was found that the crest of the EPR consists of an elongated domed structure that is about 800 m in height (from the 3,300 meter contour line), 5 km wide and 37 km in length (Fig. 7.16). Normal faulted blocks characterize the eastern and western flank of the ridge. A linear graben about 30–60 m deep and less than 600 m wide occurs on top of the rise. It is in this axial graben that most of the tectonic, volcanic and hydrothermal

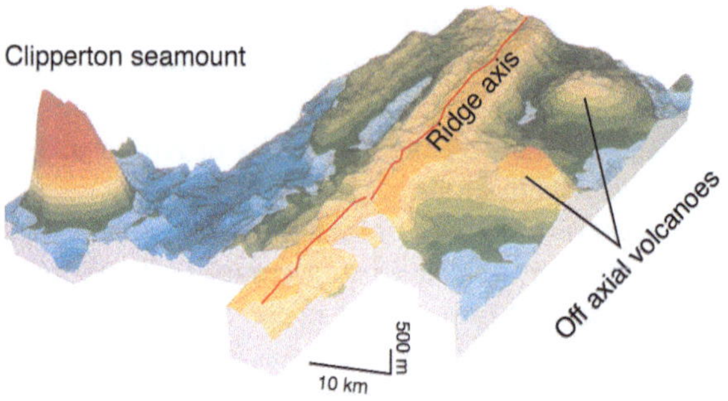

Fig. 7.16 Block diagram of the East Pacific Rise segment at 12°50′ N was reconstructed after the multibeam bathymetry survey of The N.O Jean Charcot during the Clipperton cruise. The volcanic ridge at 2600 m depth dominates more than 300 m above the surrounding sea floor. It is flanked by parallel linear faulted structures and by volcanic seamounts located between 1 and 18 km from the ridge axis. Axis consists of a narrow axial graben

activities take place. The axial graben is bounded by marginal highs with a gentle rolling topography that is often interrupted by small discontinuous fissures and normal faults. At a distance away from the axial graben (<7 km to the east) but still considered as being part of the EPR structure, we could see four small off-axial volcanic features which were located between 1 and 18 km away from the ridge axis.

The main purpose of our exploration was to find more traces of hydrothermalism. After completing our bathymetric map, and based on the topography, we then started to define the locations for our geological and hydrological (water sampling) stations. The geological stations consisted essentially in using a deep towed bottom camera system, plus dredging and sediment sampling. The deep towed camera, which gave us images from the seafloor, was primordial for observation but its limited surface coverage was not satisfactory for pinpointing active hydrothermal vents. The hydrological stations covering a much larger area would be more useful for localizing hydrothermal venting.

In order to detect hydrothermal venting areas, hydrological profiles were made by lowering plastic bottles to sample the water column at various depths on the EPR axis at 12°50′N. The experiment was headed by Liliane Merlivat (Merlivat et al. 1987) from the physical–chemistry department of the Centre d'Energie Atomique, in Saclay. She was helped by Bernard Dimon, Jean-Francois Minster, and Jacques Boulègue from the University of Paris 6 and 7, and Jean-Luc Charlou from IFREMER. The data revealed the high concentration of helium, manganese and methane at two sites along the axis at 2,500 m depth. The anomalous increase of helium and manganese observed in seawater suggested the presence of hydrothermal fluid and or volcanic discharge in the water column. The helium and manganese anomalies are found at about 200–300 meters above the active hydrothermal vents. This suggested that hydrothermal fluid circulation has enhanced the presence of these compounds in seawater.

The water sampled above the sea floor was taken to shore in order to analyze the helium content as well as other the compounds. Helium is an element found in solar nebula and is considered as being a primordial element during the formation of our planet. It is believed that the mantle is enriched in helium, which was captured during the early formation of Earth. Thus, any helium that could be measured was probably derived directly from the degassing of the Earth's mantle during volcanic and hydrothermal activity. Helium has a low mass density and does not accumulate on the surface of the Earth, in addition to being only weakly soluble in seawater. This is why helium is a good tracer for detecting hydrothermal and volcanic activity.

The CLIPPERTON cruise served as a preparation ground for other more detailed diving expeditions such as the GEOCYATHERM (1982) and the GEOCYARISE (1984) cruises. Furthermore, this area has become a natural laboratory site for hydrothermal and biological in situ experimental work during the past twenty years. The segment of the East Pacific Rise near 12°50′N is one of the world's ridge segments where the largest numbers of international submersible dives (more than 60) and surface ship operations have taken place, between 1980 and 1990.

Low Velocity Zone Underneath the EPR

During a discussion on board the RV. *JEAN CHARCOT*, Felix Avedik had the brilliant idea of using the combined SeaBeam bathymetry swath-mapping system with his seismic water (1.1 l) gun to detect any sub-crustal discontinuities existing underneath the spreading ridge axis. We could start our profile along the ridge axis as soon as the various in situ stations were finished. Since we had seen from our bathymetric coverage the capability of covering the entire 2 km wide axis of the ridge with one single-pass bathymetry swath, by combining both systems we might be able to make a radiographic subcrustal image of the material existing underneath the ridge axis. Thus, after completing our geological and hydrological work at 13°N on the EPR, in the early morning at 4:30 AM on May 21, 1981 we started the seismic profile along the ridge axis.

At the time, Felix did not know if the seismic signal would be strong enough to see any distinct subcrustal features. However soon after we started, the signals that came on board to the receivers were encouraging. For the first time, a combination of multichannel bathymetry combined with towed seismic gear was able to show deep seated (<2 km depth) subcrustal discontinuities in the form of a reflection of the seismic waves. Such a discontinuity is related to a decrease in the seismic velocity crossing a material of different density than the surrounding formation. This experiment allowed us to calculate the size of an low velocity zone (LVZ) along a stretch of the spreading center. The extent of the signal was proportional to the length of the segment in the area of our experiment, which was less than 30 km (Figs. 7.15, 7.17).

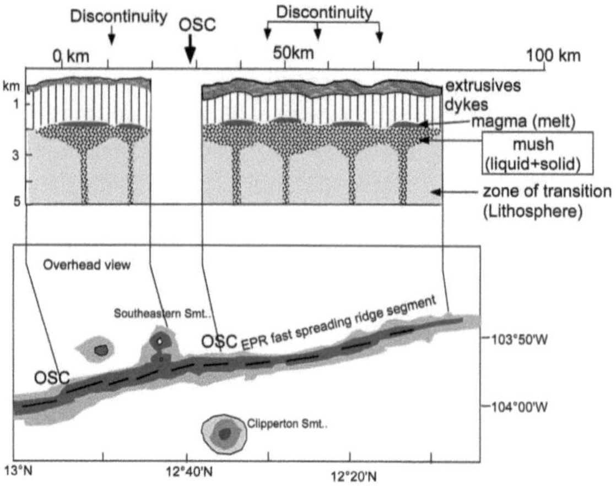

Fig. 7.17 Subcrustal section obtained from seismic exploration during the IFREMER Clipperton Cruise in 1981 by Avedik and Geli (1987) shows the distribution of magma chambers located at 1.5 km depth underneath the ridge axis of the East Pacific Rise near 13°. The volume of magma stored in the crust decreases in the vicinities of the 2° order ridge discontinuities

Such a type of discontinuity encountered within the crust is called a Low Velocity Zone (LVZ). It indicates the presence of melt that is magma and/or a mushy mixture of magma and partially solidified melt. Later on, in 1986, another such experiment with more sophisticated and efficient seismic instruments was carried out along the strike of the various EPR segments essentially located between 8°N and 14°N by U.S. scientists. This study also revealed similar results. They also interpreted the presence of an LVZ, which was observed to be discontinuous along its extent (Detrick et al. 1987; Avedik and Geli 1987).

Unfortunately, Felix waited until 1987 for his data (from the 1981 cruise) to be published, because some of his colleagues at IFREMER who had read the manuscript were skeptical about his results and did not believe that the small water-gun he had used was reliable, and thus he was worried that his interpretations were risky. Nowadays, it is well accepted that the types of sub-crustal discontinuities that can be seen on the seismic signal correspond, in most cases, to the topographical and structural offsets observed. It is also inferred that the seismic discontinuity corresponds to the top of a low velocity zone represented by magma chambers. Within an individual ridge segment, the size of this shallow level magma chamber (<2 km deep) is believed to be less than 5 km in width.

Clipperton Island

Because of bad weather conditions during the CLIPPERTON cruise, Captain Guy Paquet decided to leave the working area of the ridge axis on June 2 and go westward in order to escape from a tropical storm called "Adrian", with gale winds up to 60 kn, which was forecasted on June 1. The storm was coming from San Francisco and moving at 4 kn in a northeasterly direction. The swells started to increase in the evening and before leaving the site, we had to secure all our gear.

Fig. 7.18 South side of Clipperton Island with the Research Vessel JEAN CHARCOT in the foreground. (Photograph courtesy of Gerard Vincent 1981). To the right of the ship, the island's volcanic outcrop of trachyte is visible in the background

The transponders, which had served to make our precise map, were left on the bottom in order to be used when another mission would return to the site. We decided to go to the Island of Clipperton, which would also serve to reinforce French presence in this remote area. In 1986, the fact of our passage on Clipperton was of help when the French government wanted to construct a meteorological station in the framework of an international climate study as well as set up a seismic station on this small Island, in order to monitor earthquakes on the west coast of Mexico.

The Island of Clipperton is located at the intersection of two different structures (Fig. 7.18). To the north there is the prolongation of a volcanic alignment called the Mathematician Seamount Chain, and the other is the large East–West trending fracture zone, also called the "Clipperton Fracture Zone" (FZ). Clipperton is a volcanic island, which was active about 500,000 years ago. It is located on the northern wall of the Clipperton FZ, on a crust about 15 million years old. The island is about 300 km west of the East Pacific Rise near 10°17′N and 109°13′W at a distance more than 1,300 km from the Mexican west coast.

Clipperton Island was discovered by Ferdinand Magellan in 1521, and was later named after a British sailor called John Clipperton, during the voyage of a British explorer, William Dampler, in the Pacific Ocean in 1704. The sailor was either abandoned by the skipper of the ship or he escaped from the ship as Pierre de Latil reported in his article of June 1986 in a publication called "*Science et Connaissance Magazine*". Another story says that John Clipperton was the captain of a pirate vessel who used the island as a refuge. Two French navy ships called "La Princesse" and "La Découverte" commanded by captains Martin de Chassiron and Michel Du Bocage moored near Clipperton on April 7, 1711, and renamed the island "l'île de la Passion". This long-forgotten island is one of the smallest French territories, and covers about 2 km^2 with a circumference of 2 km. It was in 1850 that the French government took possession of the island and a company from Le Havre started to exploit the detritus of birds called "guano", made up of phosphate compounds, that was covering the volcanic, rocky outcrop.

On November 17, 1858, under Emperor Napoleon III, the French annexed Clipperton. However, in 1897, Mexico reasserted its claim over the island and established a Mexican garrison under the command of Ramon Arnaud, which remained in place for many years. The conflict between Mexico and France over the island was submitted to the arbitration of King Vittorio Emmanuel of Italy. It was in February 1930 that the king of Italy finally declared in favor of France.

Debarking on the island it is not an easy task because a coral reef and the constant swell makes any debarkation approach very risky. The N.O. *Jean Charcot* stopped about 1 mile away from the coast, and a zodiac was put over-board. A sailor piloted the zodiac and waited for us, so that I, along with the chief engineer (Jacques Guillemin) plus another engineer from the engine room were the three people who were able to visit this deserted outcrop. In the meantime, the ship made a bathymetry survey around the Island.

Because of the swell, we could not approach too close to the shore with the zodiac. Coral that could have damaged our inflatable boat covered the coastal area. The sailor stopped the engine about 150 m from the sandy beach and we had to swim the rest of the way. Both the chief engineer and his crewman from the engine room dove in first and I followed them. I had my socks on and I had laced my shoes together and hung them around my neck. It seemed that it took me forever to reach the shore. When I finally arrived safely, without encountering any sharks, I was very tired and I had lost my socks. I had to put my shoes on quickly because the beach was full of red-colored coconut crabs that were not at all afraid to come close to try to nibble our feet.

Prior to leaving the ship we had equipped ourselves with a small bag of cement and a commemorative bronze plate indicating the passage of the N.O. JEAN CHARCOT near the island. The plate was made on board and was going to be sealed into the rocky outcrop on Clipperton Island.

The island hosted thousands of birds, mostly boobies, as is the case for the Islets of St Peter and St Paul's Rocks in the Atlantic Ocean. It seems that animal life can develop extensively when humans are absent. In addition, a multitude of yellow-pink colored crabs are always waiting along the sandy beach to greet any adventurer. In 1968, when French scientists surveyed the island they estimated that eleven million land crabs had once stripped the island of all plant life except for the tall palm trees. The southern side of the Island is made-up of a 30 m high volcanic trachytic (alkali enriched) outcrop covered by guano. The first petrographic study was made by Teall (1898) who identified the rock as a trachyte. Near the rocky outcrop, there is a lake (lagoon) consisting of fresh water less than 50 m deep, but the water also contains dissolved sulfates which give off a rotten egg smell. The far northern end of the island consists of a coconut forest with an abandoned wooden house. The wooden house was the site of tragedy that happened on this island at the beginning of the twentieth century.

Around 1905 the Mexicans had annexed the island and they sent a detachment of soldiers with their wives. About 100 people were living in a Mexican army garrison and some phosphate-mining people from a British company were also living on the island, which was regularly supplied by ships from Mexico.

However, at the outbreak of World War I in 1914, which also corresponded to the Mexican Revolution, Mexican authorities forgot all about Clipperton Island.

Three years after the last supply ship had come to the island, most of the people had died from starvation or from scurvy. Captain Arnaud, head of the Mexican detachment, wanted to leave the island with the healthiest men. When they put a small boat at sea, it capsized, and the sharks did not miss out on finding them. Among the sick people left on the island, only one man, a black person, was left with four women. The man was a soldier and he had taken charge of the situation by killing the other men and by trying to enslave the women. For two years he terrorized the women until one day they managed to kill him with an axe. The next day, the U.S. Navy vessel "Yorktown" made a call to the Island and rescued the three surviving women. More details on the Island can be found in a book written by Jean-Louis Etienne (2005).

After the Clipperton cruise, a diving expedition called *Geocyatherm* (short for GEOlogy CYAna hydroTHERmalism) followed and it had two Legs. The first Leg started at Acapulco on January 10, 1982 and ended in Manzanillo (Mexico) on January 25, 1982. The second Leg left Manzanillo on January 28 and returned to Acapulco on February 10, 1982.

The beginning of this cruise was difficult because of a conflict with the submersible group and their administration called GENAVIR (GEstion NAVIRe of IFREMER), concerning the demand of an additional shore-day in port. Usually the crew and the technical group have the right to a period of 2–3 days off at shore depending on the length of the cruise during the port calls. The threatened strike could have postponed our departure and jeopardized the mission, therefore I took the initiative to shorten our time on the site by one day in order for the crew to have an extra day at shore. In the end, this decision really paid off because the crew and the submersible group put in extra time for maintenance and repair, which helped to save one of our dives during the cruise.

Ship refueling incident and scientific party. Another incident that marked the cruise took place while we were in the port of Acapulco to refuel the N.O. *LE SUROIT* on January 5, 1982. The trucks that were supposed to come to the dock to refuel the ship did not show up. Apparently there was a problem between the agent and the port authority. I remember that I spent all day long sitting in the agent's office until he finally decided to take care of our business. In the meantime, I had to call the French embassy in Washington where I knew the scientific attaché, Jean Jarry, in order to ask them to intervene with the Mexican Oil Energy Ministry's office via the French embassy in Mexico City. This was efficiently handled and in a few hours we had obtained the authorization to refuel. However, another problem arose with the agent because he was in no hurry and he was taking his time to react. Probably he was waiting to have an under the table tip, "bakshish". At this point we had already lost one full day of our cruise time just waiting around. So I decided to spend another sitting day in the agent's office until he would finally give us satisfaction and order the trucks of fuel needed for our ship. At last, the agent got tired of seeing me complaining in his office. So, an unmarked fuel truck pulled up along the side of the ship in the late afternoon of January 9 to deliver the fuel.

Finally the vessel left the port of Acapulco on January 10, 1982, at midnight and headed out for the diving site on the EPR at 12°50′N. During the first leg of the *Cyatherm cruise,* the scientific party included Jean Francheteau, Bob Ballard, Vincent Renard, Jean-Luc Charlou, Pierre Choukroune, Francis Albarede and me. The first Leg (January 9th–January 25th 1982) comprised the first 20 Cyana dives of the cruise. The second Leg (January 28th–February 10th 1982) included 11 dives and the last dive number was Cy 82-32 (Cy = Cyana, 82 = year 1982 and 32 = dive number). The participants on the second Leg were mostly the same scientists as the first Leg except for two new members: B. Marty, a fluid geochemist, and Jean-François Minster, a geochemist from the Institute de Physique du Globe de Paris (IPGP), who joined the ship in Manzanillo. Jean-François Minster later became one of the directors of IFREMER for the period of 1999–2004.

The person responsible for the submersible team was Jean Roux, a tall, thin man who called me his cousin because he was half-Armenian in origin, on his mother's side. After talking about our families and their stories, we figured out that his grandmother was probably from the same town, Kharpert, as was my mother's family, before the genocide of 1915–1920 perpetrated by the Ottoman government, which killed more than 1,500,000 Armenians.

The use of submersibles and deep towed bottom camera coverage had enabled us to find more than a hundred (149 counted by Fouquet et al. 1996) hydrothermal deposits dispersed along the strike of the EPR in the 13°N axial graben over a distance of 31 km. These deposits vary in size between 2 and 15 m in height and are less than 50 m in diameter. They are distributed in a sequence at a distance of about every 150 m along the EPR axis. All the hydrothermal deposits encountered in the axial graben consist of moderately to extremely porous zinc sulfide, iron–zinc sulfide and copper–zinc sulfide material. Most of the active hydrothermal venting is taking place on a narrow band of less than 50 m long, and about 25 m wide along the en-echelon linear fissures of the axial graben. The fissures have variable lengths and could be as long as 7 km.

Cyana Dives on the EPR at 12°50′N

The diving operations with *Cyana* on the EPR segment at 12°50′N was the first time we would try to use this submersible for conducting detailed geological exploration such as contour line mapping, sampling and deploying complex instruments such as water bottle collectors, markers for site recognition, and baskets for biological sampling on the sea floor. The geological exploration was mainly focused on finding and sampling hydrothermal fields. We already had some indications about our exploration targets from previous surface ship surveys due to hydrological water sampling data and deep-towed bottom photographic coverage.

Because of the limited space available on the *Cyana*'s sample storage rack, a special system was built on board at the last minute in order to be able to store

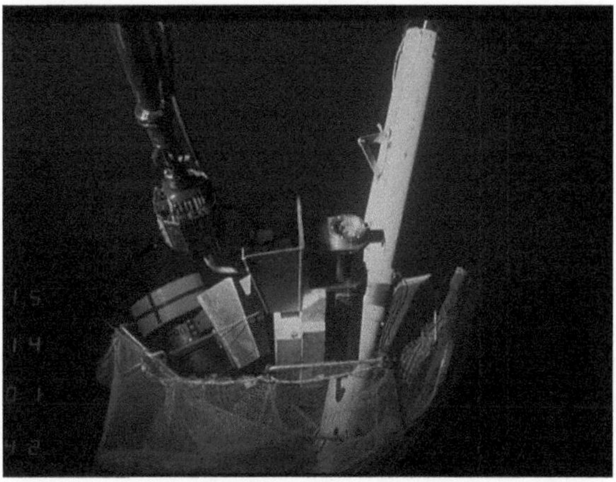

Fig. 7.19 The storage device invented for the IFREMER Cyatherm cruise of 1982 consisted of a plastic basket, also called an umbrella basket, made with a net about 2 m in diameter and moored 20 m above the sea floor, equipped with a transponder (12.5 KhZ), plus a weight to hold the basket on the sea floor bottom. The basket's weight could be released from the surface, and the basket and its contents would be recovered by the ship. (Photo courtesy IFREMER, Cyatherm 1982, Dive Cy82-17)

more instruments and all the sampled material collected. This was a clever way to maximize our data collection during each dive. The storage device consisted of a plastic basket (umbrella basket) with nets of about 2 m in diameter moored at 20 m above the sea floor, equipped with a transponder (12.5 KhZ), plus a weight to hold the basket contraption on the bottom, and another plastic rope tied to a large, light-weight foam float which could raise the basket to the surface when it was full (Fig. 7.19). The transponder had a release mechanism and could be located by the submersible and the ship. The entire system was called the "ascenseur" (meaning "lift/elevator", in French) because it could be released when full and dropped back down again to the sea floor, with a new weight to hold it on the bottom, after it had been emptied.

Water bottles about 40–50 cm in length were made of a titanium alloy and carbon. They were used for sampling the fluids from the hydrothermal vents. The bottles or any other instruments were inside their holders and ready to be picked by the submersible arm once they were on the sea floor. Once used, they could be returned by the submersible to the "lift" and put back inside the umbrella net.

The hydrological work on board was taken care of by B. Marty and Jean-François Minster both from the CNRS, since they were experts on the study of water and hydrothermal systems. At the time, Jean-François was a young and bright scientist who had graduated from a prestigious *Polytechnic* school in Paris and was working at IPG (Institut de physique du Globe). He was eager to see the hot smokers as was everyone else on board. His dive took place on the ridge axis at 2608 meters depth (dive Cy82-26) on February 20, 1982. Marty dove (Cy82-32) on February 26, at the "Bryce Canyon" site where he sampled hydrothermal fluid from an active black smoker.

During the two legs of the cruise, 32 submersible dives were made plus there were other surface-ship stations at night.

During the second Leg of the cruise we met up with the German vessel, the F.S. *SONNE*, which had arrived in the area around February 2, to conduct geological exploration from the surface. I communicated with the chief scientist, Harold Bäcker from Preussag, a mining company in Germany, and he asked me if it would be possible to participate on one of our dives. Knowing Harold's reputation as a conscientious and internationally known marine geologist, I accepted his request after consulting with the captain and Jean Roux who was responsible for the submersible group.

Harold arrived on board and after an evening's briefing he was ready to dive the next morning. It was during dive Cy 82-16 on January 4, 1982, that Harold Bäcker with the pilots Arnoux and Pottier, discovered an important field of several active hydrothermal chimneys at the foot of a lava pond in the axial graben near 12°50′N. This dive was the first of a series of dives planed to carry out a systematic visual survey across the spreading axis in order to make the first geological map of the seafloor using a submersible. We intended to make several transects, at least six from wall to wall and perpendicular to the spreading axis where the freshest lava and the most active hydrothermal vents occurred.

Harold and the pilot and co-pilot arrived on the sea floor at 10:15 at a depth of 2610 m at the foot of the east graben wall. After few minutes for positioning the submarine with respect the transponder, they took a westerly course (270°). They crossed a lava pond 30 m away from the landing site and collected their first rock sample coated with rusty colored hydrothermal deposits. Above the lava pond, they found a hydrothermal chimney spewing white, low temperature (47 °C) fluid (a "white smoker"). A few other inactive chimneys were crossed and then they returned to make another crossing after reaching the west wall's first scarp. After the third crossing from wall to wall, they arrived on a very active hydrothermal field spewing out black, high-temperature fluid. When the temperature probe was put at the interface between the solid metallic orifice of the chimney and the exiting dark fluid, the temperature reading reached 232 °C. A sulfide sample was taken near the active chimney at 14h52 at 2,616 m depth. After that, the diving party spent about one hour on the site and at 15h02 they left to terminate their crossing and reached the western wall's first scarp about 15 min later. At this point, it was time to leave the bottom.

At that time, the divers did not realize that they had found a hydrothermal field that would become a natural laboratory where geochemical and biological instruments could be deployed during several future cruises. Harold Bäcker, for whom this was his first submersible dive, was very enthusiastic about what he saw and was extremely grateful for having been able to observe the sea floor for the first time with his own eyes.

The next three dives (Cy 82-17, Cy 82-18 and Cy 82-19 on February 5th, 6th and 7th) were also planned to use the same transponder network which had been deployed on the sea floor in order to continue the transect crossing of the graben. Jean Roux, head of the submersible team, was very eager to dive on a hydrothermal field. I found it normal that someone who had been spending so much energy and time to direct each dive would also like to see how the sea floor looks in this particular active spreading ridge segment. Also, this would be an opportunity for Jean to observe the high quality performance of his group in handling the *Cyana*. Hence dive Cy 82-19 was made with Jean Roux as the science observer together with Guy Sciarrone and Arnoux.

At 09:51 *Cyana* was in the water and ready to dive. The team reached the bottom at 11:04 at 2,600 m depth in the axial graben near the eastern wall. The purpose of the first part of this dive was to perform water sampling, which meant going to pick up a bottle made of titanium/carbon that was stored in the "lift". So, first the submersible had to find the basket and pick up one of the two water sample bottles that were inside. The basket was about 80 meters from the submersible as shown on the radar screen and by the transponder beacon beeping its signal. At 11h32, after localizing the "lift", Cyana went to pick up a water bottle and then went back to the target. At 11h48 they arrived on the "black smoker" and started to sample the hot fluid exiting from the chimney. Using the starboard arm to open the small valve of the bottle that was held by the port side arm, the pilot, Guy Sciarrone, completed this delicate operation despite the small size of the valve, about 3 cm long.

Bottle filling required a pilot to show a great deal of skill to successfully complete the maneuver. The submarine had to be stabilized correctly in neutral buoyancy without moving. The observer and the co-pilot held their breath, and no motion was allowed on board. This could jeopardize the operation and even be dangerous for the safety of the submersible because of the high temperature of the exiting fluid (about 300 °C). At 13h19 the bottle was on top of the chimney and the valve was opened so the hot fluid could pour in. It took only a couple minutes to fill the bottle, however, the time spent for the entire operation lasted about 2 ½ hours. After taking a few more pictures of the site, they left to return the full bottle to the "lift" and then followed their course at 080° towards the eastern wall at a distance of 250 m, so they could make another visual crossing. Then, they went back again towards the west wall of the graben where the dive terminated at 16:15 at a depth of 2,623 m.

Since the fluid sampling was successful, a geological dive (Cy82-24) was made in the same area by the geochemist, Francis Albarede, on February 12, along with Guy Sciaronne as pilot and Pottier as a co-pilot. As long as I have known Francis, he has always had a mustache, with a gentle smiling face. As a young specialist in rock geochemistry working at Institut de Physique du Globe de Paris (IPGP), he had become very interested in hydrothermal fluid circulation.

The *Cyana* left the surface at 09:28 and reached the sea floor at 10h44 at 2,595 m depth. They landed on the first step of the eastern wall and they went to the graben floor before starting their profile towards the western wall at the hydrothermal site. The aim of this dive, in addition to doing the geological profile of the spreading axis, was to take fluid samples on two black smokers and, if time permitted, to take an additional water sample on a low temperature white smoker that had already been discovered on the previous dives. Francis described the site with four chimneys in the most detail: three black and one white smoker. On the east side, a small chimney looking like a “Chinese hat” also occurred at the foot of the main edifice.

The main edifice is about 10 m in diameter located on the western edge of the lava pond along a fault scarp oriented at 340° corresponding roughly to the direction of the spreading axis. An animal community made up of Galathea crabs, dandelions and vent fish were seen. When going further northwest towards the western wall, the terrain became older and the sediment dusting on top of the lava flow increased. A field of at least three dead chimneys 2.5 m tall was observed at less than 50 m from the previous active black smoker site. This field was accidentally found while going back to the “lift” in order to pick-up the fluid bottles. After each sampling, the submersible had to go back to the “lift” in order to put the full bottle back into the storage rack. At the end of a day’s sampling, the “lift” was released from its weight and rose to the surface where it was located and recovered by the ship’s crew.

At 16:05, at a depth of 2,608 m, pilot Guy Sciarrone activated the vertical propeller and took *Cyana* above the black plume of the hydrothermal fluid, which was being diluted in the seawater. The temperature at the edge of the plume, as registered by the outside sensors, was about 3.8 °C which is more than the twice the temperature of the ambient seawater. Above about 25 m over the hydrothermal chimney, the black smoker fluid had decreased in intensity and was almost totally diluted in the seawater. The dive ended at 16h15, above the hydrothermal plume.

Guy Sciarrone is a reserved man who is a very conscientious engineer and an experienced pilot. When at sea, he spent most of his time taking care of the submersible as if it were his “baby”. He showed his concern during and after each dive, being worried that the submarine might come back damaged after its voyage to the bottom.

It was after dive Cy82-19 on the EPR “*Chain site*” hydrothermal field that Guy Sciarone came to me because he was concerned. He told me:

> Roger, I had difficulty recognizing the site this time. I am worried that we might have missed the previous vent and that we may have sampled the wrong one.” He added: “I am really puzzled, how can it be possible that the aspect of the site has changed so much since the last time we saw it, which was on dive Cy82-16, only 4 days ago?

This was a serious piece of information. Geologists are used to observing long-term changes on the order of hundreds or thousands of years. A change of geological form in a time lapse of just a few days seemed highly unexpected. However, then some doubts came to my mind. If what Guy said was true, and if it were really possible for a site to change its appearance in only a few days, then we could be dealing with a time span that is closer to the time span on a human scale. Could this be possible?

I knew Guy was an experienced pilot and a thorough navigator, so he could not have made a mistake. If he questioned his ability to recognize the site, this meant

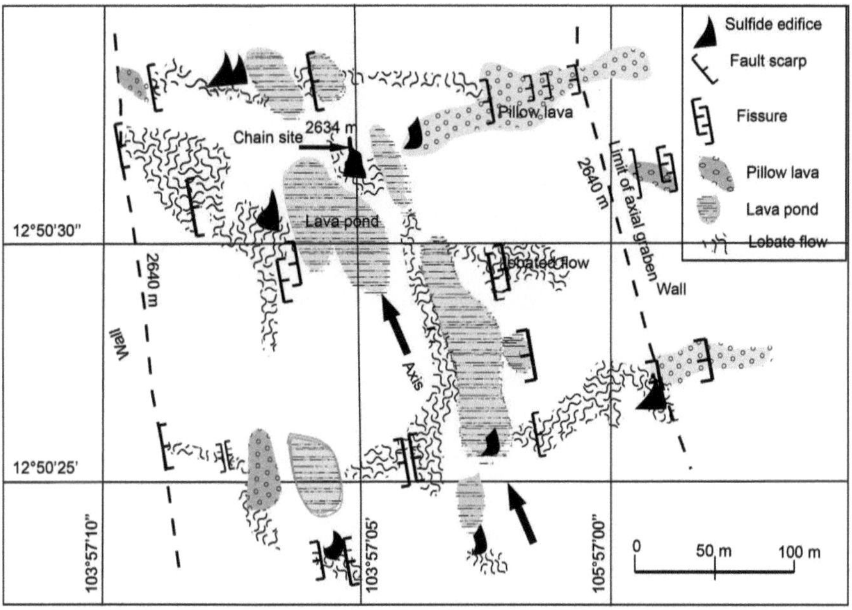

Fig. 7.20 Geological map made during six dives that criss-crossed the axial graben from wall to wall (EPR at 12°50′ N at 2550–2650 m depth) (See also Fig. 7.15, ridge axis area). The lava morphology is indicated: *grey* with *open circles* = pillow lava formation, *grey* with *hashed lines* = lava pond depression, *squiggle lines* = lobated sheet flows. The active hydrothermal edifices are shown by black *irregular triangles*

that something was happening. We had to find out. If it were the case that the site looked different, then this could mean that the morphological landscape of the ridge was changing rapidly and the geological time scale, where we were used to counting in thousands and millions of years, was no longer valid.

Then I decided, along with my scientific colleagues, to make several other dives to verify if rapid changes in the growth of the hydrothermal edifices were in fact taking place so quickly. Thus, we sent another dive (CY82-24) with Guy Sciarrone, Pottier and Francis Albarède to continue our systematic geological field mapping of the spreading axis but at the same time, they were asked to take a closer look at the particular hydrothermal site the pilot had visited shortly before.

By now, Guy was familiar with the terrain, so after crossing the lava pond depression, he found the site on the East side of graben axis and he recognized the active chimney. During his communication to the surface, he told us that, according to his estimation, the same chimney that he had seen last time had grown about 20 cm (approximately 1 foot). This was astonishing news, if true.

The Chain site. The most striking experiment carried out during this diving cruise was the measurement of the growth rate of a hydrothermal chimney. This was performed during several dives on the same hydrothermal field, which permitted us to sketch a geological map of the area (Fig. 7.20).

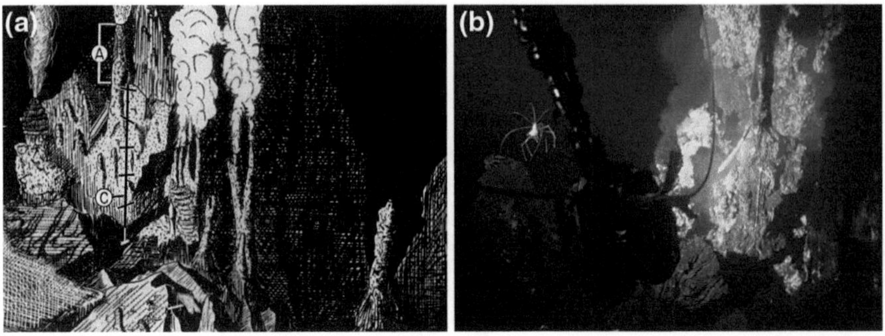

Fig. 7.21 **a** Sketch of the graduated chain device made to measure the growth of a hydrothermal chimney on the sea floor. **b** On February 19, 1982, a graduated chain device was laid on top of a "black smoker" chimney to visualize the growth rate of a hydrothermal chimney on the axis of the EPR near 12°50′N at 2634 m depth. Five days later the graduate chain was sealed by the growth of a chimney. (Photo courtesy *IFREMER Geocyatherm cruise, Cyana dive Cy82-25)*

Dive 82-25 with Nivaggioli, Normand and Vincent Renard as the scientist, had the objective of measuring the growth of the active chimney found during dive CY82-16, when Harold Bäcker was on board, and revisited during dives CY82-19 and CY82-24. Now that the site was well localized, with a sonar buoy left as a marker as well as the detailed survey map that had been made, we were able to organize other dives to return to the exact same area.

In order to measure the growth rate of the newly formed chimney, a device consisting of a chain of titanium alloy was cleverly constructed by one of the engineers under the supervision of Guy Sciarrone. The chain consisted of 10 cm long horizontal bars welded every 20 cm (Fig. 7.21a, b). This would permit the observers from inside the *Cyana* (dive 25) to evaluate the dimensions of the target.

On February 19, 1982, during dive Cy 82-25, Pilot Nivaggioli, Co-pilot Normand and Vincent Renard reached their target at 13h25 so they could place the device for our measuring experiment. Nivaggioli carefully approached the chimney that was spewing hot, black hydrothermal liquid like a locomotive. The temperature probe inserted on top of the orifice gave a reading of 275 °C. After replacing the temperature probe in *Cyana*'s sampling basket, Niva broke off about 25 cm of the black smoker chimney. After securing the piece of chimney inside the sampling basket of *Cyana*, he took the graduated chain and laid it on top of the active orifice. Then he communicated to the surface to share the success of the operation, which made me very happy. Now they had another 2 h on the bottom to continue their geological observations and sampling and at 16:23 they returned to the surface.

February 24, only 5 days later, another dive (Cy 82-30) was programmed to return to the same site with the same team as on dive Cy 82-25. At 11:00 they reached the site and it was then that the divers were able to witness the growth of the chimneys at 2630 m depth with their own eyes. As reported in the diving report, the co-pilot Arnoux wrote :"All the chimneys seen five days ago have grown". On the edifice where the chain had been placed, a fragile, few millimeters thin and 40 cm high tubular conduit had been built on top of the "chain measuring device". The temperature probe inserted into the black smoker gave a reading of 314 °C. Two sulfide samples were collected from where the chain was located. At 11h27, after sampling at the foot of the edifice and recording the observations of the site on videotape, *Cyana* left the site. Four hours later, when the submersible returned once again to the same hydrothermal field, the divers witnessed a regrowth of the active vent. A thin, golden yellow metallic coating, which was probably chalcopyrite or pyrite minerals, had been built, but it was so fragile, it floated away in the water column as soon as the submersible's mechanical arm tried to grab it.

Hence, the first quantitative measurements were made on the growth of a hydrothermal edifice on the ocean floor at the "Chain site". Based on the observations made during the *Cyana* dives, it was now possible to evaluate the growth rate of a sulfide edifice. A cylindrical shaped chimney of 3 cm (internal diameter) and 10 cm (external diameter) with an average density of 2.9 gms/cm^3 will increase its mass by about 1.6 kg per day. It is known that sulfide edifices are formed from the coalescence of several chimneys. Thus, edifices averaging about six meters in height and three meters wide, weighing about 41 tons, could be built over a period of seventy years. This is the average duration of a human's lifetime.

Since then, the "Chain Site" has been used as a reference site by several other expeditions, and has become an area where biologists and geochemists have also conducted various in situ experiments.

More sophisticated methods of calculating growth rates based on the age of the sulfide deposits were later performed by Claude Lalou (Lalou et al. 1985) in Saclay at the Centre d'Energie Atomique (CEA). Several short-life radioactive elements such as thorium (Th^{238} = 75,000 yr), uranium as well as lead (Pb) and radium (Ra) were used to determine the young age of the sulfide deposits. The method is based on the fact that no Th^{238} is present in the samples when they are first formed, therefore any measurable thorium (Th^{238}) must have been the result of the decay of radium (Ra^{238}). From the U/Th disequilibrium method (Lalou et al. 1985), it was calculated that the age of the gossans material and of the massive Fe-sulfide deposit found on the "Southeastern seamount" (near 12°43′N-103°50′W, EPR) is about 30,000 years old. This age is much younger than that of the basement on which the seamount was formed (130000 years old, as predicted by the magnetic anomalies). Other isotopic age-dating studies on samples from near 21°N on the EPR show that massive sulfide deposits located at less than 10 km from the axis are 2,000–5,000 years old. The oldest samples, consisting essentially of Fe-oxyhydroxides as products of sulfide chimney alteration, have an age of about 36,000 years old and were collected at a distance of 10 km from the ridge axis.

Hydrothermal activities are limited in time. For instance, in the Trans Atlantic Geothermal (TAG) area on the MAR, intermittent hydrothermal activities have lasted only during the past 40,000–50,000 years (Lalou et al. 1993).

Bryce canyon site. Further investigations along the spreading axis about 3 km south of the Chain Site took place during dive Cy82-31 on February 25, 1982, with

Jean-Louis Cheminée, Nivaggioli and Normand. They discovered another hydrothermal field near 13°47′N-103°56′W at 2629 m depth and they called this area the "Bryce Canyon" site since it reminded Jean-Louis of the landscape at a place with this name in Utah (USA). This hydrothermal site, like the "Chain site", was built at the western edge of a lava pond along a fissure oriented N165°. Active and dead hydrothermal mounds overlaid by chimneys reaching 15–20 m high were observed. One of them was a very active chimney with three orifices spewing black fluids at the foot of the tall mound. Temperature probes gave a value of 283 °C in the most active and taller edifice, while a reading of 330 °C was obtained on a smaller and thinner 10 cm wide chimney.

After the dive we had a briefing, where we were able to watch the videocassette and hear a description of the dive by Jean-Louis. We decided to make our last dive on the same site in order to collect samples of hydrothermal fluid and hydrothermal sediment precipitated on the surroundings near the edifice. It was decided that Marty, a fluid chemist from IPG, was going to make the last dive with the pilots Guy Sciarrone and J-M Nivaggioli. On February 26, 1982 *Cyana* was launched from the support ship at 9h15 and at 9h24 the submersible was diving. At 10h29, the divers reached the sea floor near the sample storage basket moored on the sea floor. It took only about 35 min between their landing and the first bottle sampling. The submarine left the bottom at 16h26 terminating the last dive in the area.

After the submersible was on board, the navigation team started to call back the three transponders that we had left on the bottom, which it took until one o'clock in the morning. Soon after we had caught all three transponders, we left the site on February 27, to transit towards Manzanillo, Mexico. The night of the 27–28th Vincent Renard, Echardour, Jean-Luc Charlou, Jean-Louis Cheminée, Francis Albarede and Jean Francois Minster took turns on watch. We arrived in Manzanillo in the afternoon of February 28. The cruise was over and everybody was pleased with the results. All of us were eager to return to our labs to make detailed analyses on all the samples we had collected.

After the Searise (1980), Clipperton (1981) and Cyatherm (1982) cruises, a fourth scientific expedition called "Geocyarise" took place. This project had 3 Legs on the EPR at 13°N and took place in 1984 with the submersible *Nautile* and the support ship N.O. *LE NADIR*. The first and third Legs of the cruise were devoted to performing morphological and structural observations and sampling. The second leg was dedicated to a geochemical approach for studying the water chemistry of the hydrothermal vents. I participated on leg 1 and was the chief scientist on Leg 3.

Geocyrise Leg 3 began in Manzanillo on February 18, 1984, at 8:00 AM and returned to Manzanillo on March 5. We left Manzanillo at a speed of about 9 kn but we had to slow down our speed in order to let the electric conducting cable out. This cable was important for the geo-electrical experiment that Felix Avedik, geophysicist, wanted to perform during the cruise. Indeed, we had to verify the state of the cable before we arrived on the site, and the only way to do it was to lay out the cable at sea and verify if the electrical connections for the experiment were operational. Felix and an electronic engineer, Perron, verified the damaged cable and fixed the portions that were defective. In the morning at 7:30, after working all night, Felix and Perron with the help of François, the boatswain, pulled back and rolled-up the cable on the winch and then we resumed our voyage at full speed to get to our target zone.

We arrived on the site of the dives in the evening of February 20th. The weather was sunny and the sea calm. The normal routine of finding the ridge axis and preparing the next morning's diving target started with the navigation team who deployed the three bottom transponders on the sea floor and then calibrated them with respect to geographic coordinates obtained by a Global Positioning System (GPS). This would take a few hours while the vessel circled around each transponder in order to have sufficient, coherent satellite fixes for the accuracy of positioning.

The main objective of the diving program was to perform a detailed geological survey on the four off-axial seamounts that had been discovered on the previous "Clipperton" cruise. A total of 11 dives were performed. Most of the dives were made on the off-axis seamounts and only three took place on the ridge axis. The three dives that took place in the axial graben of the ridge axis were to localize and evaluate the degree of hydrothermal alteration on the rocks during fluid circulation. It was found that the eastern wall of the axial graben was the site of intense hydrothermal alteration. Greenish gray faulted scarps with abundant talus were seen. Dead hydrothermal chimneys that had been partially fractured and dislocated were found associated with staircase-faulted scarps.

The off-axis seamounts visited are located between 4 and 18 km from the ridge axis. The two nearest seamounts to the ridge axis are located near 12°43′N (called the Southeastern seamount) and near 12°50′N (called the Northeastern seamount). A third seamount was further away, about 10 km to the west of the ridge axis, and was named the Clipperton seamount (Fig. 7.15).

On Leg 3 of this cruise, we had the privilege of having James Franklin, a specialist of ore deposits from the Geological Survey of Canada, on board. Jim had spent most of his life in the field exploring and finding ore deposits in Canada and other places in the World. His knowledge, as a land-trained mining geologist, was helpful in order to make some comparisons with land deposits. From the dives conducted on the Southeastern seamount, where we discovered one of the most extensive hydrothermal deposits, we learned that several conditions were necessary for such deposits to exist: (1) the path followed by the hydrothermal fluid to carry the metals in solution to the surface must be canalized through a narrow zone, (2) it helps if the area of hydrothermal circulation is associated with faults and fissures, and (3) it is also better if the hydrothermal activity is moderate and constant rather than very high and rapid.

Off-Axial Seamounts on the EPR at 12°50′N and 11°20′N

When spreading takes place at the ridge axis, it sometimes happens that the lithosphere, during its motion away from the axis, could also carry the magma away from the main ridge-centered upwelling zone due to lateral subsurface flows through preferential pathways. This displaced magma could form seamounts at some distance from the axis, which can be found on the flanks (<500 km) and/or away (>2,000 km) from ridge axis in deep-seated basins (>4,000 m deep). Off-axis seamounts could be formed during axial and/or off-axial volcanism or in deep-sea basins within the rigid plates during hotspot activities (See Chap. 9). Evidence of volcanic activity at a distance away from accreting plate boundary regions, which are not true intraplate areas, but are rather situated on the flank of ridge segments, are found along the flanks of the EPR. These off-axial structures differ from volcanic hotspot chains such as the Society, Austral, Pitcairn and the Hawaiian volcanoes by their morphology, structure and geological settings. They do not necessarily form lines of volcanoes, but rather they form isolated edifices as well as short (<100 km in length) small, ridge segments. They also often differ from intraplate structures by the presence of a crater and/or caldera at their summit, which are attributed to the collapse of their magmatic reservoirs when they have been carried away from the ridge axial-magma supply zone.

The distribution of volcanic edifices across the various segments of the East Pacific Rise is often related to axial discontinuities (Macdonald 1989). They are either found in the prolongations of small transforms and/or on overlapping spreading centers (OSC). They are interpreted as being the remains of volcanic edifices formed at the tips of accreting ridge segments and which are then moved away during spreading (Macdonald 1989). Isolated volcanic edifices, which are found at less than 50 km away from the ridge axis, are formed from axial magmatism lasting for several hundreds of thousand years. Many off-axis seamounts are generated near the ridge axis and have the same age as the crust (Batiza 1981; Hekinian et al. 1983) on which they are formed. The rate of seamount formation decreases with the age of the crust, due to the decrease of magmatic production (Batiza 1989).

Four off-axis volcanic structures discovered near 12°50′N and 11°26′N were investigated between 1980 and 1984 during the three cruises known as "Clipperton", "Geocyatherm" and "Geocyarise" with the surface ships N.O. *JEAN CHARCOT*, N.O. *LE SUROIT* and N.O. *NADIR* (Figs. 7.15, 7.22). Two off-axis volcanic seamounts are located on the eastern flank at 12°43′N (the Southeastern Smt) and near 12°50′N (Northeastern Smt) and the third is near 11°30′N on the western flank (Western Smt) of the EPR axis. Both eastern seamounts are less than 7 km from the axial graben (Figs. 7.15, 7.16). These two volcanic edifices are structurally similar; they are both relatively shallow in relief (<350 m in height), have about the same basal diameter (6–7 km in diameter), and both terminate with peaks composed essentially of hydrothermal deposits. No central craters were observed and their southern flanks are characterized by the occurrence of volcanic aprons resembling funnel shaped protuberances extending down the summit for a distance of about 1 km. This type of apron structure, also observed on the summits of some Northwest Pacific Basin seamounts, was interpreted as being due to flank eruptions injected laterally from shallow magmatic chambers (Vogt and Smoot 1984).

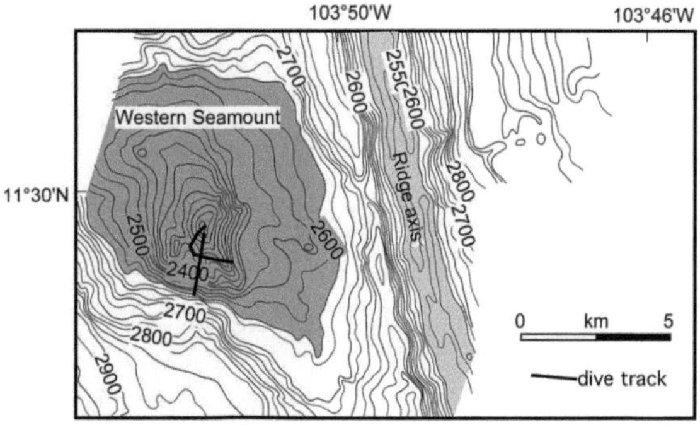

Fig. 7.22 Bathymetric map of the Western seamount located at 7 km from the adjacent spreading ridge axis near 11°30′N and 103°50′W

The Northeastern seamount near 12°50′N-102°53′W was visited during a dive in order to find hydrothermal deposits. The seamount had the same shape as the Southeastern seamount where extensive sulfide deposits had been found, so I thought that this would also be the case for this seamount. However, dive Cy84-28 made by Yves Fouquet, as the scientific observer, with Arnoux and Nivagioli as the pilot and engineer, failed to see any hydrothermal activity. This dive started at the foot of the seamount and never reached the top. Nevertheless, my intuition to dive on the Northeastern seamount was revealed to be correct because a couple years later, during another expedition conducted by Yves Fouquet, hydrothermal activity in the form of "black smokers" was indeed found on the top of this seamount.

Another of these seamounts, called the Southeastern Seamount, revealed the presence of large deposits of metallic sulfides. A dredge was lowered early in the morning on May 17, 1981, at 07h07 on the southern flank of the seamount that we named the Southeastern Seamount. The dredge recovered 700 k of massive metallic sulfides from a depth of 2,100 m. When the dredge was pulled up by the crane and emptied on the back deck, we had an unexpected surprise. Golden colored crystals of the pyrite rich material were shining on the deck of the ship, to the astonishment of the scientists and the crew. In addition to the massive Fe-rich sulfides, hydrated and oxidized iron and massive silica-rich hydrothermal material was lying in front of us (see Chap. 6). The presence of metallic sulfides on an off-axial seamount was an important discovery and opened a new horizon for looking at metal ore deposits on the sea floor.

Returning to the site the following year, in 1982, during the *Geocyatherm cruise* with the submersible Cyana, we had confirmation of our findings. However, it was only in 1984 during the Geocyarise Leg 3 that we were able to do a detailed survey, almost as a mining geologist would have done on land. Jim Franklin, a mining specialist, was the best qualified to conduct this survey and I was eager to have his input. I was anxious all day waiting for him to come back to the surface and tell us what he thought about the site. Finally, at 6 PM the chief of the submersible group gave the order for Cyana to start to come back up. It was only at about 7:30 PM that Jim Franklin could give us a briefing after getting out from the

Fig. 7.23 Massive sulfide deposit at dive site at 2562 m deep on the south flank of the Southeastern off-axial seamount near 12°43′N on the EPR. (Copyright IFREMER *Geocyatherm cruise, Cyana dive* Cy82-14)

Cyana and having had the traditional "initiation ceremony" that all new divers received with the water hose prior to a cocktail party. Finally, Jim was able to talk about the dive and he said he was surprised by the type of massive sulfides, which reminded him of some land-based deposits in Canada. Also, he was certain that these sulfides were deposited along a faulted elongated terrain on the southern flank of the seamount, just as he had often observed on land. So our discovery was an important landmark for the future exploration of massive and extensive deposits that had been formed underwater.

Thus, after four dives (Cy 82-14, Cy 84-25, Cy 84-23, Cy 84-31) and one deep-towed camera station (RAIE 3), we were able to have a much better idea about the geology of the Southeast seamount (Fig. 7.23). Our survey showed that the hydrothermal deposit was confined roughly within a triangular shaped surface with its apex situated at the top of the seamount and its base marked approximately by the 2,600 m bathymetric contour line. The outcrops of massive sulfides were associated with step faults, which are the pathways through which hydrothermal fluids circulate and precipitate sulfides at the foot of fault scarps. The sulfide fields lie along the fault scarps on top of pillow lava flows. These fault scarps located on the southern flank of the seamount have exposed up to about 20 m of massive sulfide deposits. Some of the structures recognized appear to be relics of ancient chimney mounds. One massive tubular fragment of a sulfide edifice of about 2 m in diameter was seen lying on the edge of a step-faulted structure bounded by a vertical wall about 10 m high. Powdery, yellowish-red gossan material consisting of the low temperature powdery iron–manganese oxyhydroxide was observed on the top of massive sulfides as a result of the sulfides' alteration. Opal and idiomorphic quartz were often associated with this Fe–Mn oxyhydroxide in the recovered dredges.

Although geo-electrical measurements are usually carried out in Ocean Drilling operations and in land-based surveys for the search of metalliferous deposits, this type of measurement was never done before 1984 on the ocean floor. During a

Cyana dive on the Southeastern seamount of the EPR near 12°42′N, geo-electric measurements were used to detect the presence of sulfide deposits. Felix Avedik from IFREMER and Tim Francis from the Institute of Oceanographic Sciences, in Wormely, Godalming, Surrey in the UK, carried out the experiment during the 1984 Geocyarise diving expedition. Both men were well known geophysicists interested in the seismicity of oceanic structures and they were very imaginative in developing techniques for exploration at sea. Tim and Felix contributed a great deal to the survey of the hydrothermal deposit in evaluating its extent and volume.

Felix Avedik was involved in measuring the distribution of the iron rich deposit by lowering a conducting cable from the ship to near the sea floor and then measuring the differences in electrical potential emanating from the oxidation of the sulfide deposits. The first and the second attempt to measure the electrical potential from the surface was conducted on the Southeastern Seamount at 2,415 m depth and in the axial graben at 2,650 m depth respectively. An anomaly of 6–8 mV was observed to occur deeper than 2,500 m. This was interpreted to be due to hydrothermal events taking place in the area. Such experiments were usually done at night in order that they would not interfere with the daytime diving operations.

The geo-electric measurements of Tim Francis were more constraining. The theory is that by sending an electrical impulse between connecting electrodes along a cable where a voltage reading could be observed on the potential electrodes, it would then be possible to measure the resistivity of the sea floor. This needed some preparation for passing the conducting cable inside the submersible's sphere. The cable was composed of three electrodes of silver chloride attached to a battery of 12 v through which the electric current flowed. A 50 m-long cable was made and attached to a small winch and put into the basket of Cyana. A 5 kg weight was put on the end of the cable and held by the claw of the manipulating arm. It was hoped that this configuration would help in the deployment of the cable on the sea floor. When reaching the sea floor, the deployment of the cable began by putting the weighted cable on the sea floor, which was essentially covered by a light-brown colored sulfide deposit. The pilot had to move the submersible backward and at the same time keep a watch on the cable. This was a risky task because if the Cyana was backing up on slope, the pilot was not fully in control of the engine. The unknown rugged environment was a major obstacle and he was afraid to inadvertently hit a rocky outcrop. The pilot had to maintain the tension on the cable by moving at constant speed in order to lay the cable in a straight-line. The experiment in laying the cable on the sea floor took half-an-hour and only a few minutes were needed to make the electrical measurement. The measurements and the readings of resistivity were made directly inside the sphere and the result was communicated directly to the surface during normal, radio communication.

When the first measurement was completed, *Cyana* rose about hundred meters upward with its vertical propeller in order to move the cable from the sea floor and transport it to the next site for further measurements. From previous dives, it was found that further to the South, the sea floor was flatter and less rugged and consisted of pillow lava. It would be important for comparison to have a geo-electrical measurement on a volcanic surface. When the Cyana reached the area where the contact zone between the sulfide deposit and lava flow is presumed to occur, then a second deployment of the cable was made by lowering the submersible to where it had visual contact with the sea floor. This contact between the sulfide and the lava flow was found thanks to the fully automated acoustic navigation network deployed on the seamount and relayed to the ship and to the Transit satellite navigation system. During two dives (Cy84-25 and Cy 84-31) conducted in the area of the seamount containing a sulfide deposit, four geo-electrical measurements were made. At the beginning of the last dive, after two measurements, it was decided to cut the cable and leave it on the bottom in order to free the submersible from the constraints of the cable and so it would be better able to continue exploring the sea floor.

The results of the geo-electrical measurements under the supervision of Tim Francis carried out by submersible (Cyana) at four different locations on the southern flank of the Southeastern Seamount helped in evaluating the extent of the deposit. The results were astonishing since, for the first time, we could determine the differences between the various types of formations through these measurements. Now the task was to compare the in situ readings with that obtained on the different samples collected from the area and measured in the laboratory. When matching the laboratory and the in situ results, these measurements gave an average apparent resistivity for the sulfide of 0.214 Ω/m while that of the pillow basalt was 0.552 Ω/m. In addition, it was noticed that the sulfide generates a self-potential. Comparing the resistivity measurements on a Fe-sulfide sample of known dimensions permitted us to estimate the thickness of the deposit to vary from about 8 m up to 15 m thick (Francis 1985).

Using previous submersible and deep-towed camera stations, plus in situ observations on the Southeastern seamount on the EPR near 12°43′N, enabled us to assess the size of the sulfide deposit. Assuming that sulfide deposits cover the entire surface of the explored triangle with a base of 900 m length, sides of 800 and 700 m respectively, and an average thickness of 5 m when assuming a sulfide density of 2.9 gr/cm^3, it is estimated that the deposit is less than 3.8 million tons. This is a conservative estimate since the entire surface within the triangle was not entirely explored and, as shown from the survey tracks, only a portion of the entire seamount was covered (Fig. 7.23). In terms of its size, this deposit is similar to some of the small exploitable deposits found in Cyprus of 15,000 to 15 million tones (Searle 1972) and if it were emerged, the seamount would be exploitable for ore.

The theory is that circular cones, because of their shape, are better suited to canalize ore enriched fluids than along large fissures where a dispersion of matter and energy is more prominent, such as on the spreading ridge axis. Indeed, when considering the geometry of a seamount (an inverted funnel-shaped structure), the concentration of energy sources is likely to be more efficient with centrally located conduits, than that of the accreting plate boundary regions where the energy source (heat release) to drive the system would seem to be more dissipated due to extensive linear fissuring. Although most off axis seamounts are not volcanically active, some structures that are located up to 50 km (i.e. at 21°N) from the ridge axis show hydrothermal activities and sulfide mineralization. Up to now, only a few cases of extensive hydrothermal fields have been reported in association with seamounts on the ocean floor. Some of these consist of massive sulfide deposits capped by a mixture of Fe-oxyhydroxide and manganese crusts. It has been shown that part of this capped material is the result of low temperature fluid precipitation giving rise to a laminated crust of Fe-oxide-hydroxide, and a green and reddish-yellow clay called nontronite.

The off-axis Clipperton Seamount is the tallest volcano in the area of study. It is about 1000 m high above the surrounding floor and located 18 km west of the EPR axis near 12°36.507′N-104°02.02′W. The seamount was found on May 15, 1981 during the CLIPPERTON cruise of the NO. *JEAN CHARCOT*, and its name was given after the cruise of the same name. This seamount, about 4 km in diameter, has a summit that culminates at 2,200 m depth. It is formed by four volcanic cones, which are located on top of a plateau that is shallower than 2,200 m. Three of the cones are less than 60 m tall and the major cone is about 150 m high above the plateau. This seamount is of particular interest because it was constructed about 200,000 yr ago near the ridge axis, and since then, it has been moved to its present location. It is not known whether or not the Clipperton seamount it is still volcanically active.

I decided to do a dive on this seamount and the goal was to make a survey by reaching the flank of the seamount near the summit and then explore the top to look for recent volcanism. The dive with *Cyana* (CY82-29) took place on February 23, 1982, with Nivaggioli as pilot and Normand as co-pilot. We started our dive at 8:42 and reached the sea floor 11:06 at 2182 meters depth. When looking through the porthole, I saw a few scattered pillow lavas that were partially buried by sediment. I had the track of the dive by my side and told Niva to take the course to point "B" towards one of the transponders that served as a reference for navigation. This location was near one of the small volcanic cones located on the southeast corner of the seamount and we had to travel for about 2 km in order to reach the top. The sediment blanket on top of what seemed to me to be a flat flow surprised me. The sediment cover was thin (millimeter scale) and thus the flow must have been relatively young (<1000 yr). At 11h52 we climbed on the flank of the volcanic cone and stopped to collect a sample in a sheet flow type of terrain in the form of "drapery". It was then, while the pilot had extended the starboard mechanical arm for grabbing the rock, that suddenly we noticed droplets of oil leaking through the fore-arm branch. Niva did not stop the sampling. He took the rock and put it into the basket. However the leak had now extended to the level of the claw. But this did not stop Niva, who had not yet reported the anomaly to the surface ship.

We continued our ascension towards the top of the volcanic cone following a heading of 345°N. Some organic material (gorgonians) covered the 5-m high scarp at 2,158 m depth. We reached the summit of the volcanic cone at 2,162 m depth at 12h11 and stopped again to sample. When the arm was pulled out we still could see the droplets of the oil leak. This meant that water must be coming into the system. There was no rush because this type of situation had happened before, so

Fig. 7.24 Domed shaped structure or tumulus feature observed in a crater during at 2,179 m depth on top of the Clipperton Seamount located near 13°30′N 18 km west of the EPR axis. The brittle crust has cracked during cooling of the lava flow. (Copyright IFREMER, *Geocyarise cruise 1984, Cyana dive Cy84-29*)

we continued our work as planned. After our second sampling station, we progressed in the same direction to NNE (course 347°) towards our target, which was on the second volcanic cone located on the northeast corner of the seamount. We went down slope from the first cone and reached a flattish sedimented bottom. At 13h07 we arrived on a domed shaped feature or *"tumulus"* about three meters high made up of sheet flows with a corrugated surface. We explored this structure, which was pounded into a depression. It was probably a crater. "Tumuli" such as I observed are formed during the flowing of hot lava underneath a chilled crust that swells upward (Fig. 7.24). The brittle crust could buckle as the underlying hot lava continues to flow, in order to accommodate the hot lava flow and thereby form a central crack along the length of the tumulus.

As we proceeded with our dive, and continued looking through the portholes, both Niva and I noticed that droplets of oil were continuing to flow from the leak. Now the leak was more substantial and Niva called the surface ship to inform Jean Roux, head of the *Cyana* team. Niva told the shipboard crew about our problem and waited a couple minutes for an answer. Jean said we should be vigilant about controlling the pressure gage of the hydraulic circuit. In the meantime, I was busy seeing that the volcanic terrain had changed its aspect. We were now on another large domed lava structure with a fissure on top that had busted open during the cooling of the lava flow. This "cooling crack" was large enough to show the internal part of the dome that now could be seen as being an empty cavity. The flowing magma inside the conduit had almost completely drained leaving only a cooling ledge about 29 cm thick. Further away on the same structure, the broken up chilled crust formed irregular *"clinker"* structures due to the occasional bursting out of the hot lava that had flowed underneath. In fact, we were now following a river of hardened lava that had flowed on top of the volcanic cone after its extrusion.

By now, the hydraulic leak had become serious because Niva turned towards me and said that we must leave the bottom at soon as possible. Then he took the phone and called the surface to ask permission to come up. It was 17h30 when Niva tried to switch the button on the control panel on the port side. Suddenly a red light flashed and the electro–magnetic device that releases the iron pellets from its container did not open. In order for Cyana to come up, it was necessary to "drop some of its weight" by releasing the iron pellets it carried. Because of the hydraulic leak in the system, we could not use the "electrically commanded valve release". Normand, the co-pilot sitting behind us, tried to release the weight with an emergency hand-manipulator, but that did not work either. The manual emergency device was stuck, so it did not move at all. There we were, sitting on the domed-shaped volcanic cone where I was trying to keep cool and continue my description as well as take some pictures of the same outcrop at different angles, while the submersible Cyana was slightly moving and turning by inertia. In the meantime, both the pilot and co-pilot were trying to switch on several devices as well as using the emergency handle for releasing the weights. In such a case, several options are possible for releasing weight and making the Cyana lighter. One is the release of the mechanical arm and eventually the rack with our collected samples. This would have been disastrous because we would have lost all our dive data.

Eventually, the weight pellets were finally released so we could float back up to the surface. We usually have a tendency to try and forget about any dangers that we have experienced on the bottom once we are back on board. These are the kind of events that are not discussed, except for the need of making any technical repairs.

Two years later in 1984, during the Geocyarise cruise, another dive was scheduled to take place on the same off-axis volcano, called the Clipperton Seamount. The purpose was to verify if any changes had occurred on the summit of the volcano. This time the best person to explore this type of edifice was Vladimir Nesteroff, a marine geologist specialized in sediment transport generated from volcanic edifices and from the slumping of continental margins. Vladimir's dive was programmed with George Arnoux, the pilot, and J-M. Nivaggioli, co-pilot, to explore the top of the seamount. The objective was to study the volcanic structure and its landscape. It was found that the summit of the seamount represented a collapsed crater with its western wall still intact rising 100 m from the summit. It was in this crater that we had observed the tumulus feature described in the previous dive. A concluding observation was that the collapsed nature of the crater is due to the retrieval of magma during eruption without the magma chamber being replenished.

The off-axis Western seamount, also found during the Clipperton cruise of 1981, is another off-axis volcano 670 m high above the surrounding sea floor and located at less than 6 km from the adjacent ridge axis at 11°30′n and 103°50′W (Fig. 7.22). In addition to having scoria-like flows (resembling "aa" in subaerial volcanism) and sheet flows, the Western seamount is characterized by hydrothermal Fe–SI oxyhydroxide edifices (Fig. 7.25a) and hyaloclastite slabs. Recent

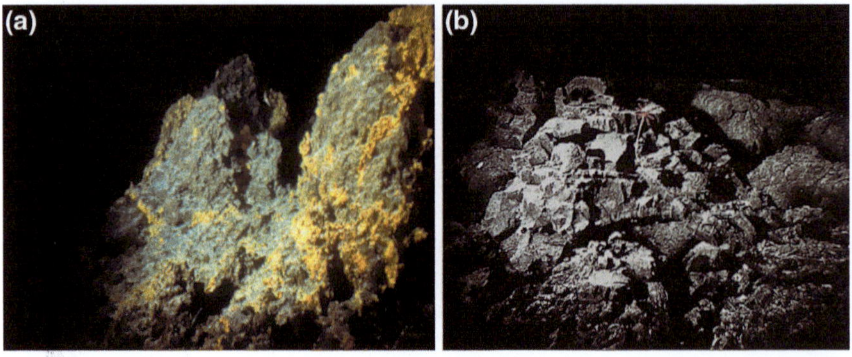

Fig. 7.25 (**a**, **b**) *Bottom* photographs (Copyright IFREMER, *Geocyarise cruise 1984, Cyana dive Cy84-22*) on the Western seamount of the East Pacific Rise showing a) a hydrothermal chimney made of iron oxyhydroxide at 2945 m depth and **b** a fragmented scoria-like lava flow at 2356 m depth

pillow lava flows, "spiny aa", and traces of a collapsed lava pond were observed near the summit of this volcano (Fig. 7.25b).

A series of *Alvin* dives on which I participated took place in March 1988 along the EPR further south near 10–12°N and were aimed at studying volcanic and hydrothermal activities, north of the Clipperton fracture zone. The scientific cruise of the R.V. *ATLANTIS II* was headed by Geoff Thomson and Bill Bryan. The captain of the vessel was Reuben Baker and the chief pilot and expedition leader of *Alvin* was Ralph Hollis.

The vessel left Manzanillo on March 4, 1988, at 22:00 and was on site on March 6 at 14h30. The task of the cruise was to test the model of a ridge centered magma supply associated with topographic volcanic highs along a spreading ridge axis between 10°15′N and 11°53′N. The section of this ridge system was chosen because previous work with bathymetry and bottom imaging in the area had showed the presence of recent volcanism less than 1,000 years old, super-imposed on a 7,000 year old sediment-laden pillow lava terrain.

Twenty dives were programmed during this expedition along the spreading ridge. I participated in three of them. The first one (*Alvin* dive # 1985) took place the next day after our arrival, with the two "veterans", Bill Bryan and me, as scientific observers and the pilot. On March 11, 1988, I also did a second dive (dive # 1989) with Susan Humphries from WHOI, a geochemist interested in hydrothermal deposits and a former student of Geoff Thompson. Susan is a very bright and imaginative scientist and she is also pleasant to communicate with.

On March 13, 1988, I made a third dive (# 1991) with Kathy Gillis, at the time a PhD student at WHOI. The dive took place on a segment of the EPR at 11°27.2′N in the deepest part of the axial graben at depths of 2520–2540 m. I must say that she seemed very much at ease for a person performing her first diving experience.

Both dives (# 1989 and # 1991) took place in the axial graben near 11°06′N and 11°29′N at 2,520–2,550 m depth. We landed on the first steps of the graben's western wall and went in a northeasterly direction until we crossed the ridge axis. On both sides of the axial graben, the walls are made up of pillow lava but as soon as we approached the axis of the ridge, flat lava (sheet flow) and collapsed lava ponds were observed and the flows were younger in appearance with a lack of interstitial sediment. This is an indication that we were approaching the most recently active volcanic zone. Several lava ponds with depressions (20–50 m deep holes about 15 m wide) were found aligned towards the strike of the ridge axis. We were attentive to look for the area with the freshest lava flow. We made a stop at the interior of a lava pond, to sample the most recent lava flow having a shiny appearance. After sampling, we continued the dive and rose above the depression leaving the lava pond. Progressing northward (11°07.40′N), the structures of the ridge axis changed and became more chaotic with rock debris and collapsed features due to extensive fracturing of the crust. Abundant fissures (dive # 1891) cutting through sheet flows were filled with fragmented pillow lava suggesting that we had come across a tectonically active area. Such type of activity is due to the separation of two plates. Indeed, we crossed the limit where the Mexican plate at the east is separated from the Pacific plate to the west.

It was during this expedition on the EPR axis that my colleague Bill Bryan from the Woods Hole Oceanographic Institution and I had the opportunity and the privilege of making the 2000th dive of ALVIN on March 22, 1988. Ralph Hollis was our pilot during the dive and we had the pleasure of opening a bottle of wine at 2300 m depth to inaugurate the event. The submarine was launched at 9 AM at 11°50′05′' N at about 2 km west of the ridge axis. The purpose of the dive was to document the eruptive cycle, the nature of the ridge-axis segmentation and the type of faulting. The Alvin landed on a pillow lava constructional hill at about 2,700 m depth. Before starting our exploration, Ralph, the pilot, took a plastic marker with two flags, French and American, and laid it on top of the pillow lava mound.

Several of these volcanic constructions were crossed on our way towards the ridge axis, however, no really fresh lava was noticed. The degree of lava freshness is noticeable to the experienced diver by the black and "oily" or "shiny" type of coloration on the surface of the flow. Although we did not really observe shiny lava flow, we saw partially buried fissures and "pillars" of lava ponds along the ridge axis. The fact that the delicate features such as pillars are still intact suggested that since the last eruptive events no tectonic event due to spreading had taken place in this particular portion of the ridge segment. The spatial and temporal relationship between the pillow lava and the younger sheet flow indicates that magmatic activity is episodic and discontinuous along the spreading axis. Thus magmatism follows a cyclical pattern, progressing from an initial eruption of pillow lava to the extrusion of sheet flows to a waning phase of pillow extrusion and finally to pelagic sedimentation. If there is a rapid rate of expansion, sheet flow will be the main type of flow extruded. If magma is depleted, or if the emplacement is slow, then pillow lava will dominate.

When the dive ended, we were pleased to return on board the *Atlantis II* and had the surprise to find that a reception was prepared. The captain handed the pilot a telex from Washington D.C. It was a message of congratulations from Massachusetts Senator *Ted Kennedy.*

Hess Deep: One Million Year Old EPR Crust Exposed

About one million years ago, the spreading of the EPR created a rifted structure bounded by a depression with depths of more than 4,000 m, called the Hess Deep, named after the famous scientist Harry Hess who did pioneering work in marine geology in the 50–60 s (see Chap. 1). This area is located in a region near 2°14′N–101°30′W which is North of the Galapagos Island in the Equatorial Pacific. It consists of oceanic lithosphere created on the axis of the fast spreading East Pacific Rise that was exposed during crustal uplift and faulting (Fig. 7.26a). Early works in the 70's by Russian scientists have reported that this EPR section has exposed the lower crust and upper mantle rocks of peridotite and gabbros. In 1988, another cruise called the *Hymas II* cruise organized by the University of Karlsruhe (Germany) on the R.V. *SONNE* (Leg 60) also revealed the existence of similar rocks in the trough of the Hess Deep.

This section of the EPR in the area of the Hess Deep represents a rifted structure about 8 km wide and 25 km long oriented East–West, parallel to a more recent spreading ridge called the Cocos-Nazca ridge. The Cocos-Nazca spreading ridge extends from the Middle America trench (Central-eastern Pacific) westward. This ridge system propagates towards the East Pacific Rise with a total spreading rate of about 6 cm a year. The Cocos-Nazca ridge propagates westward and becomes less significant and decreases in its volcanic activity. The northern wall of the Hess Deep is the shallowest part of the EPR ridge, also called an Intra-Rift, and is 2,900–3,000 m deep, while the deeper trough is found at 3,750–4,400 m depth. This Intra-Rift is an uplifted structure that has exposed deep-seated lower crust and upper mantle peridotite-gabbros complexes (Fig. 7.26b). The other northern wall bordering the Cocos-Nazca ridge propagator reaches shallower depths (2700 m) and consists essentially of volcanic rocks, dykes and basaltic flows.

The first diving cruise in the Hess Deep took place with the *Nautile* and its support ship *Le NADIR* in October–November 1988, during the Nazcopac (achronym for NAZca-COcos PACific) cruise.

Uncomfortable research vessels. For many years, until 1996, the N.O. *Le NADIR* was the main support ship for the *Nautile.* This vessel of 56 m long, 1,875 tons, commissioned in 1974, could carry 15 crewmembers and 25 scientists. When compared to other ocean going vessels, it already looked old at the time it was built (Fig. 7.27). The *NADIR* has a flat keel, rides low on the sea, there is limited space and it could be very uncomfortable to be on board. It rolls up to 35–40° during average weather conditions. The ship is not adapted to spend more than 30 days at sea. The accommodation quarters and water supply were

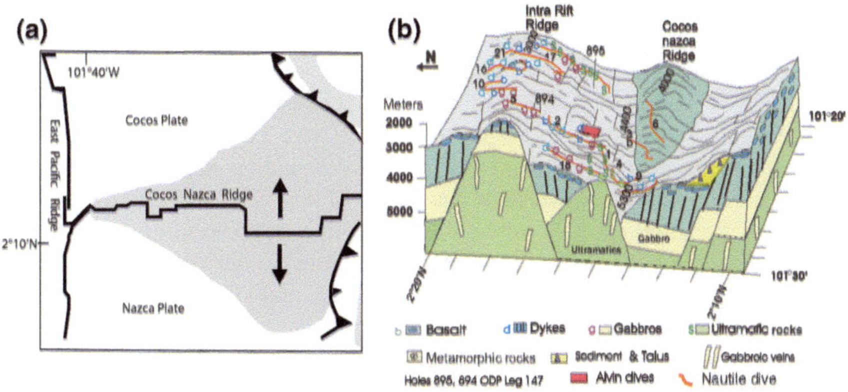

Fig. 7.26 General location map (**a**), and a Block 3D diagram (**b**) of the Hess Deep in the Equatorial Pacific shows the Cocos-Nazca spreading center separating the ancient, 1 million year old East Pacific Rise lithosphere-crust. The submersible dive tracks are indicated with the corresponding dive numbers 1–21. A drilling site 895 is shown

Fig. 7.27 The Research Vessel N.O.Nadir in the South Pacific Ocean (photo courtesy of IFREMER, by Victor Chaperon). The submersible Nautile hangs on the "A" frame that supports 12 tons. The N.O. Nadir with a weight of 1857 tons is 55.75 m long and 11.91 m wide. The vessel was built in 1974 and can carry 15 crewmembers and up to 25 scientists

limited on board. Indeed, the number of cabins for the scientific party did not exceed three and accommodated a total of eight scientists in each room. The engineers and technicians of the submersible, and the navigation team, occupied the rest of the cabin space. Later on, they increased the ship's capacity by adding two cabins, and they managed to stack five persons in each one. The scientific team on board was always the least favored for accommodations, but we scientists tolerated this since we are volunteers and usually spend less time at sea than the crew and technicians. When the ship was full, it was very uncomfortable to live on board the *Nadir*. I always marveled since this type of scientific vessel was in total contrast to the highly sophisticated submersible that it carried.

The Nazcopac cruise was proposed and conducted by Jean Francheteau. At this period Jean was still at IFREMER before moving to Paris to work at the *Institute de Physique du Globe* (IPG). The diving team included scientists from several disciplines who were familiar with the geology and the structure of the seafloor. The goal of the project was to obtain continuous observations and sample the exposed lower and upper mantle complex in the context of a fast (6 cm/year total rate) spreading ridge system. Twenty-one dives were conducted on the Cocos-Nazca spreading center and the Intra-rift ridge bounding the young spreading ridge. The Intra-Rift ridge represents the section of the EPR over hanging the Hess Deep. More than 400 samples were collected, with rock compositions that ranged from basalts, dolerites, cumulates (olivine gabbros), non-cumulates (isotropic gabbros, containing clinopyroxene, olivine and plagioclase), to ultramafics (dunites and peridotite).

For the first time in the Pacific Ocean, a nearly complete section of a relatively young crust and upper mantle representing a portion of a spreading ridge were observed and served as a model of comparison with other land-based ophiolite complexes (Fig. 7.26a, b).

After the *Nautile* dives, two other international cruises with the *Alvin* submersible investigated the upper crust (lavas and dykes) and uppermost gabbro formations exposed along the northern rift valley scarp crust (Karson et al. 1992) in the same general area. Furthermore, the Ocean Drilling Program (ODP) Leg 147 drilled gabbroic rocks at the crest of the intra-Rift ridge (Site 894 and 895) and has recovered peridotites intruded by numerous gabbroic veins, at site 395, located about 10 km southeast of Site 894 (Gillis et al. 1993).

I would like to mention in the following section a particular dive that took place in the Hess Deep in order to show how unpredictable underwater exploration could be even after thorough and detailed preparations.

The Intra-Rift Ridge Dive NZ08 in the Hess Deep

Dive NZ08 (standing for NaZcopac dive # 08 of the Nazcopac (NZ) cruise) took place on the *Nautile* with Triger, LeBigot and myself. We started at 9:00 o'clock in the morning of October 21, 1988. We reached the bottom at 11h08 and landed at the foot of south facing steep slope at 5391 m in the Hess Deep near 02°15′N-101°33′W in the equatorial Pacific. The depression forming the Hess Deep is covered by sand sized debris and pelagic sediment. As usual we looked for the first sample before starting to explore. As the submersible moved upward along the slope of the Intra-Rift north wall we encountered several faulted steps showing green and white colored peridotite outcrops (Fig. 7.26a, b, 7.28) crisscrossed by variable sized massive gabbroic dykes. Slumped rocks and debris of variable size going from cobble size to large blocks up to the size of the submersible were lying on the slope. This was not very reassuring because these large blocks of massive gabbros are unstable and could move at anytime during our ascension.

As we progressed upward, the large unstable blocks decreased and were replaced by smaller debris (coarse sand and boulder size) of basaltic and dolerite

Fig. 7.28 Hydrothermal deposits are seen partially covering a serpentinized peridotite outcrop. Photo taken at 4837 m depth in the Hess Deep (Copyright IFREMER, *Nazcopac Cruise 1988, Nautile dive NZ08*)

dykes. To our surprise, at 14h43 at a depth of 4,850 m, we encountered an outcrop of reddish-brown and greenish white hydrothermal deposit, with a small pit that might have been a collapsed hydrothermal vent that was now inactive (Fig. 7.26b). This was an unusual situation since, up to now, most hydrothermal products to be discovered were associated with basaltic basement along spreading ridges. Here we were for the first time looking at a polymetallic deposit sitting on an ultramafic peridotite basement. Serpentinized peridotite was the major constituent exposed along the north-facing slope of the intra-Rift segment. The volcanic rocks encountered were debris slumped above the 4300 m contour line.

I was very pleased with the dive and the sampling that we had conducted. I already had in mind the laboratory work that my colleagues and I could perform on such valuable, well-classified and recorded samples. Prior to putting the samples into the basket, the pilot and I made sure to photograph each individual piece of rock that had been taken. Also, I was careful to take pictures before and after sampling the outcrop for our records. It was important to identify the geological formation at each site.

However, when we reached the surface, there was an incident. The sea was unusually rough and the submersible rolled and pitched. I felt uncomfortable due to being tossed in the waves. The pilot was in radio communication with the ship and I heard the chief of the submersible group saying, "We can see you and we are heading toward the submersible". In the meantime the wind and swell increased and we were rolling and pitching in the waves, like a cork. Suddenly, the basket that was carrying our precious samples got loose from the front of the submersible and through the porthole I could see all the material that we had collected sink back down into the abysses. We had lost all our precious samples. I was discouraged and felt depressed. I guess even at sea it is wise to remember the old saying, "Don't count your chickens before they're hatched."

From the dive observations and sampling, the scientific team was able to draw a stratigraphic sequential exposure of the lithosphere in the Hess Deep. The dyke and lava are exposed on both sides of the Cocos-Nazca spreading center. It was also observed that the transition between dyke intrusions and erupted basalt is gradual and interlocking. This mixed transitional zone of dykes and erupted lava was observed throughout all the crossings and reaches up to 500 m in thickness. In the mixed zones, the lava complex is massive and sub-horizontal, probably fed from the underlying dykes (Francheteau et al. 1992). A significant observation was made during dive NZ20 where the observers saw the contact between the dykes and the gabbroic unit, which was partially buried by talus at 3046 m depth.

Combining the dive results, we were able to extrapolate the thickness of the dyke complex to be about 1,200 m. On the other hand, the thickness of the volcanic units, including basalt and mixed dyke-basalt overlying the dykes, is only 150–200 m. The overall sequence of the volcanic rocks lies on top of a gabbroic unit representing a fossil magma chamber exposed on the 1 million year old lithosphere of the ancient EPR. The lower and upper mantle material on which this gabbroic and volcanic complex lies consists of residual peridotite that had accumulated after a partial melting of the mantle giving rise to the volcanic rocks.

The composition of the rocks in relation to their lithological sequences revealed the presence of several types of gabbroic rocks going from cumulates (heavier, olivine enriched) up to lighter ferrogabbros and isotropic (clinopyroxene plagioclase enriched) gabbros (Hekinian et al. 1993; Coogan et al. 2002). This rock sequence suggested that crystal liquid fractionation took place in a confined environment. Thus, for the first time, direct field observation on the sea floor confirmed the existence of a fossil magma chamber on a fast spreading ridge system.

South East Pacific Rise

The ultrafast spreading segments in the South Pacific (SEPR, South East Pacific Rise) were first investigated by the N.O. *JEAN CHARCOT* (Searise cruise) and by the F.S. *SONNE*, both of which used multibeam echosounder bathymetric systems (also called SeaBeam and Hydrosweep) in 1979–1980 to map the sea floor. The scientific interest in studying the South East Pacific Rise was to find out how the ridge's tectonic activity and associated magmatism contributes to shape ultra-fast spreading ridge systems. The axial region of the SEPR is interrupted (<40 km, in length) by small ridge-discontinuities, which are small transforms or overlapping spreading centers and/or kinks (Macdonald 1989). These spreading ridge segments are located south of the Garrett transform (13°26′S) and extend for about 1000 km up to the northern boundary of the Easter Microplate (23°S) (Fig. 7.29). The full spreading rates of these various segments vary between 141 and 162 mm/yr (DeMets et al. 1990). The general axial depth of the ridge decreases gradually towards the south and varies between 2590 and 2850 m depths.

The ultrafast South Pacific Spreading Ridge (SEPR) segments are characterized by volcanic domes. They are the closest structural settings on the sea floor to what is observed on subaerial shield volcanoes. The bulging of the domed shaped ridge

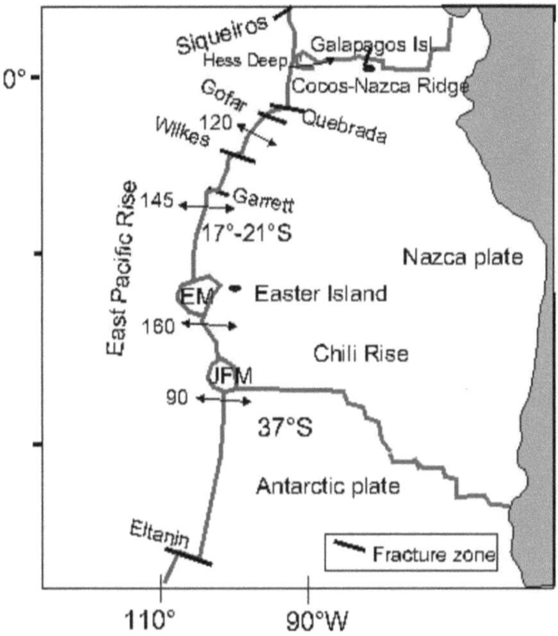

Fig. 7.29 General map of the South East Pacific Rise segments with the transform faults and fracture zones (heavy dark lines crossing the spreading ridge) and the rate of spreading (total rate with arrows) of the various segments. The Easter Microplate (*EM*) and the Juan de Fuca Microplate (*JFM*) are indicated

axis is dominated by the upwelling of a large amount of magmatic input by dyke and sill injection. These loci of magmatic upwelling could flow along the ridge strike or retract down into the lithosphere during the periods when there was a lack of magmatic upwelling. White et al. (2000) reported domed structures during the *Sojourn cruise*, of the R.V. *MELVILLE* in1996. On the larger domes of 2–3 km wide and up to 6 km long, lava lakes and sheet lava were observed at 17°52.5′-17°53.3′S on the south EPR. These lava domes are associated with third order short segments averaging about 10 km in length on ultra fast spreading ridges (full rate >140 km/million years) (White et al. 2000). Also in the Mantle Electromagnetic and Tomography (MELT) experiment area of the South East Pacific Rise (SEPR) near 17°26′S, the thickness of the crust is about 7 km at mid-segment and decreases to 5 km at the segment-ends (White et al. 2000).

The presence of "rift zones" and the distribution of small volcanic cones indicate that there is sub-crustal magma circulation along the spreading ridge axis.

A diving program in the south Pacific Ocean was initiated and conducted by scientists from the Geosciences department of IFREMER in collaboration with other French, German and US Institutions to explore, for the first time by submersible, the ultra-fast spreading ridge segments between 17°S and 21°S in January 1984 during the GEOCYARISE cruise Leg 1. This initiative of diving in the South Pacific Ocean opened the way for other surface ship and submersible expeditions carried out by international Institutions in Japan and the USA between 1993 and 2004. The Germans (Bäker et al. 1985), Americans (Scheider et al. 1993), French (Renard et al.1985; Hekinian et al. 1995; Auzende et al. 1994; Jollivet et al. 2004) and Japanese, have used their manned submersibles *Nautile*, *Alvin* and *Shankai* respectively for this endeavor.

The first submersible cruise along the ultrafast EPR segment took place with *Nautile* and its support ship N.O. *LE NADIR*. The cruise was organized and conducted by scientists from IFREMER as part of the project Geocyatherm (short for GEOlogy CYAna hydroTHERmalism). This cruise was divided into three legs that were programmed on the previously surveyed area of the EPR where traces of hydrothermal deposits were discovered. This was an ambitious French program to understand and evaluate the metalliferous ore deposits along an ultrafast spreading ridge segment.

The first Leg of Geocyarise Cruise headed by Vincent Renard as chief scientist left Easter Island in January 1984 heading northward following the East Pacific rise axis to the next port of call, which was Manzanillo in Mexico.

On board the N.O. Le Nadir there were scientists from French and US institutions. Dr. R. Ballard from WHOI and Emery Christoff from the "National Geographic Magazine" were invited in the framework of the French-American collaboration on projects FAMOUS and CYAMEX. In addition, Dr. Harold Backer from the German mining company "*Preussag*", who had previously explored and mapped the area with the R.V. *SONNE* in 1979 was invited. He was also the first to discover signs of hydrothermalism on this southern East Pacific Rise.

Field observations on ultrafast spreading centers in the south Pacific have shown that some ridge segments are at different stages of development and were also the sites of various types of tectonic, magmatic and hydrothermal activity. This was confirmed during the first Leg of the GEOCYATHERM cruise submersible dives on four different short segments of the spreading ridge. The ultrafast, short (3–15 km in length) spreading ridge segments were investigated during six submersible dives. Three main types of ridge segments having a different structural evolution were chosen for diving (Renard et al. 1985) (Fig. 7.30): (1) Two rifted segments with a relatively deep gashed (>50 m relief) graben at 21°26′S and 18°30′S), (2) moderately-rifted (20–40 m depth) segments at 20°S and (3) a non-rifted segment at 17°30′S. The purpose was to understand the mode of magmatic and hydrothermal fluid circulation in relation to the tectonics of spreading. Furthermore the exploration helped to understand the cyclicity of magmatic and tectonic events involved during ridge accretion as shown below.

Magmatic and Hydrothermal Cyclicity at Different Ridge Segments

Magmato-tectonic processes in the lithosphere are responsible for the different stages of evolution of the magmatic and hydrothermal activities on spreading ridge segments. The compositional and the morphological changes observed in the extrusion of lava are related to the dynamics of magma supply and residence time

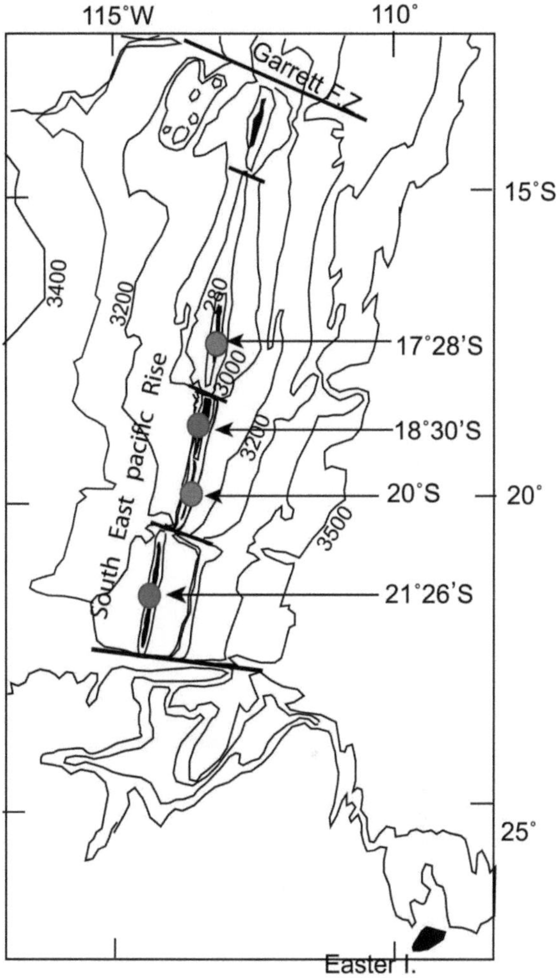

Fig. 7.30 The south East Pacific Rise segments showing the submersible diving sites (*grey circles*) between 17 and 21°S

in a magma reservoir. Also, the tectonics of spreading might control magmatic and hydrothermal activities at the ridge axis.

The evolution of spreading ridge axial volcanism at a fast spreading center in the South Pacific indicates that hydrothermalism is intimately related to the tectonic and magmatic processes.

A generalized overview of crustal extrusion and the temporal evolution of a ridge segment (over a few tens of thousand years) reconstructed from sea floor topography, lava morphology and the extent of hydrothermal activities was obtained from submersible observations on three ridge segments, which were at different stages of formation, and which are reported below (Fig. 7.31).

(1) *Magmatic stage* is marked by voluminous eruptions of mafic lava, forming narrow shallow-depth grabens with domed shape structures, and large magma

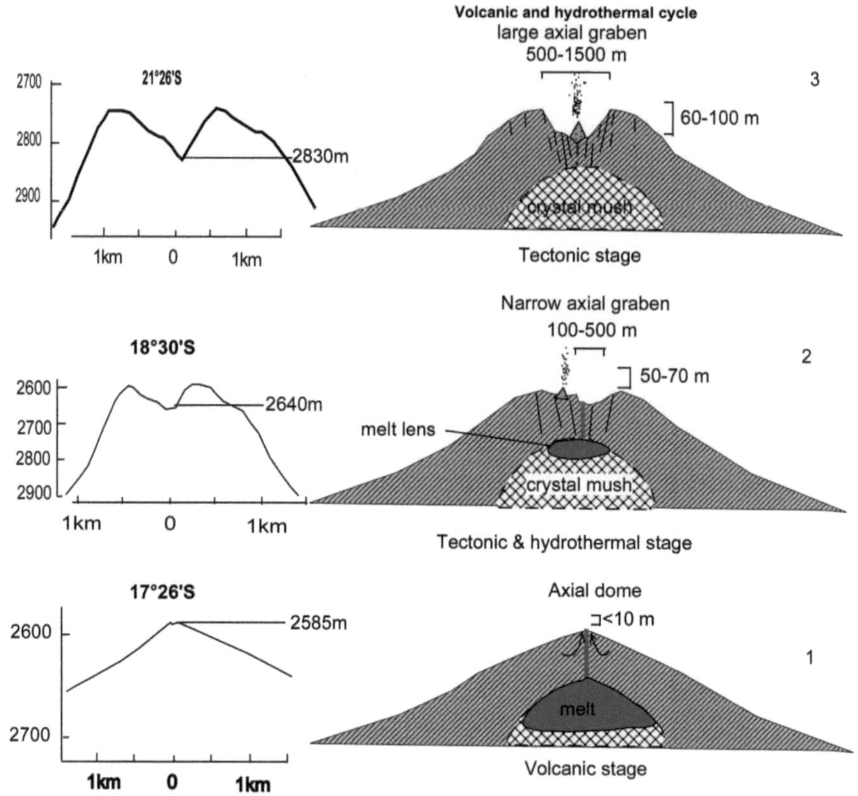

Fig. 7.31 Reading from the *bottom* to the *top* of this figure, we can see three stages of tectonic, magmatic and hydrothermal activity along ultrafast spreading ridge segments at 17°26′S-21°26′S. (*1*) Stage of intense volcanic activity shows a domed shaped structure. High thermal regime and lack of tectonic activity is accompanied by eruption of fluidal lava and diffuse hydrothermalism. (2) Stage of lithosphere cooling accompanied by tectonic activity. The formation of an axial graben due to magma retrieval and extrusion of sheet flow and pillow lava. Hydrothermal fluid circulation and metallic discharge influenced by cooling and fracturing of the lithosphere. (*3*) Stage of tectonic activity and lack of volcanism accompanied by intense hydrothermal events. When tectonic activity due to spreading increases, there could be a renewal of magma ascent underneath the axial graben, so stage 1 of intense volcanism will repeat itself

lenses located at relatively shallow depths (<1.5 km). High sheet flow/pillow ratios and episodic dyke injection occur. Very little hydrothermal activity except for shimmering of low temperature fluids with low metal content and gaseous phases (H_2S) takes place. The chimney constructions are in an immature stage; they consist essentially of thin veneers of Zn-sulfides, anhydrite and amorphous silica. For example, the domed shaped ridge segment similar to a Hawaiian "shield volcano" near 17°26′S at 2,585 m depth has been recently in activity for a length of about 800 m along the ridge axis.

(2) *Rift failure stage,* which is responsible for decreasing magmatic eruption and enhancing the cooling of the sub-axial magma lens, which leads to the eruption of more evolved lava and increases the crystal-liquid fractionation process in magma reservoirs. This stage is observed on the ridge segments at 18°30′S at 2640 m depth with a small graben 50 m wide and 60–100 m depth, showing tectonic activity with fissured and faulted crust associated with declining hydrothermal events. Also, it is shortly after each eruption that low temperature (<60 °C) fluids emanated from the interstices of fractured fresh lava flows. Prior to the main eruption, bacteria and tube worms live on the warm pillow crusts associated with diffuse venting.

(3) *Amagmatic stage* occurs when magma upwelling is replaced by tectonic activity responsible for faulting and fissuring with the formation of a larger graben 150–200 m deep and 1 km wide as shown on the 21°26′S EPR segment at 2,830 m depth. This ridge segment is in a stage of tectonic extension and characterized by intense hydrothermal activity with black smokers and several chimneys up to 20 m high. Both low (30–60 °C) and high temperature hydrothermal vents (>200 °C) are observed. This is due to the retrieval of deeper lying dykes marked by the decrease/or absence of magmatic input. Because of the large amount of crustal fissuring, the concentration of larger hydrothermal fields due to more intense fluid circulation is increased. Lobated flow and sporadic pillow lavas form the floor of the valley.

In summary, during crystal-liquid fractionation a zoned magma chamber will favor episodic volcanic eruptions underneath a spreading ridge. The domed structure and the fresh aspect of the lava flows suggested an intense magmatism. When this magmatic activity declines, hydrothermalism takes over and becomes the prominent event. The reduction of magmatic activity is related to a cooling of the magma reservoir underlying the ridge segment. The domed shape structure along the ridge will eventually collapse and form a linear graben or rift valley. It is at this stage that the faulting and fissuring increases and opens the path for hydrothermal fluid circulation giving rise to active hydrothermal chimneys. This will last until new magma upwelling occurs, to replenish the magma reservoir.

Another expedition called NAUDUR (anachronism for NAUtile Dive Ultrafast Ridge) took place between the 2nd and 30th of December 1993, in the same area of the East pacific Rise. The cruise was headed by Jean-Marie Auzende from the Geoscience department of IFREMER and included petrologists and geochemists (Daniel Bideau, Yves Fouquet, Rodey Batiza, John Sinton), structural geologists (Yves Lagabrielle, Jean-Marie Auzende, Marie-Helene Cormier, Valerie Ballu and Piera Spadea) and a biologist (Patrick Geistdoerfer). The invited American scientists were from the Universities of Hawaii (Batiza and Sinton) and from the University of California in Santa Barbara (Cormier) who had previously worked in the area. The other scientists were from different French institutions such as IFREMER in Brest (Auzende, Bideau, Fouquet), the University of Bretagne Occidentale in Brest (Lagabrielle), University of Paris 6 (Ballu), the Museum of

Natural History in Paris (Geistdoerfer) and the University of Udine in Italy (Spadea).

The purpose of the Naudur cruise was to make detailed observations on the various segments of the south EPR between 17°S and 19°S and do more extensive sampling of the hydrothermal fields discovered previously during the Geocyatherm cruise (Fouquet et al. 1996). Indeed, the Geocyatherm cruise spent only 7 days in the area. The major objective of Naudur's 30 day cruise was to find out the extent of fresh lava on the domed shaped regions, also called the "Hump" area by our American colleagues (Cormier et al. 1996), which marks a shallow regional depth with respect other segments of the EPR.

Between the two cruises (*Geocyatherm* Leg 1 and *Naudur*), a total of twenty-three dives were made and about 50 hydrothermal sites were visited. The main observation made during these cruises was that the domed shaped topographic highs of the different spreading segments visited are topped with fresh lava. New hydrothermal fields in their early stage of formation were also discovered.

Losing the deep-towed camera system in a lava pond. The name SCAMPI is an acronym in French for "Système de CAMera Ponctuel Intéractif". The device is an interactive deep towed camera system that was constructed at IFREMER in Brest and used during the Geocyatherm cruise (Leg 1) in 1984 on the South East Pacific Rise between 17°S and 21°S. This deep towed instrument consists of a metal frame of 2 × 0.8 x 1.2 m in size weighing about 700 k. Large probe lights, photographic and television cameras are attached on the sled (See Chap. 1). SCAMPI is towed by an armored coaxial cable, which sends real-time information to the surface and has a depth capability of 6,000 m. In order to be able to view and recognize the structures, it has to be at an altitude of less than 5–8 m above the sea floor. While we were surveying a rough terrain along the axis of the SEPR with deep towed instruments, an accident happened. Suddenly a blurry signal was received from the electric conducting wire and the total weight on the cable dropped down suddenly. It was clear that we had lost the engine. Indeed, when we rewound the cable, nothing was hanging off the end of it. It was quickly decided that we should try to recover the instrument using the submersible.

The submersible group was familiar with rescue operations conducted on other occasions. They had been involved in several recovery operations for airplane wreckage, missiles and other material lost at sea. In the morning, the group set a thick rope of about 10 cm in diameter aligned carefully in several rows along the back deck. The chief of the submersible group, Jean Roux, gave permission to go ahead and try to recover the lost engine with the Cyana. The total length of the rope had to be longer than the depth of the ocean floor, that is to say, more than 2600 m in length. The portion of the rope laid on the deck was rolled-up on a winch, which is used to pull material back up from the depths. The goal was to deploy the rope down to the seafloor with a transponder that the submarine would interrogate in order to find it. A hook was attached to the end of the rope, and then the *Cyana* was supposed to grasp it and carry to the site of the lost engine so it could be hooked to the rope. Then the ship would have to pull on the rope. This appeared to be quite simple, but it required a lot of extensive preparation, which lasted all morning.

During dive Cy84-08 taking place on the ridge axis at 17°26′S Vincent Renard (chief scientist) asked me to dive with Nivaggioli to the site where the deep towed engine had been lost in order to try to recover it. At 10:20 AM on January 12, 1984, we arrived on the seafloor and stopped at 2,554 m depth to calibrate the gyrocompass of the submersible for about 5 min prior to going to look for the lost vehicle. However, we noticed that we had a problem with Cyana, because we were light weighted so we had a problem to stay in neutral buoyancy, meaning that Cyana tended to leave the bottom and rise on its own. We had risen 29 m, and this ascension could be irreversible if we could not find a way to come down again. Niva (short for Nivaggioli) was now using the engine propeller to take us back down to the sea floor again, knowing that this would use precious energy from the battery. The seafloor was shiny with a freshly quenched glassy crust from a recent eruption.

There was no sign of the SCAMPI. It seemed that a new lava flow had covered all signs of faults or fissures on the older lava that was previously seen in 1982. Indeed, we could distinguish two generations of lava flows by their relative degree of freshness; the older flow had a dull looking appearance either underneath or at contact with the new shinier flow. We localized the transponder that gave us the signal of the location for the lost engine and we started to go towards the target (course 270°). At 10h49 we were on top of a small volcanic mound with a small collapsed pit at the top. Than suddenly we found ourselves at the edge of a lava pond at 2,561 m depth. A large lava pool had drained out, leaving a collapsed roof and chaotic fragments of rocks and broken pillars. We saw the broken and twisted wire attached to the lost engine, which was half planted in the wall of the lava pond about 17 m below us. The scenario was impressive; rock fragments were partially covering the lost vehicle, which stayed in equilibrium between the wall and a pillar (Fig. 7.32). The seafloor underneath was about 5 m below the engine.

Then at 10:58 we left the site and moved towards the rope that had been lowered down to the seafloor to serve in recovering the instrument. At 11h01 we saw the rope lying on the seafloor. With the mechanical arm we picked up the line and returned to the lava pond. In the meantime Niva said "We are still too light weighted and I should stop to get another rock in order to increase the weight of the submersible". We stopped and Niva took another sample and put it into the basket. When we reached the target and as soon as we were going to clamp the hook on the engine, suddenly we felt ourselves being pulled upward. Even after trying to use the vertical engine to reverse our ascent, it was impossible to come down again. At 11h09 the pilot, Niva, contacted the surface ship telling them that we had left the bottom and we were heading up. That was the end of the dive, which terminated with our failure to recover the deep-towed camera sledge. As far as I know, the deep-towed Scampi is still there at latitude 17°26′30″S and longitude of 113°12′30″W at 2,567 m depth.

In 1993, another French diving expedition conducted by Jean Marie Auzende (Naudur cruise) took place in the same area where the Scampi had been lost

Fig. 7.32 A deep-towed camera (SCAMPI) was lost during the *Geocyatherm Leg 1 cruise,* organized by *IFREMER* in January 1984 in a lava pond at the ridge axis near 17°26′30′S–113°12′30′W. Photograph of the lost SCAMPI was taken by the submersible Cyana at 2567 m depth. Note the basalt fragments that fell on *top* of the deep-towed frame when it hit the lava pond's wall. (Copyright IFREMER, *Geocytherm cruise 1984, Leg 1, Cyana dive Cy84-06*)

9 years earlier near 17°25′ S on the EPR axis. The exact satellite location where the engine was lost was communicated to the head of the diving team, Guy Sciarrone, but the divers could no longer find the SCAMPI. Instead, they saw a new fresh lava field that covered the depression inside the lava pond, where the engine had first been first lost and now was buried. Since then, several visual observations during diving on marked targets have enabled us to determine that the recurrence of volcanic eruptions on certain ridge segments could be on the order of weeks or months.

The Pacific–Antarctic Ridge at 36–42°S

The Pacific-Antarctic Ridge (PAR) looks like a continuance of the East Pacific Rise in the southern hemisphere. It is located between Australia and Antarctica and it continues up to the southern boundary of the Juan Fernandez Microplate near 35°S-111°W. It extends for about 5,000 km in length.

An area of the PAR was visited during three sea-going expeditions shared between French, German and American institutes in 1995, 2001 and 2004. Although, the Pacific-Antarctic Ridge between 35°S and 56°S was previously surveyed (Lonsdale 1994), it was in 1995 and 1997, during the French-German cooperative program, that the most detailed work was carried out with the R.V. *SONNE* (Leg 100, coordinated by the University of Kiel in Germany) and R.V. *L'ATALANTE*. In June-July 2001, the *FS SONNE* (Leg 157) with the newly installed SIMRAD EM120 multibeam bathymetry system returned to the area. The

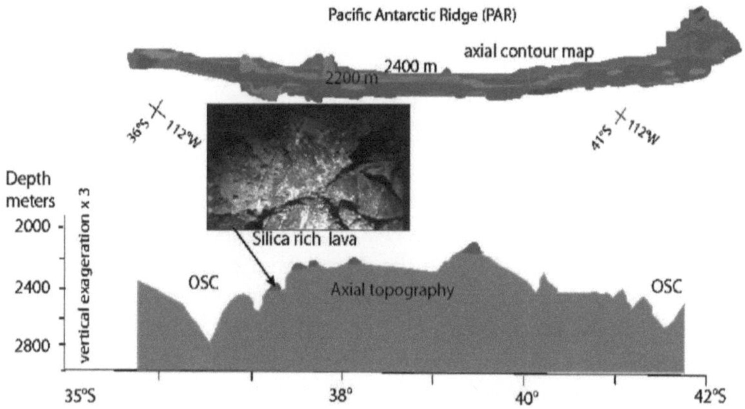

Fig. 7.33 The Pacific Antarctic Ridge (*PAR*) segment about 555 km long and 5–8 km wide at the 2500 m contour line is interrupted by two major discontinuities (Overlapping Spreading Centers, *OSC*). The sea floor photograph representing a silica enriched lava flow was taken with a deep-towed camera system (*OFOS*) during the SONNE 1995 cruise

goal of *SONNE*'s 157th Leg, which was also organized by the University of Kiel in Germany, was to investigate magmatic and hydrothermal processes at a spreading ridge axis influenced by the hotspot-controlled Foundation Volcanic Chain that reaches the PAR.

For evaluating the influence of the hotspot on the ridge axis, it was planned to survey the PAR axis extending for about 555 km in length, with a width between 5 and 8 km, at the 2,500 m contour line located south of the Juan Fernandez microplate between 36°55′S and 41°35′S in 2001 (Fig. 7.33). Morphological and tectonic reconnaissance and rock sampling was done in order to study the changes in the style of eruption in relation to the compositional variability of lavas and the distribution of hydrothermal activities in relation to their setting. Since the bathymetric coverage had shown details about ridge segment orientation and axial topography, this information was used to constrain the structural and magmatic evolution of the various spreading ridge segments. This approach helped to define the area influenced by the hotspot.

From the detailed bathymetry survey, we were able to further sub-divide the PAR segment into shorter en-echelon segments also called third order discontinuities. These small discontinuities or DEVALs (DEViation from Axial Linearity) occur at 38°10′S with no ridge offset but with a change in the general direction of the ridge strike from 012°N for the northern segment to 009°N for the southern segment. This area is marked by the presence of a plateau (2,250 m depth) with an off-centered seamount culminating at about 2,180 m depth and having domed shaped axial topography.

The various spreading ridge segments are characterized by elongated topographic highs with domed shaped features about 3 km long and 1 km wide that are gouged by summit depressions or grabens of less than 50 m deep between two elongated marginal highs. These domed-shaped axial highs are particularly prevalent south of latitude 38°10′S at 39°35′S and 39°45′S (dredge station SO63DS) where detailed studies were conducted. The northern segment (north of 38°10′S, at 37°50′S on the ridge) is of particular interest since it represents the area where the hotspot of the Foundation Volcanic Chain meets the spreading PAR axis. Deep-towed camera stations (OFOS and SCAMPI), dredging and TVG (television-grab) stations were conducted and revealed the presence of numerous active and inactive hydrothermal fields along the northern (north of 38°20′N) ridge segment. Recently erupted lava flow was detected with the deep towed camera station (OFOS) in a small graben located at the top of a dome shaped feature near dredge station 86DS (37°39′S) made by the R.V. *SONNE* in 1995 (Leg 100). This recent volcanism was revealed when comparing previous (N.O. *L'ATALANTE*) deep towed television (SCAMPI # 2) coverage carried out in February 1997 in the same graben, which showed hydrothermal deposits. Returning to the same spot during R.V. *SONNE's* Leg 157 in 2001, we could not find the same features. The hydrothermal material was buried by newly erupted silicic lava. Silicic lavas (andesite and dacite) associated with sulfide deposits were recovered from the axial grabens as well as from other topographic highs forming at least four volcanic domes along the various segments. This was a major discovery because silicic domed shaped features, although common in subaerial silicic volcanic events, were never seen to occur on the seafloor at depths of more than 2,000 m. The silica-rich lava consisting of andesite and dacite occurred at 37°39′S (SO100-86DS), at 37°50′S, at 38°05′S (station SO100-91DS), near 39°35′S and near 39°45′S. These silicic lavas are associated with basaltic tubes and pillow lavas, which have covered the lower slopes of the "domes" (see Chap. 5). The silicic "domes" are narrow, about 3 km long and 1 km wide, and are gouged by small summit depressions or grabens that are a few tens of meters deep, at depths of 2,100–2,200 m. Silicic lavas are highly viscous and they give rise to explosive volcanism.

Further to the south of latitude 39°35′S, the ridge segment shows a smooth topography with an almost constant depth of about 2,150 m and a continued and gentle slope deepening southwards, reaching about 2,300 m depth near 39°56′S (Fig. 7.33). This segment-end near 40°S consists of small discontinuities (DEVAL) marked by a depression (2,300 m) between the two incipient overlapping segments. Thus, in the vicinities of this DEVAL tip near 39°56′S, the general pattern of the axial orientation is changing more to a northerly (005°N) direction. At least four Non-Transform-Offsets (NTO) propagating ridge segments are observed along this nearly N–S oriented ridge segment. The segment-ends of the NTO occur at 39°56′S, 40°14′S, 40°37′S, 40°50′S and 41°13′ and show an increase of intervening depths from 2,300 m to 2,600 m between the segment-ends southwards, before reaching the southern overlapping spreading center (OSC) near 41°13′S. The segments migrate southwards and show a general deepening of the topography. Also, from the nature of the collected samples (degree of rock alteration and manganese hydroxide coating), it was inferred that the volcanic activity of the northern ridge segment had decreased concomitantly southwards towards the OSC. Thus the southern segment moving away from the influence of the Foundation Volcanic Chain hotspot could become a "normal spreading center".

The last area surveyed during Leg 157 of the R.V. *SONNE* was the southern OSC region near 41°35′S, which shows an overlap basin (depression) of about 2,800 m separating the two ridge segment propagators (Fig. 7.33). In this region, the eastern and the western ends of the two major spreading ridge segments north and south of the 41°35′S OSC are en echelon and curving towards each other with small offsets less than 2 km wide. A tall seamount about 2,210 m in height rises above the seafloor at 2,420 m deep, and it is about 200 m shallower than the surrounding topography of the ridge propagator's tips. The seamount is asymmetrical in shape. Since the base on the eastern (2,600 m) flank propagator is shallower than the one to the west (2,820 m), it is believed that magma was channeled from the eastern end (tip) of the OSC propagator to construct the seamount.

Interaction between off-axis hotspot and spreading ridge magmatism

As mentioned in another chapter (see Chap. 9), upwelling of magmatic flows could be either transported laterally at the lithosphere-asthenosphere boundary or come close to the sea floor underneath spreading ridge segments as well between diverging plates giving rise to hotspot types of activity.

Two main hypotheses are proposed to explain the volcanic activities giving rise to the off-axis volcanoes in the vicinities and on the axis of the PAR:

1. The first hypothesis implies that two separate mantle upwelling zones occur: one deep seated mantle plume underneath the lithosphere (>70 km depths) giving rise to the hotspot volcanoes, and a second shallower upwelling magmatic zone within the lithosphere (<70 km depth) giving rise to spreading ridge lava. In this case, the mixing between the two types of magma will give a hybrid composition during subcrustal magma channeling. This could explain the compositional variation of the lava erupted such as the alkali enriched lavas found at a distance from the spreading ridge and the MORB type of lava erupted on the ridge axis. An interaction between the mantle plume of the intraplate region and the spreading ridge system itself has been suggested by several authors including Zhang and Tanimoto (1992), Schilling (1991), Hekinian et al. (1999). For example, it was calculated (Ito and Lin 1995) that excess mantle temperature anomalies of 50–250 °C could influence hotspot-ridge interaction, up to a maximum distance of 500 km. Thus the Foundation Volcanic Chain magmatism, that gives rise to seamounts, has interacted with an ancient (20–23 Ma) Failed Rift Propagator of the Selkirk microplate and the more recent (<5 Ma) ridge type magmatism to produce the off-axis linear

volcanic ridges and volcanic cones observed on off-axis volcanoes forming the linear volcanic ridges (Fig. 7.33).

2. An alternative hypothesis implies that hotspot-ridge interaction is mainly the result of magma flow differentiation during the hot plume-derived melting process (Niu et al. 1999; Hekinian 2004). In this case, the plume that gives rise to hotspot volcanism is the main source for driving volcanic activity on the ridge axis. Niu et al. (1999) suggested that much of the material needed to create ocean crust and oceanic lithosphere must be supplied by lateral sub-lithospheric flow through the Low Velocity Zone (LVZ) at 670 km depth. When this mantle material reaches the base of the lithosphere (70–100 km depth), it flows laterally following the plate motion and the path of least resistance, such as faults and fissures. When the plates diverge during spreading at the ridge axis, it will drain the magma through the fissured and fractured lithosphere. The ridge-ward flowing and melting plume material is progressively depleted in its enriched components as it approaches the ridge.

According to this hypothesis, the commonly interpreted geochemical "mixing" between compositionally distinct enriched-plume material and a depleted "MORB source" in the context of a plume-ridge interaction is misleading. Hence, the apparent "mixing relationship" observed in geochemistry could in fact be the consequence of a partial melting of two different mantle sources.

References

Auzende J-M, Sinton J, Ballu V, Batiza R, Bideau D, Fouquet, Y, Cormier M-H, Geistdoerfer P, Lagabrielle Y, and Piera Spadea P (1994) NAUDUR explorers discover recent volcanic activity along the East Pacific Rise. Eos Trans Am Geophys Union 75(51):601, 604–605

Avedik F, Geli L (1987) Single-channel seismic reflection data from the East Pacific Rise axis between latitude 11°50′ and 12°54′N. Geology 15:857–860

Bäcker H, Lange H, Marchig V (1985) Hydrothermal activity and sulfide formation in axial valley of the East Pacific Rise crest between 18°S and 22°S. Earth, Planet Sci Lett 72:9–22

Ballard RD (1983) Exploring our living Planet. National Geographical Society Washington D.C ., p 366

Ballard RD, McConnell M (1995) Explorations: a life of underwater adventure. Hyperion, New York, p 407

Barberi F, Varet J (1970a) The Erta Ale volcanic range (Danakil depression northern Afar, Ethiopia). Bull Volcanol tome XXXIV, Fascicule 4:848–891

Batiza R (1981) Quantitative data on the abundance and distribution of seamounts in the Pacific Ocean, McDonnell Center for Space Sciences Publ. MCP 498, p 23

Batiza R (1989) Petrology and geochemistry of eastern Pacific spreading centers. In: Winterer EL, Hussong DM, Decker RW (eds) The eastern pacific ocean and Hawaii: the geology of North America. Geological Society America, Boulder, pp 145–160

Barclay AH, Toomey DR, Solomon SC (1998) Seismic structure and crustal magmatism at the Mid-Atlantic Ridge, 35°N. J Geophys Res 103:17827–17844

Barberi F, Varet J (1970b) The Erta Ale volcanic range. Bull Volcanol Fascicule 4:848–917

Bideau D, Hekinian R, Sichler B, Gracia E, Bollinger C, Constantin M, Guivel C (1998) Contrasting volcanic-tectonic processes during the past 2 Ma on the Mid-Atlantic Ridge: submersible mapping, petrological and magnetic results at lat. 34°52′N and 33°55′N. Marine Geophys Res 20:425–458

Choukroune P, Francheteau J, Hekinian R (1984) Tectonics of the East Pacific Rise near 13°N from field observations. Earth Planet Sci Lett 68:115–127

Cormier M-H, Scheirer DS, Macdonald KC (1996) Evolution of the East Pacific Rise at 17-19°S since 5 Ma: splitting of rapid migration of axial discontinuities. Marine Geophys Res 18:53–84

Coogan LA, Gillis KM, MacLeod CJ, Thompson JM, Hekinian R (2002) Petrology and geochemistry of lower crust formed at the East Pacific rise and exposed at Hess Deep: a synthesis and new results. Electronic J Earth Sci, Geochem Geophys Geosyst G3, Am Geophys Union 3(11):1–30

DeMets C, Gordon RG, Argus DF, Stein S (1990) Current plate motions. Geophys Int 101:425–478

Detrick RSP, Buhl P, Vern E, Mutter J, Orcutt J, Madsen J, Brocher T (1987) Multi-channel imaging of a crustal magma chamber along the East Pacific Rise. Nature 326:35–41

Detrick RS, Needham HD, Renard V (1995) Gravity anomalies and crustal thickness variation along the Mid-Atlantic ridge between 13°N and 14°N. J Gephys Res 100:3767–3787

Eissen JP, Bideau D, Juteau T (1981) Présence de basalts porphyritiques dans les zones de fractures de la dorsale Est-Pacifique. Compte Rendu Acad Sc Paris 293:61–66

Etienne J-L (2005) Clipperton- l'Atoll du bout du monde. Edition Septième continent, PAO Editions du Seuil Normandie ROTO Impression SAS Lonrain, France, p 208

Février M (1981) Hydrothermalism and mineralisations sur la dorsale Est Pacifique à 21°N. PhD Thesis, Université de Bretagne Occidentale, Troisième cycle : Structure et évolution géologique des Océans, p 270

Flueh ER, Michael J, Fisher A, Bialas J, Childs JR, Klaeschen D, Kukowski N, Parsons T, Scholl DW, Brink U, Trehu A-M, Vidal N (1998) New seismic images of the Cascadia subduction zone from Cruise SO108 ORWELL. Tectonophysics 293:69–84

Fouquet Y, Knott R, Cambon P, Fallick A, Rickard D, Desbruyeres D (1996) Formation of large sulfide mineral deposits along fast spreading ridges. Example from off-axial deposits at 12°43′N on the East Pacific Rise. Earth Planet Sci Lett 144(1):147–162

Francheteau J, Armijo R, Cheminée J-L, Hekinian R, Lonsdale P, Blum N (1992) Dyke complex of the East Pacific Rise exposed in the walls of Hess Deep and the structure of the upper oceanic crust. Earth Planet Sci Lett 111:109–121

Francheteau J, Ballard RD (1983) The East Pacific Rise near 21° N, 13°N and 20°S—inferences for along-strike variability of axial processes of the mid-ocean ridge. Earth Planet Sci Lett 64:93–116

Francis TJG (1985) Resistivity measurements of an ocean sulfide deposit from the submersible Cyana. Marine Geophys Res 7(3):419–437

Fujioka K, Massoth G, de Ronde C, Tanaka-Nakano A, Gendron J, Kinoshita M, Nishizawa A, Okano O, Paradis G, Row K, Shinohara M, Shitashima K, Tsunogai U, Ytow N (1995) Active hydrothermal systems of a super-fast spreading ridge, Southern East Pacific Rise (13°58′S to 18°26′S). InterRidge News 4:7–10

Gillis KM, Mével C, Allan J, Leg 147 Scientific Party (1993) Hydrothermal evolution of EPR lower crust exposed at the Hess Deep. EOS, Trans Am Geophys Union 74(Suppl.):653 (Abstract)

Gracia E, Parson M, Bideau D, Hekinian R (1998) Volcano-tectonic variability along segments of Mid-Atlantic ridge between the Azores platform and Hayes fracture zone: evidence from submersible and high-resolution sidescan sonar data. In: Mills R, Harrison K (eds) Modern ocean floor processes and the geological record, vol 148. Geological Society, Special Publications, London, pp 1–15

Haymon RM (1996) The response of ridge crest hydrothermal system to segmented, episodic magma supply. In: MacLeod CJ, Tyler PA, Walker CL (eds) Tectonic magmatic, hydrothermal and biological segmentation of Mid-Ocean Ridges, vol 118. Geological Society, Special Publications, London, pp 157–168

Heezen BC, Tharp M (1968) Physiographic diagram of the North Atlantic Ocean. Map Edition, Geological Society America, New York

Hekinian R (2004) Petrology of young Submarine Hotspot Lava: Composition and Classification. In: Hekinian R, Stoffers P, Cheminée J-L (eds.) Oceanic Hotspots. Springer-Verlag, Heidelberg (Germany), pp 431–459

Hekinian R, Fevrier M, Avedik F, Cambon P, Charlou JL, Needham H, Raillard J, Boulegue J, Merlivat L, Moinet A, Manganini S, Lange J (1983) East Pacific rise near 13°N : geology of new hydrothermal fields. Science 219:1321–1324

Hekinian R, Auzende JM, Francheteau J, Gente P, Ryan WBF, Kappel ES (1985) Offset spreading centers near 12°53′N on the EPR. Marine Geophys Res 7:359–377

Hekinian R, Thompson G, Bideau D (1989) Axial and off-axial heterogeneity of basaltic rocks from the East Pacific Rise at 12°38′N–12°51′ N and 11°26′N–11°30′N. J Geophys Res 94:17437–17463

Hekinian R, Bideau D, Francheteau J, Cheminée J-L, Armijo R, Lonsdale P, Blum N (1993) Petrology of the East Pacific-Rise crust and upper mantle exposed in the Hess Deep (Eastern Equatorial Pacific). J Geophys Res 98:B5 8069–8094

Hekinian R, Bideau D, Hébert R, Niu Y (1995) Magmatism in the Garrett transform fault. J Geophys Res 100:10163–10185

Hekinian R, Stoffers P, Ackermand D, Révillon S, Maia M, Bohn M (1999) Ridge-hotspot interaction: the Pacific-Antarctic Ridge and the Foundation seamounts. Marine Geol 160:199–223

Hekinian R, Pineau F, Shilobreeva S, Bideau D, Gracia E, Javoy M (2000) Deep sea explosive activity on the Mid-Atlantic ridge near 34°50′N: magma composition, vesicularity and volatile content. J Volcanol Geotherm Res 98:49–77

Hekinian R, Binard N (2008) Le Feu des abysses, Editions Quae, Planète Terre, RD 10 F-78026 Versailles, p 175

Ito GT, Lin J (1995) Mantle temperature anomalies along the present and paleo-axes of the Galapagos spreading center as inferred from gravity analyses. J Geophys Res 100(B3)L3733–3745

Jollivet D, Lallier FH, Barnay AS et al (2004) The BIOSPEEDO cruise: a new survey of hydrothermal vents along the South East Pacific Rise from 7°24′S to 21°33′S. 13. Inter Ridge News 13:20–26

Karson JA, Hurst SD, Lonsdale P (1992) Tectonic rotations of dikes in fast-spreading oceanic crust exposed near Hess Deep. Geology 20:685–688

Lalou C, Brichet E, Hekinian R (1985) Age dating of sulfide deposits from axial and off-axial structures on the East Pacific Rise near 12°50′N. Earth Planet Sci Lett 75:59–71

Lalou C, Reyss J-L, Brichet E, Arnold M, Thompson G, Fouquet Y, Rona PA (1993) New age data for Mid-Atlantic Ridge hydrothermal sites: TAG and Snake Pit Chronology revised. J Geol Soc London, Special Publ 148:1–15

LePichon X, Hekinian R, Francheteau J, Carré D (1975) Submersible study of lower continental slope abyssal plain contact. Deep sea Res 22:667–670

Lipman PW, Prostka HJ, Chistiansen HL (1971) Evolving subduction zones in the western United States, as interpreted from igneous Rocks. Sciences 148:821–825

Lonsdale P (1994) Geomorphology and structural segmentation of the crest of the Southern (Pacific-Antarctic) East Pacific Rise. J Geophys Res 99:4683–4702

Macdonald KC, Fox PJ, Perram LJ, Eisen MF, Haymon RF, Miller M, Carbotte SP, Cormier M-H, Shor AN (1988) A new view of the mid-ocean ridge from the behavior of ridge axis discontinuities. Nature 335:217–225

Macdonald KC (1989) Tectonic and magmatic processes on the East Pacific Rise. In: Winterer EL, Hussong DM, Decker RW (eds) The geology of North America: the Eastern Pacific Ocean and Hawaii, v. Geological Society of America, Boulder, pp 93–110

Macdonald KC, Fox PJ (1998) La dorsale Médo-océanique. Dossier hors serie Pour la Science, Scientific American, les humeurs de l'Océan, pp 28–34

Merlivat L, Pineau P, Javoy M (1987) Hydrothermal vent water at 13° N on the East Pacific Rise isotopic composition and gas concentration. Earth Planet Sci Lett 84:100–108

Needham H, Voisset M, Renard V, Bougault H (1992) Structure and volcanic features of the Mid-Atlantic Rift zone between 40°N and 33°N. EOS Trans AGU, fall Meeting, 552 (Abstract)

Niu Y, Collerson KD, Batiza R (1999) Origin of enriched type of Mid-Ocean Ridge basalt at ridges far from mantle plumes: the East Pacific Rise at 11°20′N. J Geophys Res 104(B4):7067–7089

Renard V, Hekinian R, Francheteau J, Ballard RD, Bäcker H (1985) Submersible observations at the axis of the ultra-fast-spreading East Pacific Rise (17°30′ to 21°30′S). Earth Planet Sci Lett 75:339–353

Scheider DS, Macdonald KC, Forsyth DW, Scientific party (1993) Wide spread volcanic and tectonic activity on the southern East Pacific Rise and its flanks. Ridge Newsletter 4(1–2):8–10

Schilling J-G (1991) Fluxes and excess temperature of mantle plumes inferred from their interaction with migrating mid –ocean ridges. Nature 352:397–403

Searle DL (1972) Mode of occurrence of the cupriferous pyrite deposits of Cyprus. vol 81. Institute of Mining and Metallurgy, Transaction, section B, London, pp B189–B197

Teall HJJ (1898) A phosphotorized trachyte from Clipperton Atoll (Northern Pacific). Quart J Geol Soc London 54:230

Vogt PR, Smoot NC (1984) The geisha Guyots multibeam bathymetry and morphometric interpretations. J Geophys Res 89:11085–11107

White SM, Macdonald KC, and Haymon R (2000) Basaltic domes, lava lakes, and volcanic segmentation on the southern East Pacific Rise. J Geophys Res 105:23519–23536

Wright DJ, Haymon RM, Fornari DJ (1995) Crustal fissuring and its relationship to magmatic and hydrothermal processes on the East Pacific Rise crest (9°12′–54′N). J Geophys Res 100:6097–6120

Zhang Y-S, Tanimoto T (1992) Ridge hotspots and their interaction as observed in seismic velocity maps. Nature 355:45–49

Chapter 8
Fracture Zones and Transform Faults

Abstract Fracture zones are major discontinuities disrupting the linearity of spreading ridge segments. They form depressions separating two regions of plates that slip in opposite directions. The motion of the plates has gashed the lithosphere down to more than 4000 m deep, often exposing deep mantle material. Fracture zones vary in length from a few tens to hundreds of kilometers and are narrow, no more than a few tens of kilometers wide. The seismically and tectonically active portion of a fracture zone is also called a "transform fault". The area of the Equatorial Atlantic between the African and South American continents located between 5°N and 5°S corresponds to the highest concentration of closely spaced fracture zones in the World's ocean. That is where the St. Peter and St. Paul's Rocks Fracture Zone crosses the entire Atlantic Ocean from the Amazon basin at the west through the coast of Liberia to the east. Most transforms are regions of intense tectonic events but are deprived of volcanic activity, except for a few transform faults associated with the fast spreading ridge systems in the Pacific Ocean where recent volcanic activity has been detected.

Introduction

Fracture zones are oriented perpendicularly to the spreading ridge axis and represent large topographic irregularities, which interrupt the spreading ridge system along its entire length. The fractures, extending across the ocean basins away from the ridge axes, were called *transcurrent faults* by Tuzo Wilson (1965) and are more commonly referred to as "fracture zones". The seismically active area of a fracture zone displacing the immediate vicinity of the spreading ridge segments is called a *transform fault.* The area that is more distant from the displaced spreading ridge segment and which is seismically inactive and older is called a *fracture zone (FZ).*

One particularity of a fracture zone is to reveal a sliced portion of the oceanic lithosphere, thereby creating steep and unstable scarps. Hence, one of the characteristics of a fracture zone is the abundance of broken rocks forming talus and

R. Hekinian, *Sea Floor Exploration*, Springer Oceanography,
DOI: 10.1007/978-3-319-03203-0_8,

avalanche debris, which accumulate along steep slopes. Because of their faulted nature, scientists believe that Fracture Zones are an "open window" into deeper layers of the oceanic lithosphere. The depths measured in a fracture zone are roughly equivalent to the amount of vertical displacement of the FZ area. Nevertheless, caution has to be exercised in interpreting any data acquired from the apparent exposition of so called deep-seated material, which is believed to represent in situ exposed mantle material such as peridotite.

Francheteau et al. (1976) had questioned the assumption that deep-seated layers (representing deep sections of the crust-lithosphere) are exposed during fracture. Several factors need to be taken into consideration. For example, staircase faulting along the wall of a fracture zone might never reach deep-seated material because of its small vertical relief displacement. Indeed, at least two major processes must be operating in order to expose Earth's deeper layers. One process takes place without appreciable vertical displacement but, instead, is due to plate extension producing the cracks that expose Earth's deeper sections. The other process involves the forceful injection of deep-seated material during the tectonic activity of faulting. For example, a diapir intrusion of altered (serpentinized) peridotite could be squeezed through faulted blocks like "tooth paste".

In order to expose deep portions of the lithosphere, it would be necessary to have a relatively large vertical motion of the fault plane surface. Most dives in the FAMOUS area near 37°N on the Mid-Atlantic Ridge or in the Garrett transform of the Pacific, have shown that the zones of faulting have small vertical scarps (<5 m throw). However, the presence of numerous, small fault scarps would hardly be able to expose layers from 2–3 km deep. In order to do this, it would be necessary to be in a special situation with large vertical and quasi-uninterrupted uplift, which is preferably associated with magma-starved regions (*see* Chap. 4). Areas that have exposed deep-seated layers of the lower-crust or upper-mantle are found in several locations such as the Vema FZ, the Saint Peter and St. Paul's Rocks FZ, or the Romanche and Terevaka (south Pacific) transform faults, which show uninterrupted sections of peridotite-gabbro-dyke complexes. These regions have undergone the special tectonic mechanism known as thrust-folding giving rise to uplifted slivers of transverse ridges (TR) and/or to the diapiric upwelling of buoyant material (due to buoyancy forces) (i.e. serpentinization) (*see* Chap. 4).

Other fracture zones explored in the Atlantic Ocean (Campsie et al. 1973; Fleming et al. 1970; Hekinian and Aumento 1973; Rusby 1993; Detrick et al. 1995; Lagabrielle et al. 1992a,b) have recovered basalt, plus mafic and ultramafic intrusive rocks. The fracture zones displacing slow spreading ridge segments of the Atlantic Ocean are the most privileged sites for exposing ultramafic rocks compared to the Pacific's fast spreading ridge segments.

Transform faults are generally not privileged sites for oceanic crust creation. Nevertheless, scientists have accumulated evidence that some transform faults are also sites of magmatic upwelling producing volcanic activity. This happens when the transform domain is influenced by the mechanism of extension during a change in direction of the moving plates. In other words, a transform fault could become the site of crustal extension, much like a spreading ridge segment, should it become the site of volcanic activity. In such cases, it will be called a "leaky" transform fault. Several transform domains have been recognized as being magmatically active. Examples of these types of "leaky transforms" are found in the equatorial Pacific at 3°–8°S, near 13°S (Garrett F.Z.) and 15°N (Siqueiros F.Z.) (Lonsdale 1978; Searle 1983; Fox and Gallo 1989; Hekinian et al. 1992).

Nomenclature of a fracture zone:

A seismically active transform fault consists of several tectonic domains: (1) The Transform Domain (TD) limits the width of the main transform fault bounded by two walls forming the displaced lithospheric plates between two spreading segments. (2) The active tectonic zone (ATZ) consists of the region where the most recent sheering and faulting is taking place in the Transform Domain, also called the transform fault zone (Arcyana 1975). *The ATZ runs parallel to the strike of the transform. (3) The Transform Valley (TV) is the floor of the transform bounded by the Active Tectonic Zone (ATZ). (4) The Transverse Ridge (TR), when present, is an uplifted structure that is shallower than the surrounding displaced lithospheric plate and forms the exterior walls of the transform fault. The TR is the result of large vertical uplift during compression caused by the tectonics of thrust faulting within the transform domain (TD).*

Garrett Transform Fault

The Garrett transform is the fastest-slipping transform in the World with a lateral slip rate of about 145 km/million years. The transform separates two segments of the ultra-fast EPR near 13°28′S–112°W by a distance of about 130 km (Fig. 8.1). Bathymetry and structural data have shown that the Garrett Transform is composed of a deep (>4000 m) *transform valley*, 24 km wide from wall to wall at the 3000 m contour interval. Within this transform valley, there is a *transform domain*, which is about 10 km wide and deeper than 3000 m. This is where most of the tectonic and volcanic activity takes place. Another peculiarity of the Garrett Transform Fault is that it is one of a few fracture zones where recent volcanic activity has been reported but never observed in its geological context prior to the Garrett cruise. Other transforms with evidence of recent volcanism were reported from the Siqueiros transform (15°N) (Natland 1989; Casey et al. 1991; Perfit et al. 1994) and from a region comprising several transforms between 2°S and 8°S along the East Pacific Rise (Lonsdale 1989).

After preliminary surface-ship exploration of the Garrett Transform was carried out by the international scientific community, the area was then ready to be explored in more detail by submersible. It was important for the success of our mission that we had been able to obtain the SeaBeam bathymetry coverage and

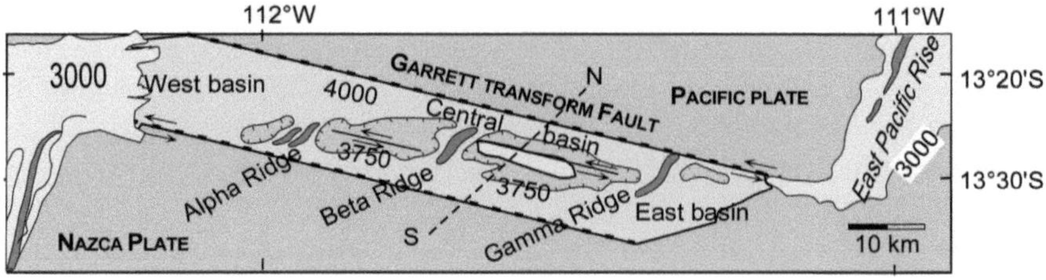

Fig. 8.1 Sketched map of Garrett transform fault. The transform fault is 130 km long, 15 km wide and has three oblique volcanic ridges called Alpha, Beta and Gamma bounded by four depressions forming the central basin, which marks the Active Tectonic Zone (ATZ). The trace of a north–south profile is shown on Fig. 8.2

some deep towed SEAMARK imagery from America. Jeff Fox from LGO (Lamont-Doherty Geological Observatory, New York) had kindly provided us with this information prior to the preparation of our cruise. In addition, other more localized topographic data on samples dredged from previous U.S and German cruises (FS *SONNE* in 1980, Geometep cruise) were made available to us. I was very thankful to my colleagues John Sinton (University of Hawaii) and Harold Bäcker from Germany with whom I had shared fruitful discussions on the geology of the area.

Rejean Hébert from the University of Laval in Canada was also one of my PhD students at UBO-IFREMER. In 1983, he started to study the samples obtained from Harold Bäcker and published his results in an international scientific journal prior to our cruise. The result of his work showed that the ultramafic rocks had undergone deformation that took place in the lower crust and upper mantle. His findings strongly suggested that we were in the presence of an uplifted lithosphere accompanied by plastic deformation revealing a layered rock structure. However, it appeared that the data obtained from surface ships was not going to be sufficient for understanding what was happening inside the deep trough lying at a depth of 3000–5000 m.

Some important questions remained to be answered such as: Is there evidence of recent volcanism? How much? Where and in what form is any recent volcanism located? Can we find evidence of predicted recent volcanic activity on the basis of the reconstruction of plate motion simply by using the data obtained during bathymetric coverage, or must we obtain in situ observations to confirm the presence of this volcanism?

I first submitted our project for the exploration of the Garrett Fracture Zone in 1987, however it took until 1991 before the cruise was finally accepted and scheduled by the IFREMER administration. At the time, I recall that this project of exploring a remote area of the south Pacific did not have the support of our scientific director in Paris. During a departmental meeting, I remember hearing him say: "*Do you really think this project Garrett is so important?*" My answer was obvious: "*I certainly do, and if I did not think it was important, I would not ask to go*".

Indeed this was to be the first diving cruise at more than 5000 m depth in a zone where we had evidence that deep-seated upper-mantle/lower-crust had been exposed. At that time, this exploration was the first of its kind to look at the Earth's upper mantle on an ultrafast spreading ridge segment of the Pacific Ocean.

Once the cruise was accepted and planned, we started our journey by plane to the town of Callao in Peru on January 2, 1991. Arriving at the airport, we discovered that a policeman, after stopping Rejean Hébert, had kept his wallet with all his money and there was nothing that he or we could do about this. Little adventures dealing with human behavior have been additional learning experiences in my life as a scientific explorer.

We continued our trip by flying to the small fishing port of Talara, located further north at about 4°34′N-81°16′W. While I was passing the embarking gate, I saw a sign at the airport indicating an air flight for Garrett, probably a town with the same name as that of the oceanic fracture zone that we were heading for.

The town of Talara is located in a barren region of the Andes and it looks like a western style "cowboy town". We had to spend the night in a shabby hotel. The ship was scheduled to arrive the next day at noon. It was very hot and we were dehydrated because we did not dare to drink any local water. Of course, there was beer but one can only drink so much, and beer does not always quench one's thirst.

From Talara, we were going to board the N.O. *NADIR*, the support ship of the submersible *Nautile*. The *NADIR* was coming from Fort de France (Martinique) via the Panama Canal. Captain Guénhaël Thébaud and Norbert Compagnot, head of the submersible group, were already on board. They moored the ship outside the harbor of Talara, and remained there only long enough to embark the scientific team.

The journey from the Peruvian port of Talara to the Garrett transform diving site lasted 7 days and covered a distance of 1867 nautical miles from Talara to the site of 13°20S-130°W. During this transit time, we prepared and coordinated our strategy with all the team members. I organized briefings with the Captain, the Chief Engineer, the head of the submersible team, and the scientific party. The meetings took place in the science container (a sort of box car), located on the starboard side of the deck, just above the engineering container for the *Nautile*. The access was somewhat risky because we had to cross a gangway that was unstable, so it swayed with the ship's motion. The science-container was also the place where the scientific team had to view their videocassettes and write up their daily reports after each dive. During the first "briefing meeting", we decided to use part of the evenings, after the dives, and part of the early mornings, before the submersible was launched, to do extra work with the deep towed television camera in order to have additional bottom observations to implement the diving data. Since all the dives had to take place during the day for security reasons, in order to optimize our time at sea, we all agreed to conduct additional exploration after the *Nautile* was safely secured on board.

As soon as we left Talara, the magnetometer (used for measuring the Earth's magnetic field) was put overboard, once we had left the Peruvian economic zone, but unfortunately, our magnetometer failed to work and was pulled back on board by hand because the winch was broken. The connecting wire had traces of shark teeth bites and was quite damaged. Luckily, the electronic engineer on board was able to fix the cable. During our transit, the weather was sunny and the sea was calm. We used our travel time to become better acquainted with the *NADIR* crew and the members of the submersible team.

I was pleased to see some familiar faces from the submersible group, who were involved in this expedition. For example, I saw Christian LeGuern, a mechanical engineer and frogman, who did the same work as Charlie LeGuenn. Both men were very handy and reliable, and could fix anything on board.

Before arriving on site we had another science meeting to discuss the operational orientation of the dives. The meeting was supposed to define our scientific priorities and the order of the scientists for the first series of dives. Usually, the chief scientist makes the first dive of a cruise. The reason for this is because the first dive after the maintenance of the submersible is considered as being a test-dive and things could go wrong. There are chances that the dive might be shorter due to technical problems. For this reason, I was the person who was expected to make the first dive on the presumed site of the exposed deep mantle peridotite.

Our first priority was to make detailed observations on the lower-crust/upper-mantle formation of peridotite in one of the areas of the fastest spreading regions in the World. Another important aspect was to determine if the transform fault was the site of recent volcanic activity. Therefore, another target was to dive on the flank and the top of one of the ellipsoidal shaped ridges crossing the transform valley. This dive was assigned to Daniel Bideau whose competence in volcanology was well known. The dive on the Median Ridge located at the center of the Garrett transform where we were expecting to find outcrops of mantle material (peridotite) was assigned to Mathilde Cannat and Rejean Hebert, who were respectively a structural geologist and a petrologist. The fourth dive was assigned to Jean Francheteau, a veteran of sea floor landscape and an expert in the geophysical study of submarine structures. Jean was to dive on the northern wall of the transform fault in order to determine the succession of tectonic events that had constructed the 3000 m high wall. The fifth dive was scheduled to explore the south wall of the Transform and its intersection with the transform valley in order to find out if there was any symmetry between the north and south walls as far as the type of lava formation was concerned. The second series of dives would be scheduled during the middle of the cruise, after the first results were obtained.

We arrived on site on January 12, 1991. The weather conditions were good and the crew and the navigators started to prepare the transponders for the following day's dive. The choice of the first dive's site was where dredge samples containing peridotites had first been taken by the FS *SONNE* in 1980. The samples had been dredged from the deepest (>5000 m depth) area of the transform, an area where tectonic activity was most prominent.

However, when arriving on the site we encountered a serious problem. The satellite Global Positioning System (GPS) was not available all the time so on the morning of the first dive, the bridge could not use this system. The reason for this was that the Gulf War of 1991 was just beginning, therefore the US military had limited the time for satellite use by civilians. The window of opportunity for having good satellite fixes was only about 6 h a day and it was most important to calibrate the transponder network set on the seafloor in order for the submersible to navigate. On the morning of the first dive, the transponders were in place, but the person responsible for the submersible's navigation did not have good fixes and he had to rely on a classical system of navigation, which was less accurate, with a distance discrepancy of up to 1 km on the sea floor.

In addition to the failure of the GPS, we were confronted with another technical difficulty. The depth recorder, which was a single channel echo sounder system called "EDO" that was supposed to record the depths, failed to work. This was an old-fashioned type of depth recorder that I remembered using during my first cruises back in 1964. This meant that for Dive 1 of the Garrett mission, we had to give up the possibility of having our depth recorded during our approach to the diving target. I asked the electronic engineer from IFREMER what was wrong with the sounder and if he could fix it before making a final decision to launch the submersible. His answer was that there was a piece that needed to be replaced, however he did not have any spare-parts on board for such old equipment.

As we approached the diving site, I knew we badly needed to have precise information about the depths, in order to follow the contour lines on a bathymetric chart. I wanted to start the dive in the deepest area of the transform valley in order to have the best chance of finding the exposed upper mantle peridotite. This was a critical time, and I was eager to find a solution to help us choose the right spot.

The officer on the bridge was also desperate; he was trying very hard to give us a precise estimate of the ship's position. Time was running out, and the submersible group, the crew and the navigation team were all waiting for a decision. Finally we managed to speculate on our location by positioning the ship using the last depth recorded, which had been given 1 h earlier by our defective sounder, and the last satellite fix. We just had to hope that in the meantime our vessel had not drifted too far from this point. We did not want to lose any more diving time by waiting for the calibration of the transponder network, so we decided to just go ahead and dive. We were going to land in one of the deepest areas of the transform fault's active tectonic zone.

The Transform Fault's Active Tectonic Zone

What characterizes a transform fault is the tectonic stress field that breaks the lithosphere and dismantles the geological formations. The strike-slip motion of the tectonic plates generates numerous fractures and fissures, while crushing the rocks into fragments and causing numerous landslides.

Where the dislocation and fragmentation of the geological formation is most intense is called the Active Tectonic Zone (ATZ). The ATZ is characterized by steep fault scarps (>10 m relief) and by fractures within semi-consolidated pelagic sediment. An abundance of fresh gullies and the presence of striations and fresh,

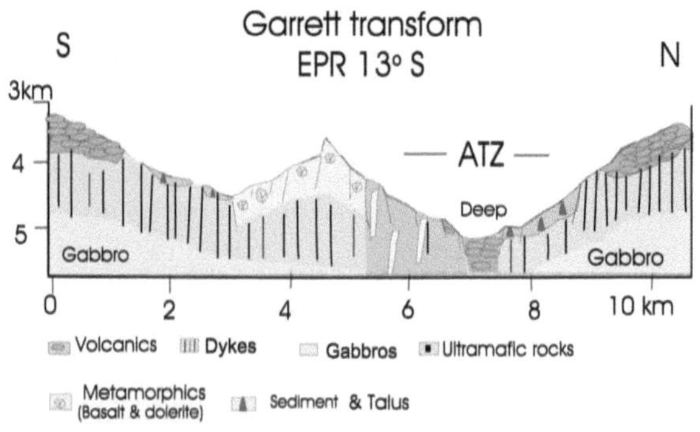

Fig. 8.2 North-South profile (See Fig. 8.1) representing a geological section of the Garrett transform fault at 13°30'S-112 W of the south East Pacific Rise (EPR). Most recent tectonic motion occurs in the Active Transform Zone (ATZ) deeper than 4000 m

Fig. 8.3 Landslide of heterogeneous basalt and dolerite fragments observed during a submersible dive at 3802 m depth on the northern wall of the Garrett Transform. (Copyright IFREMER, *Garrett cruise 1991, Nautile dive GN16*)

sediment-deprived outcrops are observed. Linear scarps and fresh talus indicates the trace of landslides (Figs. 8.2, 8.3, and 8.4). These criteria are used to define an active tectonic zone. The most dislocated area of the Garrett Transform occurs in the depths, at the base of the south facing transform wall, and on the Median Ridge's southern slope. The ATZ has a variable width varying from 200 m up to 2500 m and trends in an east–west direction (N 095°-97°). The ATZ intersects the tips of the EPR on both sides of the transform fault. The intersection is narrow (<200 m wide) and reveals mixed lava flows and fragmented volcanic rocks. The northern wall of the transform in the ATZ shows small scarps covering an area less than 600 m wide between 3650 m and 3900 m.

The Garrett Transform is associated with a shallow and elongated transverse ridge located between the two major transform fault walls. This structure is called the Median Ridge (Fig. 8.1). The ATZ on the other side, on the north-facing wall of the Median Ridge, was observed by submersible up to near the wall's summit. Below 3800 m depth, deformed and dislocated (mylonitized) peridotite underlies a brecciated and re-cemented dolerite-basalt formation. The brecciated volcanics are

Fig. 8.4 Landslide of large dolerite and basalt blocks on the slope of the Garrett Transform's northern wall at 3602 m depth observed from the submersible. Note a *polygonal block* of a dyke in the center of photo (Copyright IFREMER, *Garrett cruise 1991, Nautile dive GN16*)

chloritized and show veins of epidote, quartz and sulfides. Fault planes with subhorizontal slickenside (oriented N 95°-110°) were observed on the outcrops of the deformed breccia formation.

Diving in the Active Tectonic Zone

The first dive took place in the deepest area of the Garrett Transform's Active Tectonic Zone (ATZ). *Nautile* was put into the sea at 9 AM on January 13, 1991. Because of the uncertainty of the location, I was anxious about doing this first dive, along with the pilot, Max Dubois, and the co-pilot, Edmond Toussaint. The moment of truth was coming soon. We were not sure about whether or not we would be able to land the submersible on the spot where the R.V. SONNE had dredged and collected peridotite in 1980. During our communication with the bridge, we were informed that the current had put us 250 m away from our initial diving site in the direction 096°N. At 9:09 we started to go down for one of the longest journeys of my life. Indeed, if you hope to arrive at 5000 m depth, it will take about 3 h to go down. During our descent we listened to some music and had had one last briefing before arriving at our destination. When we landed I didn't know what I would see through the porthole. Would we be able to re-locate the dredge site of the FS *SONNE* and find the exposed peridotite?

At 11h17, at 4933 m depth, we arrived on the sea floor and I started to look out of the porthole. Max turned the *Nautile* around in a circle to check the surroundings. At the same time, our eyes were looking on the radar screen to see if there were any structures in the immediate vicinity or further away. We tested what we saw on the radar by looking outside through the portholes, first at a distance of 50 and 100 m, then further to a larger scale of 500 m. There were some

Fig. 8.5 Garrett transform at 13°30'S-111°W during the first submersible dive in the transform (Copyright IFREMER *Garrett cruise 1991 Nautile dive* GN01) at 4880–5056 m depth near the base of the Median Ridge in a sedimented depression in the Active Tectonic Zone (ATZ) (deep depression) between the Median Ridge and the north wall of the transform fault. We observed ultramafic outcrops in the deepest part of the ATZ (5056 m depth) with *white* and *dark colored* hydrothermal sediment and avalanche debris

echoes that were making noises and yellow spots on the radar screen. Apparently we were surrounded by hills. We decided to take our first sample and move on to try to find the deepest area.

The *Nautile* started to move down a gentle slope on the sedimented bottom. We reached the deep part of the bottom at 12:11 at a depth of 5052 m. We then arrived in the deepest part at 5056 m depths, in a small depression at the foot of a north-facing scarp. At 12h24 I had the submersible stop because there were small (<50 cm high) mounds of black colored Fe–Mn material at the intersection between the scarp and the sedimented depression. As the *Nautile* was turning, we noticed other black colored mounds aligned along the strike of the scarp, which was oriented East–West (N106° direction). Max stopped the *Nautile* and used the manipulating arm on the starboard side to take a small coring tube from the sample basket, which he then pushed into one of the mounds for sampling the soft looking material (sample GN01-02).

At 12h30, at a depth of 5054 m, I actually saw the tracks of a sled (dredge), which had been pulled across the deep ocean sediment. Hard as this was to believe, it appeared that we had managed to land our submersible in exactly the right place. The marks, which had remained on the sedimented sea floor, were another proof that we had come to precisely where we wanted to be. Seeing these traces of the dredge, which had been pulled on the sea floor 11 years earlier by the R.V. SONNE, was an incredible experience.

Rock debris, which had slumped down from the scarp and was composed of serpentinized peridotite, lay on the sedimented floor. When I saw the material recovered, I was absolutely sure that we were in the right area, the area previously sampled in 1980 by the R.V. SONNE. The first peridotite mixed with altered

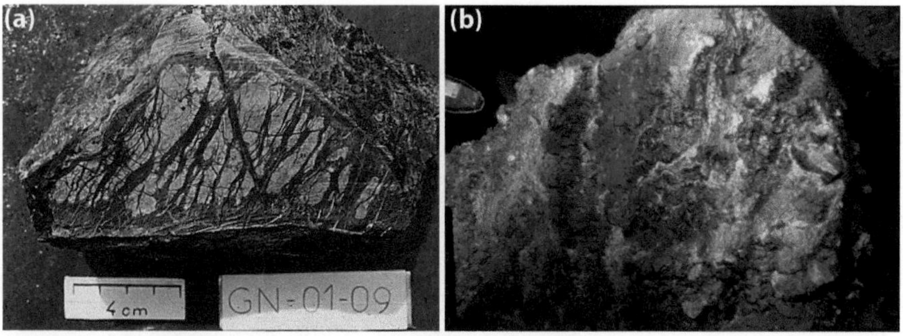

Fig. 8.6 **a** Sample of altered (mylonitized) peridotite with dark striations of granulated peridotite alternated with light colored talc bands collected near the foot of the Median Ridge at 4880 m depths. **b** Bottom photograph (copyright IFREMER, *Garrett cruise 1991, Nautile dive GN01-09*) of the exposed massive mylonitized peridotite that was sampled during the dive (see Fig. 8.6a)

dolerite and amphibolites was recovered in a talus pile between 12h40 to 12h51 at a depth of 5050–4985 m (Fig. 8.5). This debris was detached from outcrops that were probably located higher up along the slope.

We continued our ascension and encountered near vertical scarps at 4974 m depth. These scarps had parallel East–West oriented striations. It almost looked as if a gigantic paw had scraped the surface of the rock. After having crossed several scarps with talus at their feet, the lithology began to change as we continued on our way up. At 13h23 at 4804 m, we reached a 50° steep slope of basaltic rock fragments forming talus. The talus was a mixture of dolerite intrusive rocks and abundantly altered pillow lava (sample sites GN01-11, -12).

Near a depth of 4790 m, there was a break in the slope marking a transition between the peridotite and the volcanic rocks. This is where we observed a vertical scarp and a small depression marking a faulted area, probably due to the forceful intrusion of the peridotite body. At 13h38 we stopped our submersible on some pillow lava in order to eat a bite of lunch.

We had confirmation of the observed faulting when we traveled further up on the scarp. From the depths of 4804 m to about 4448 m, we encountered large talus piles that had slumped down from the major outcrops along the slope. A major tectonic event must have occurred in order to have fragmented and dislocated the entire scarp. The fragmented material consisted of pillow lava and lobated flow formations.

At 14h36, at a depth of 4598 m, we arrived at the foot of a vertical scarp. It was then that we suddenly heard a noise above the submarine. When looking through the porthole we saw rock debris rolling along the slope. We must have touched the outcrop with the submersible and caused an avalanche of rocks, which started to move along the wall. Max backed the *Nautile* away from the wall and continued to ascend while keeping visual contact with the bottom.

Fig. 8.7 Garrett Transform Fault (Copyright IFREMER Garrett cruise 1991, Nautile dive GN 17) shows an example of an isolated and almost detached block of tectonized volcanic breccia, about 2 m in diameter, standing on the southern flank of the Median Ridge at 4550 m depth

"*Was that an avalanche?!?*" Max asked. "*Yes,*" I said. "*I do not know far it extended but in any case we'd better move away from the scarp!*"

Max moved us back to a safe distance from the scarp's wall.

At 14h45 at 4550 m depth, we reached the narrow summit of the vertical scarp made up of prismatic, dislocated blocks of dolerite (dykes) and numerous fissures. The blocks were 3–5 m wide, partially detached and ready to fall (Fig. 8.7). It is likely that should such a large block have fallen on top of our small vessel, this could have put the *Nautile* in danger.

After a quick glance at the faulted and fissured structure, we started descending on the other side of the ridge into a small depression with a lot of rock debris. This was a chaotic environment indicating that a major tectonic event had taken place. Then, 10 min later at 4530 m, we encountered another vertical scarp with more slumped debris. Numerous fault blocks constituted the staircase type of terrain of a normal fault facing south. The lithology consisted of a lobated flow and flat lying ledges of volcanic rocks. At 14h55 at 4530 m depth, we reached the other side of the transform valley on the north-facing wall after passing over a depression made up of sediment and other, recent east–west oriented fissures reminding us that the area was still tectonically active. The other wall of the transform consisted of abundant rock debris of pillow lava mixed with dolerite. At that time, 15h56 at 4448 m, we received the order to come back up, so we left the bottom.

Thus, this first dive permitted us to set the ground for the other dives to be done in our study of the transverse domain, which was made up of a variety of formations that had been disturbed by recent tectonic events. It was important at this point to better define the extent of the activity that had dislocated the various formations and constructed the median ridge. One question that was important to answer was how the peridotite mantle was emplaced and what kind of influence such uplift could have had on the volcanics that were found? In order to answer such questions, we had to verify our findings with at least one other dive on the same formation, but at a distance from the location of the first dive. This would also permit us to verify the continuity of the outcrop exposing a portion of the Earth's mantle.

However before all doing this, we wanted to look closely at the data from the first dive. In the meantime, we continued to work on another important target that consisted in verifying if the oblique sigmoidal shaped ridges were really volcanic in origin (see the following section on the Garrett Transform Fault's volcanism).

Transform Fault Volcanism

Although transform volcanism was predicted by Fox and Gallo (1989) for the oblique ridges called the Alpha, Beta and Gamma ridges in the Garrett transform, proof of this volcanic activity had not been actually seen prior to our dives. In addition, it was found that transform volcanism does not only take place on the oblique ridges but it is also found in the Active Tectonic Zone (ATZ) within the Central basin (Figs. 8.1 and 8.2). All three oblique ridges were explored during dives GN02 (Daniel Bideau), GN07 (Hekinian) and GN12 (Hekinian) to determine the nature and the setting of the volcanism. These oblique ridges were not active at the same time, as was observed by the degree of alteration of the outcrops. The most eastern "Gamma ridge", which was also nearer to the northern spreading segment of the EPR, is the most active. The least active of the oblique ridges is the "Alpha ridge", located furthest to the west. It was also observed that the middle "Beta Ridge" consists of freshly erupted lava flows.

The first dive on the Oblique Ridge called "Gamma Ridge" took place on Monday, January 14, 1991, with Daniel Bideau as the observer and with Triger as pilot, and Kaioun as co-pilot. Despite some problems of navigation because the transponder network was displaced by about a mile, meaning the submersible could not get good fixes for its navigation, nevertheless, the decision was taken to hold a steady course without making any detours in order to avoid complications during the reconstruction of the dives.

At 11h06 the divers arrived on the sea floor at 3470 m depth on the east slope of the ridge and took their first sample. Then, they took a southwest direction (240°) up slope to reach the top of the ridge. On the summit, at 3170 m, they found very fresh lobated flows with an "oily" (shiny) surface appearance. Flat and ropy sheet flows similar to those found on recent EPR segments were sampled. The divers had accomplished their mission since they had found evidence of the most recent volcanism in the Garrett transform (Fig. 8.8).

The other two dives, which I made (GN07 and GN12), confirmed the findings for the other ridges (Alfa and Beta) (Fig. 8.2). These ridges differed from the Gamma Ridge because they showed mainly pillow lava and lobated flows forming small mounds and hills (<30 m high).

An example of the geology encountered on one of the oblique ridges was described during Dive GN12 on "Alpha". My dive's observations started 10h43 at 3755 m depth on the eastern flank of the oblique ridge where lobated flows were

Fig. 8.8 Garrett Transform fault's obliquely oriented volcanic "Gamma" Ridge at 3193 m depth shows freshly erupted ropy lava flows (Copyright IFREMER, *Garrett cruise 1991, Nautile dive GN 02*)

seen with abundant interstitial pelagic sediment. At 12h54 at 3543 m, we arrived on the summit of the ridge and encountered a fissure that was 3 m wide, oriented in the N 040° direction, which was also the general direction of the ridge axis. Then, we immediately changed our course and followed the shallowest depths. We crossed several faults and fissures during our exploration. At 3515 m, we descend into a fissure, which was 25 m deep, located between two vertical walls; at the fissure's bottom, we observed lava flows. Coming back to the top of this fissure, we continued our course to the southwest, and crossed several volcanic hills, which were gashed by fissures and faults with a relief of 2–5 m between 3496 and 3499 m depth. This area revealed large talus piles at the foot of the scarps. At 15h27 at 3440 m, while still on the summit of the ridge, we received the order to come up. Our dive time was already over.

Thus, the Alfa and Beta ridges showed numerous fissures less than 5 m in width cutting across the pillows and lobated flows. In addition, fault scarps with low relief (<2 m) were observed. This suggests that these ridges were in an intermediate stage of volcanism where fracturing and fissuring had replaced volcanism. Indeed, cyclic volcanic activity interrupted by the tectonic events of fissuring and faulting is a common phenomenon during magma upwelling processes.

The new discovery concerning volcanic activity in the Garrett Transform was the presence of small, deep-seated circular volcanic mounds that had been emplaced in the eastern Central Trough within the transform valley floor in the ATZ (Active Transform Zone), at the intersection of the Median Ridge's northern flank and in the deepest part of the Central Trough (4910 m). These are the sites where freshly erupted pillow lava flows (Figs. 8.1 and 8.9) were observed during several dives (GN01, -09, -13 and -15). This type of volcanism is related to the motion of the tectonic plates.

The origin of the volcanic rocks on the Garrett's transform ridges is due to the melting of a source enriched in Large Ion Lithophile Element (LILE), probably

Fig. 8.9 Pillow lava erupted in central basin within the Garrett transform fault's ATZ. Photograph taken at 4444 m depth. An octopus going upslope shows the location of pillow lava tubes. (Copyright IFREMER, *Garrett cruise 1991, Nautile dive GN 01*)

different from that giving rise to the East Pacific Rise basalt. Niu and Hekinian (1997) calculated that in order to derive the depleted type of lava that would have been responsible for the origin of samples collected from the Garrett, it would be necessary to have extensive (>20 up to 25 %) melting of residual peridotite (harzburgite type). Indeed, the harzburgites sampled in the Garrett transform fault are depleted in their more fusible mineral components such as clinopyroxene and enriched in their more refractory olivine-spinel minerals, similar to the samples that are found in the Terevaka and Hess Deep regions. This suggests that the extent of melting beneath the EPR is similar to, or even higher than, the melting beneath spreading ridges influenced by hotspots (e.g. The Azores hotspot in the Atlantic Ocean).

Median Ridge Ultramafic and Gabbroic Rocks

Within the Garrett transform domain, there is a topographic high about 1,000 m tall, 3.5 km wide and 10 km in length located between two oblique ridges (Beta and Gamma). This is called the *Median Ridge*. Since the first dive in the transform revealed the presence of peridotite at the base of the Median Ridge's northern slope, we decided to further investigate the whole structure.

Mathilde Cannat and Rejean Hébert, both experts in the study of ultramafic mantle rocks, were designated for the three dives (dives GN03, -13 and -14), which were scheduled on the eastern part of the Median Ridge (Figs. 8.1 and 8.2). The presence of peridotite mantle material was quickly confirmed during radio communication between the ship and the submersible. As soon as the submersible reached the slope, the divers observed steep scarps with a relief of 5–20 m. The scarps were extensively fractured, leading to in situ fragmentation and landslides along gullies, which suggested that the Median Ridge is an unstable structure that must have undergone vertical uplift in order to expose mantle material. In addition, the observations of more striated and schistose serpentinized peridotites and the presence of consolidated breccias indicated tectonic stress along the general east–west strike-slip motion of the Garrett Transform as well as in another direction (WNW-ESE). The dives were able to define the limits of the active tectonic zone (ATZ), which seemed at the time to be located on the northern slope of the Median Ridge. This ATZ is about 2.5 km wide to the East at 111°29′W and about 300 m wide at 111°30′W. The ATZ was responsible for the emplacement of the mantle peridotite on the northern slope of the Median Ridge. In addition, these three dives permitted us to observe the circulation of magmatic melt within exposed peridotite in the form of light colored lances and dykes, made up of plagioclase and clinopyroxene minerals. At a distance to the west from the ATZ (that is to say: to the west of dive GN15 at 13°28'S-111°34'W), the peridotite disappears and volcanic outcrops made up of basalt were observed on tectonically undisturbed terrain.

Ultramafic rocks consisting essentially of peridotite (harzburgite) and dunite plus mafic gabbroic rocks were only discovered on the north-facing slope of the Median Ridge. The first hand description of the collected samples, which was performed on board the ship, helped us classify the rocks and gave us valuable information on the composition and the mode of emplacement of the Median Ridge (Hébert et al. 1983).

Peridotite, consisting essentially of serpentinized harzburgite residues, is associated with dunites and troctolites which have textures inherited from crystal accumulation during crystal growth and metasomatism in magma-lenses. The dunites and troctolites are formed during magmatic injections into the peridotite. They show mineral lenses of clinopyroxene that are elongated to a faint foliation of serpentine in a subparalel direction. Some of the plagioclase-bearing dunites and troctolites show kink-bands and altered (saussuritized) plagioclase. The plagioclase-bearing dunites essentially contain olivine (69 %), plagioclase (28 %) and minor amounts of orthopyroxene, spinel and clinopyroxene. An important aspect of the residual harzburgite samples is that they have been affected by metasomatism (magmatic alteration) during melt circulation and their subsequent accumulation within fissures and voids. Thus, melt percolation has formed localized and discontinuous trails (<1 mm) and droplets of heterogeneous rocks (i.e. sample GN03-14) within the host peridotite as well as in larger veins and dykes (Figs. 8.10 and 8.11).

No continuous gabbroic formation, which might have suggested the presence of a magma chamber, was found. The only occurrence of gabbroic rocks is in metric sized veins, veinlets and dykes (<2 m wide) and in metric sized intrusive lenses within the serpentinized peridotites. These magmatic intrusions range from troctolite, olivine gabbro, gabbronorite and ferro-gabbros (ilmenite-bearing gabbros). From their texture and mineral composition two sets of gabbroic rocks were identified:

Fig. 8.10 A peridotite outcrop that was partially fragmented with veins of melt impregnation (*light color*) parallel to the slope was observed at 4680 m depths on the north-facing wall of the Median Ridge. (Copyright IFREMER, *Garrett cruise 1991, Nautile dive GN 03*)

Fig. 8.11 Dolerite dyke photographed at 4000 m deep on the southern flank of the uplifted Median Ridge in the Active Transform Zone of the Garrett Transform Fault. (Copyright IFREMER *Garrett cruise 1991, Nautile dive GN 06*)

(1) The gabbroic rocks consisting of olivine gabbro, normal gabbro, gabbronorite, and ilmenite-bearing gabbro. Metasomatic reactions taking place between the intrusive and the host peridotite will give rise to the transformation of the pyroxene into actinolite, probably at temperatures higher than 500 °C.
(2) The deformed and tectonized gabbroic rocks consist essentially of amphibolatized gabbros. Evidence of tectonic activity was observed on a microscopic scale. Lenses of amphibolatized and bent clinopyroxene relics, and albitized broken plagioclase set in a granulated matrix were observed in amphibolatized gabbros collected from the northern flank of the Median Ridge.

It was found that the minimum temperature of equilibrium for the Garrett FZ ultramafic–gabbro association was less than 1000 °C (Hébert et al. 1983). Such a temperature occurs in lower crust/mantle conditions, which are in agreement with the observed brittle/ductile deformation observed in the schistose or foliated, serpentinized peridotite-mylonite/cumulates (dunite), which must have occurred at the 750 °C isotherm, thought to represent the brittle/ductile transition (Phipps Morgan and Chen 1993).

In summary, our results were more than we had hoped for. Not only had we found that the oblique ridges (Alpha, Beta and Gamma) were the sites of recent volcanic activity, such as what could be observed on normal spreading ridge segments, but in addition, we were surprised to observe that other volcanic events had taken place along the deepest parts of the transform fault domain. This volcanism gave rise to smaller mounds (<20 m high), and to sporadic lava flows invading the older terrain observed in the deepest part of the transform troughs and walls (3500–5000 m), where upper-mantle and lower-crustal sequences were exposed. These observations indicated that conjugated extensional motions were responsible for the emplacement of mantle peridotite as well as for the eruption of lava.

Based on the detailed study of the geological setting in relation to the composition of the extrusive flows and the intrusive peridotite, it was possible to make comprehensive inferences on their origin. The presence of melt impregnation forming patches, dykelets and dykes throughout the exposed peridotite suggested the circulation of basaltic liquids. Some of theses impregnations forming dykelets are crossed by later examples of serpentinization (chrysotile), which suggested that there were late intrusions, which occurred after a main, plastic deformation had emplaced the ultramafic formation.

The example of the study performed in the Garrett Transform was followed and extended to other large fracture zones and to amagmatic spreading segments associated with slower spreading ridge systems. A comparative study of the Garrett FZ's volcanism and its associated deep-seated mantle rocks, when compared to other Pacific and Atlantic provinces, has been important in helping scientists create a global picture of the magmatic processes taking place in the World's oceans. Several scientific papers have been written retracing the implications of the peridotite composition and its emplacement in the Garrett transform (Edwards et al. 1996). Our cruise ended on January 31, 1991, with the last dive (GN19), which carried Jean Francheteau, Dubois and Edmond, to observe the western intersection zone of the transform fault and the western branch of the EPR. When they returned from their dive, we all enjoyed a glass of champagne and a barbecue dinner served on the main deck of the ship to mark the end of a very successful cruise.

Terevaka Transform Fault

After the Pito cruise of 1993, which was aimed at studying the Easter Microplate, I had hoped to publish a paper on the Terevaka transform fault. However, up to now, the results have not yet been published and that is why I want to dedicate this

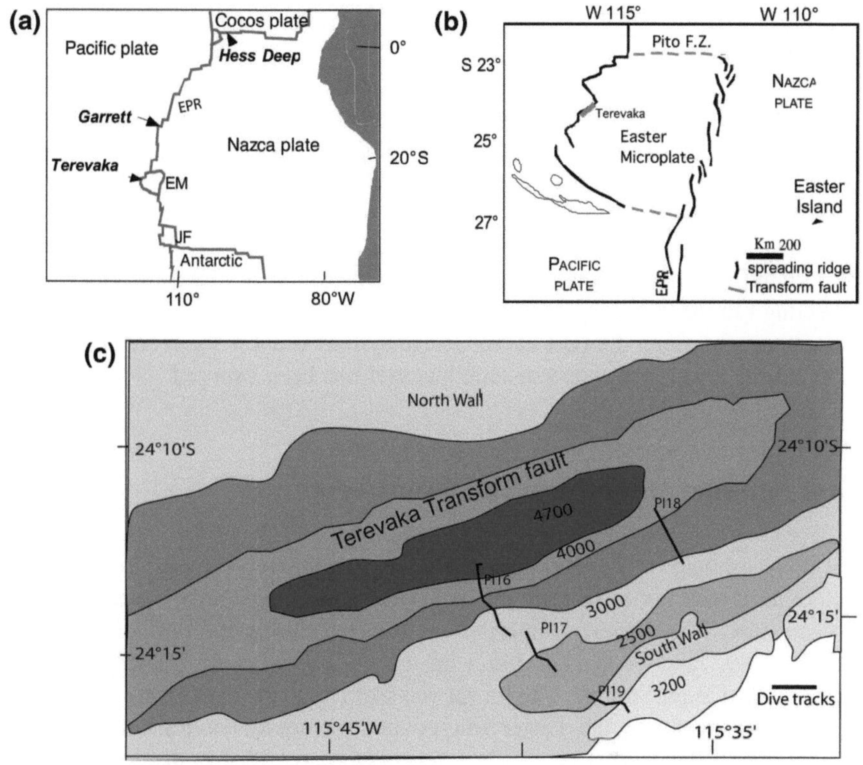

Fig. 8.12 Terevaka transform fault and geological setting, south Pacific. **a** General map of the East Pacific Rise (EPR). **b** The Easter Microplate (EM) and spreading ridge segments are shown. **c** Bathymetry of the Terevaka transform and the location of the submersible dives (*dark lines*) are shown

chapter to my colleague and good friend, Jean Francheteau, who died after a long illness on July 21, 2010. Jean was the chief scientist who conducted the diving expedition on the Easter microplate, including the work done on the Terevaka transform fault.

The Terevaka transform fault, located at 24°13′S and 115°41′W, is a left-lateral transform disrupting two segments of the western spreading ridges within the Easter Microplate (Fig. 8.12a–c). A microplate is a small, rigid plate about 300–600 km in diameter that is trapped between larger (thousands of square kilometers-sized) plates. A microplate is formed during the motion of major plate reorganization. It is characterized by two spreading ridge systems that propagate in opposite directions within the microplate, and the microplate itself is bordered by transform faults and/or fracture zones. For example, the Easter microplate is about 400 × 400 km and is located between 22°40′S and 27°S in a region where the ridge system is spreading at a high rate (total rate of 16 cm/yr) (Naar and Hey 1989; Francheteau et al. 1988;

DeMets et al. 1990). This microplate includes a major transform fault called the Terevaka Transform.

The Terevaka transform fault is made up of two parallel valleys separated by a horst, which is a type of ridge, having a total offset of about 90 km. A total offset of 100 km and a maximum depth of about 5000 m separates the EPR spreading center to the north from the West Rift spreading ridge propagator of the Easter Microplate. The spreading rates of the ridge segments, which are located to the north and to the south of the transform, were estimated to be about 118 mm/yr (118 km/million years) (Naar and Hey 1991). One of the reasons for exploring and collecting samples from the Terevaka transform is because it represents one of the deepest troughs along the East Pacific Ridge system and it is therefore likely to be a site where lower-crust/upper-mantle material has been exposed.

First Samples from Terevaka Transform

The first sampling of the transform, which suggested the presence of deep-seated peridotite, turned out to be very influential for conducting further exploration on this structure. The earliest samples were collected during the "*Midplate*" expedition (SO80, June–September 1992) on the FS *SONNE* with Captain Kull and chief scientist Peter Stoffers. The transit was from Valparaiso to Easter Island, and while transiting along the Easter Microplate, we could not resist the urge to explore one of the deepest depressions, reaching nearly 5900 m depth on the East Pacific Rise, called the Pito Deep. The Pito Deep and the Easter Microplate are associated with a Transform Fault called Terevaka.

The transform was completely mapped using the multibeam Hydrosweep system of the FS *SONNE*. As soon as the map was ready, we chose a dredging target in order to have an idea about the composition of the crust in this particular region. We were all secretly hoping to be able to sample some deep-seated mantle peridotite. This would permit us to add new information to our data set, previously obtained in the Garrett FZ and from the Hess Deep. If we could find more peridotite here, this would enable us to speculate on the emplacement and composition of the lower crust-lithosphere boundary of the Pacific Ocean. The mapping was done on July 21, 1992, and the dredging operation started the next day.

The first dredges took a long time, more than 6 h, because there were no bites. Colin Devey was in charge of the operation and he was frustrated because of the time he was spending, with no sign of success. Finally, at around 10 PM we decided to bring the dredge up and we all hoped that the dredge would contain a few rock specimens. Peter Stoffers and I left the scientific control room and went to bed leaving Colin in charge and hoping for the best.

Early in the morning, at around 6 AM when we got up, the first thing I did was to go out on the back deck and on to the sample storage room, to see if the dredge was successful. To my great disappointment, I learned from the watch that we had been unsuccessful in recovering any samples from the Transform Fault. In the meantime, Peter joined me in the lab and right away, without any hesitation, he went to the bridge and asked the officer in charge to turn the ship around and go back to our last sampling location, so we could try again to dredge until we were successful in recovering some samples. This decision turned out to be very important because we were able to obtain a considerable amount of a variety of rock types from a depth of 5100 m (Fig. 8.12c). This diversity revealed that the wall of the transform was constructed with sequential formations, which probably reflected layers of the upper-mantle and lower-crust. Later on, this finding would be confirmed by submersible observations during the "Pito cruise" (see below).

Thus, the first dredge station carried out in the Terevaka transform fault was made at the intersection of the north facing wall and the transform valley at 4850–4200 m depths during the *SONNE* cruise (leg 80) and brought up a variety of rocks consisting of peridotite, gabbro and diabase (intrusive dyke fragments). These rocks were thoroughly described by Marc Constantin, a PhD student at UBO, with the collaboration of Peter Stoffers and Dietrich Ackerman. The results were published in a special volume concerning mantle rocks edited by R.L.M. Vissers and A. Nicolas (1995).

Based on more than 250 samples, it was found that the dredge contained 8 % diabase, 15 % massive gabbros and 71 % peridotite. Among the 71 % peridotite samples, there were a large number (about 35 %) of ultramafic rocks with veins of gabbro, which indicated the percolation of magmatic melt within the peridotite formation. This is very significant when compared to other regions of the Pacific Ocean containing lower-crust/upper-mantle ultramafics, and where similar associations are observed, such as in the Garrett transform and in the Hess Deep.

The textural observations made by Marc Constantin on the minerals of these ultramafic rocks indicated that different sequences of melt circulation had taken place within the rock. By looking at the rocks' thin sections, Marc found that the first event took place during a regime of ductile deformation due to stress release when basaltic liquid circulated between fractured joints and crystal grain boundaries in the harzburgite rocks. These liquids percolated between the planes of foliation, and had crystallized clinopyroxenite. The next event is marked by melt injection from plagioclase bearing dunite during an alteration by metasomatism (transformation) of the peridotite host. Finally, the emplacement of gabbroic veins forming a reticulated network of ferrogabbros indicates the last event of melt injection (Constantin et al. 1995).

Diving in Terevaka Transform

The 1992 mission of the FS *SONNE*, headed by Peter Stoffers from the University of Kiel (Germany), provided valuable information from the dredge haul rocks. Thus, a diving expedition with the submersible *Nautile* was planned to take place in 1993 during a French expedition (Pito cruise) organized by Jean Francheteau (Francheteau et al.1994a, b).

The N.O. *LE NADIR* sailed from Hanga Roa, Easter Island, on November 1, 1993, and returned on December 22, 1993. The Captain, Guénhaël Thibault, the chief scientist, Jean Francheteau, and Norbert Compagnot, head of the submarine group, were all on board. The submarine group included technicians, engineers and pilots whom I knew from previous cruises. I was pleased be on another mission with these men with whom I had previously navigated.

I was particularly pleased to see Charlie LeGuenn, who had been a frogman in the past but who was now the person responsible for the mechanical up-keep of the submersible. Charlie no longer dove in his wet suit, but he still accompanied the divers in the zodiac when they went out to meet the submersible after it arrived on the surface at the end of a day of diving.

Charlie is a person one could count on if something should go wrong with the instruments. He was always keen to make sure that the technical gear was checked and repaired on time. Twenty-one dives were made during the Pito cruise, four of which were dedicated to exploring the deepest depression located in the Terevaka Transform Fault (Fig. 8.12c).

PI 16 (Pito dive N°16) was the first dive in the Terevaka Transform. It took place on November 22, 1993, and was made by Marc Constantin, a PhD student doing his thesis on oceanic peridotite–gabbro at IFREMER-UBO. Marc was the scientific observer accompanied by pilot Max Dubois and Henri Martinossi as co-pilot. The *Nautile* reached the bottom at 12:30 at 4910 m depth. It arrived at the intersection of the transform valley and the north-facing wall (the southern wall whose open face was directed to the north). This dive was aimed at reaching the site of the dredge haul that had recovered peridotite-gabbro during a previous cruise of the FS *SONNE* (Leg 80), and from there, the dive was supposed to start to climb the wall of the transform. The floor was sedimented with small sediment-mounds and purple colored holothurians. At 12:42, they saw two parallel grooves marked in the sediment, which were the traces of the dredge haul (DR83) made during the FS *SONNE* expedition in 1992 (Fig. 8.13). A few minutes later, the submersible stopped to have its first station and to collect a sample of rock from the same area as the dredge. However, the samples were loose and partially buried by sediment so they probably represented debris slumped from the wall. Luckily, the sample that was collected was a peridotite and therefore likely to represent a deeper portion of the exposed transform wall.

Going further up the wall, this discovery was confirmed by the presence of in situ samples of peridotites up to about 3121 m depth. It was at about 3,190 m that gabbroic veins intruding the peridotites were seen and recognized by Marc. He observed this through the porthole on an outcrop on the vertical wall. Outcrops of peridotites were seen, and in situ samples were taken up to 3190–3080 m along the wall. These veins crossed each other in two major directions (N160° and N070°). However the first veins of gabbro (10–20 cm thick) oriented in the N140° direction were seen at 4700–4690 m between sample sites #3 and #4. Other peridotite rocks veined with gabbro were observed at 4446 m and at 4304 m on sample site #5 and near sample site #7 respectively. Peridotites containing veins of finer grained rocks such as dolerites were observed at 4797 m and sampled (site #2) in a talus pile.

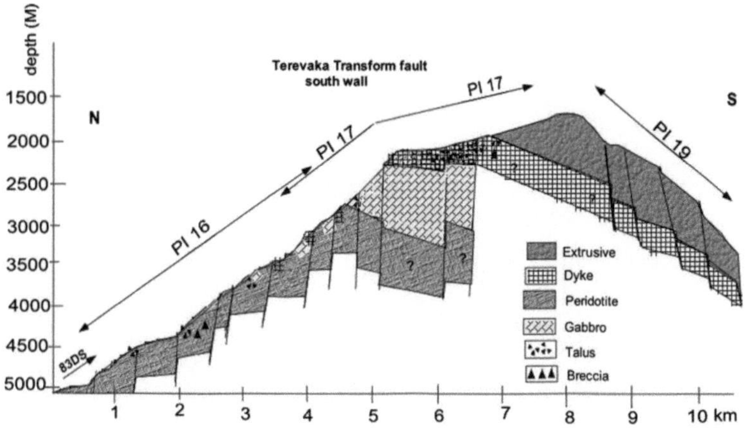

Fig. 8.13 Terevaka geological profile constructed during four Nautile submersible dives (PI 16, -17, -18 and -19) on the north-facing wall of the Terevaka Transform Fault. Each dive starts in the deepest area and goes upward

Other dolerites were also sampled in a talus pile (site #11) at 3493 m. However, the first dolerite sampled (site #14) in situ was at 3040 m depth. The sampled dolerite was associated with alternating brecciated peridotite crossed by veins of gabbro at 3004 m. The dive ended at 17:28 in this mixed variety of rock types marking a transition zone, which appears to be near the contact between the peridotite dominated outcrops and the intrusive gabbro-dolerite complex. This observation was confirmed with the next dive, which was planned to continue the ascension of the wall.

The second dive (PI 17) along the transform wall was a continuation of dive PI 16 (Fig. 8.14a–h). The scientific observer was Jacques Girardeau from the University of Nantes. His specialization on the physics of rock deformation during tectonic stress coupled with his experience in subaerial peridotite-gabbroic rock formations provided us with additional input for the project. The *Nautile* reached the site to be explored at 11:38 on November 23, 1993, at a depth of 3286 m. They found themselves close to an abundant series of serpentinized peridotite outcrops with schistose structures, which were veined by lighter colored gabbroic intrusions. After passing the peridotite-gabbro outcrops, they observed a dolerite dyke complex marked by hydrothermalism at 14:45 at a depth of 2603 m. The hydrothermal site consisted of a crowd of dead serpulite worms that were attached to a rusty-colored vertical dyke complex about 130 m high. At 15:01, Girardeau saw another area with living serpulidae worms (Fig. 8.14g) along with starfish and anemones, on another dyke outcrop 153 m high at 2408 m depths. The serpulidae worms (Neovermila species) were less than 30 cm long and formed bunches within a warm hydrothermal venting area where shimmering warm water was observed exiting from small cracks in the rock. The observation of the dyke complex on the near vertical cliff ended at 15:35 at a depth of 2305 m. At this

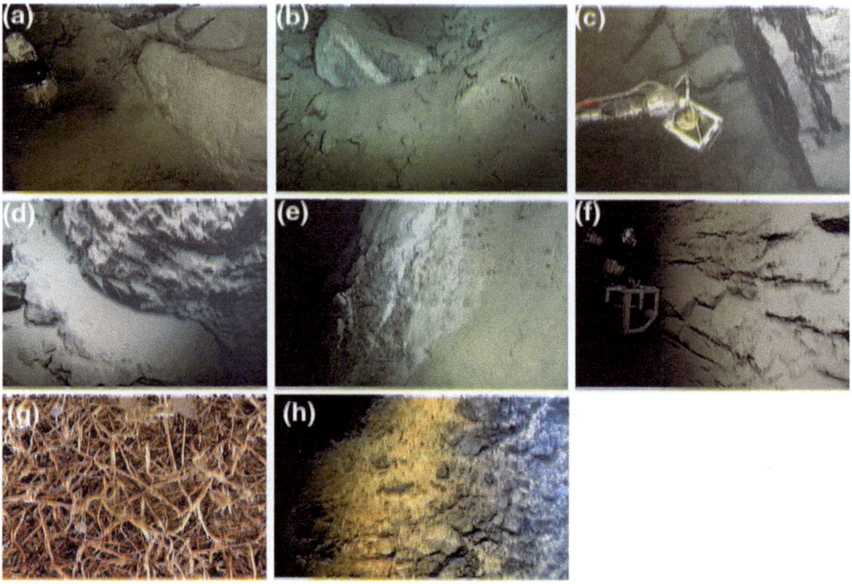

Fig. 8.14 Bottom photographs that have been extracted from a video-tape taken on the south wall of the Terevaka Transform Fault at the northwestern boundary of the Easter Microplate located near 25°S in the south Pacific. (Copyright IFREMER, Pito cruise 1993, Nautile dives PI-16, PI-17 and PI-18). The succession of pictures are arranged as the submersible moves upward along the south wall: **a** Near the base of the wall, large blocks and horizontal ledges of peridotite and talus. A fragment of dolerite containing clinopyroxene and plagioclase (Pito cruise, Sample PI 16-02) was collected from the talus at 4790 m depths. **b** A large fragment of peridotite with a vein of gabbro lies on top of partially buried ledges of peridotite at 4308 m depth. The location of the picture is near sample site PI 16-06 where serpentinized peridotite mud was taken at 4313 m depth. **c** Dyke formation showing the Nautile's mechanical arm holding the gyrocompass during sampling (Pito cruise, sample PI 18-06) at 3294 m depth. **d** Massive gabbro with horizontal fractured layering at 3241 m depth (Nautile dive PI-18). **e** Near a vertical scarp of serpentinized peridotite at 3997 m depth. Breccia and large blocks of loose peridotite occur at the base of the scarp near sample PI 16-08 which was an amphibole-bearing metagabbro (3965 m) (Nautile dive PI 16). **f** Fractured layering in massive peridotite (sample PI 17-02) at 2972 m depth. **g** Field of serpulidae tube-worms (Neovermilia species) sitting on top of hydrothermally-altered metadolerite at site of sample PI 17-08 (2408 m depth). **h** Summit of wall in a field of in situ fragmented basaltic debris (Nautile dive PI 17)

point, the slope became gentle (<10°) and was abundantly sedimented. This marked a drastic change in the geology as well as in the biology. Indeed, there was a sharp decrease in organic life and only occasional rocky outcrops were seen. At 16:03, a sample of dolerite was collected from a rare, partially buried rocky outcrop. The first basalt sample PI 17-11 (PI = Pito, 17 = dive # and 11 = sample #) was taken at 16:18 at 3035 m depth. Also, basaltic breccias and basalt were sampled at 1983 m (samples PI 17-12A and PI 17-12B)(Fig. 8.14h). During the dive, several "oriented" in situ samples were collected using a gyrocompass in order for Girardeau to study the orientation of rock deformation as well as the paleo-magnetic orientation with respect to the Earth's geode (Fig. 8.14c).

Despite the fact that Dives PI 16 and PI 17 were able to complete the profile along the north-facing wall of the transform fault, it was decided to make two additional dives in order to satisfy our questions concerning the eventual change in lithology observed along the strike of the wall.

Therefore, a third dive (PI 18) was made along the southern wall, but about 5 km to the east of the two previous dives. Jean Francheteau was the scientific observer, with Martinossi and Dubois as pilot and co-pilot. The objective of this dive was to verify whether the observed lithology on the north-facing southern wall of the fault was representative of what would be found further along. Dive PI 18 arrived on the bottom at 11:27 at 4154 m depths, which is just few meters above the level of sample site #07 for dive PI 16, corresponding to a zone of brecciated peridotite with an intrusion of gabbro. Indeed, Jean's first observation was seeing talus pile blocks of gabbro lying on sediment at 4163 m. According to Jean's description, the blocks were gigantic, twice the size of the *Nautile*, and they were associated with a stair case type of faulting (i.e. a succession of near vertical and almost flat ledges). This type of panorama lasted up to 3414 m depth where the last gabbro was sampled before coming to a different lithology. At 13:27, the dykes with their characteristic columnar jointing and planar faces were observed against a wall of gabbro. A sample (PI18-06) of dolerite was taken in situ at 3294 m (Figs. 8.12c, 8.13, 8.14, and 8.14d). The rest of the dive progressed towards a shallower depth up to 2397 m, and was in a complex of massive gabbros having a blocky appearance, sharp edges, and nearly vertical walls, when they were in place. During any tectonic motion that was strong enough to generate earthquakes, these massive outcrops could have been broken and slumped down slope to form large debris, sometimes the size of a house.

The last dive (PI 19) was done with Pierre Triger (pilot), Patrick Cheilan (co-pilot) and myself. We returned to the south-facing slope of the Terevaka transform wall to continue the observations of the previous dive, PI 18. My dive was aimed at exploring the other side (i.e. the northern slope of the southern wall), in order to find the extent of the various formations by comparing the stratigraphic sequences of both sides of this wall. The first visual contact was on the sea floor at 4797 m and the dive ended at 1767 m depth (Fig. 8.14a–h). The northern side of the wall consisted essentially of pillow lava basalts that were occasionally interrupted by dyke intrusions at 2770 and 2000 m depths. The northern flank was abundantly sedimented with avalanche debris that was partially buried by sediment. Most of the slope was broken up except for a few isolated, freshly fractured outcrops showing pillow lava sections and horizontally layered flows at depths shallower than 2940 m. The slope stayed uniform and monotonous between 2600 and 2400 m, and was characterized by sediment and partially buried lava flows. At the end of the dive near the top of the wall, at 1700–1800 m, we reached an abundantly altered outcrop with light brown and reddish ochre coloration. We encountered brecciated and incoherent looking material that was easily fragmented when touched by the submersible's mechanical arm.

Extrapolating from the diving observations made on the two sides of the southern wall, it was found that the contact between the basalt and the dyke complex is located between 2700 m (dive PI 19) and 2500 m depth (dive PI 17) (Fig. 8.14e). This finding permits us to state that the dykes are lower in the crustal section along the south-facing wall than what was observed on the north-facing wall inside the fault. Such a discrepancy is the result of the wall's tilting during the uplift of the serpentinized peridotite. Indeed, no peridotite-gabbro complexes were observed on the south-facing wall. These associations are believed to occur deeper in the crust, and are only visible on the inside of the southern wall but are not presently exposed in the area of dive PI 19, which is "outside" the transform fault.

Hence, the main results obtained by submersible observations in the Terevaka Transform are the evidence of a stratigraphic sequence exposing deep-seated ultramafic (peridotite) and mafic (gabbro, dolerite and basalt) formations. Another major observation was the evidence of melt circulation that gave rise to gabbroic and dolerite veins (dykes) <1 m thick crossing the upper mantle-lower crust peridotite. Such melt circulation and subsequent reaction with the surrounding peridotite is related to tectonic activity during plate motions, which is responsible for the fracturing and fissuring of the lithosphere. Also, the presence of hydrothermal circulation and the living biological communities observed within the dyke complex indicates a warm environment, probably related to the mechanical stress during the forceful emplacement of the peridotite-gabbro sequences within the lithosphere.

St. Peter and St. Paul's Rocks Fracture Zone (Atlantic Ocean)

The St. Peter and St. Paul's Rocks (SPPR) Fracture Zone (in the Equatorial Atlantic located at 0°40'N-25°W) is found in an area where the most prominent displacement between Africa and South America has taken place since the opening of the Atlantic Ocean 120 million years ago. The name of the fracture zone is given after that of the islets that have emerged in the area near 1°N. A large portion of the St. Peter and St. Paul's Rocks (SPPR) Fracture Zone is still tectonically active and the main spreading segments of the Mid-Atlantic Ridge have been displaced by more than 300 km to the East at about 1°N-30°20′W and 0°30′N-24°W (Fig. 8.15a, b). Along this tectonically active portion of the fracture zone, the oceanic lithosphere has been torn apart in several directions. The two main directions are East–West and North–South. The East–West direction corresponds to that of the major tectonic motions of fracturing and shearing while the North–South tectonic direction is mainly related to the volcanic events of spreading.

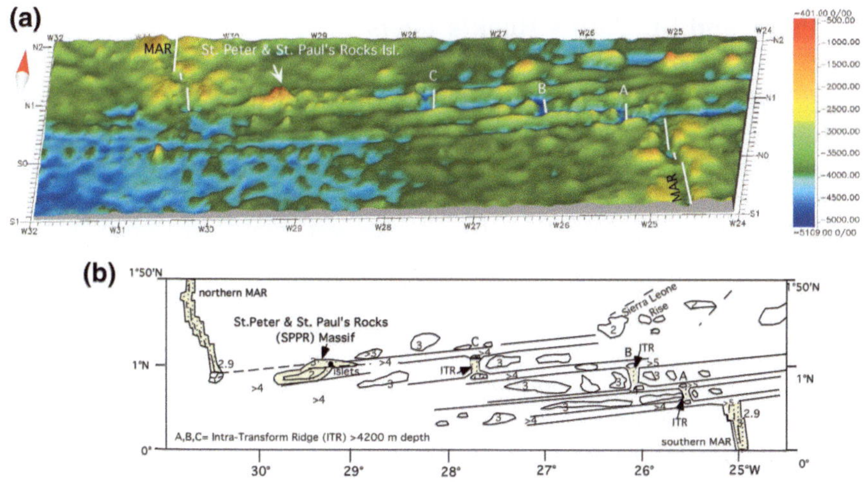

Fig. 8.15 The St. Peter and St. Paul's Rocks Fracture Zone in the Equatorial Atlantic located at 0°40'N-25°W is a multiple transform fault disrupting several short spreading ridge segments (A, B, C). The St. Peter and St. Paul's Rock's Islets are exposed on the northeastern wall of the transform fault. **a** The altimetry map shows the location of the ITR spreading segments A, B and C where the submersible dives took place. **b** Sketched structural setting based on satellite altimetry map data

The concentration of seismic epicenters with magnitudes higher than 4 indicates an "en echelon" structure (Rusby 1993) related to fracturing in the vicinity of the Ridge-Transform-Intersection within the SPPR fracture zone. Earthquake locations and focal mechanisms along the equatorial Atlantic spreading ridge segments are concentrated in a narrow band less than 5 km in width, which coincide with the recent spreading centers as well as with the E-W shear zone related to the transform. The SPPR is a multiple transform fault disrupted by at least six "en echelon" relay zones of spreading segments less than 30 km in length. These relay zones form short Intra-Transform Ridge (ITR) segments (10–30 km long) that can link the *en echelon* transform faults together. The ITR segments are somewhat comparable to the Garrett transform's sigmoidal volcanic ridges. The most prominent ITR segments with deep (>4300 m) rift valleys are located near 25°37′W, 26°10′W and 27°42′W and represent spreading ridges with recent active volcanic centers (Hekinian et al. 2000). Thus the Saint Peter and St Paul's Rocks (SPPR) Fracture Zone is a multiple transform made up of Intra-Transform-Ridge (ITR) spreading segments and the St. Peter and St. Paul's Rocks (SPPR) massif located at 0°50′N and 29°20′W.

Before discovering the existence of the Equatorial Atlantic fracture zones, the islets of St. Peter and St. Paul's Rocks, located in the middle of the Atlantic Ocean, were already known and had been visited by a number of sailing ships.

St. Peter and St. Paul's Rock's Islets

The St. Peter and St. Paul Rock's (SPPR) constitute rocky outcrops of three small islets, with a surface less than 0.01 km^2, precisely located at 0°56′N–29°22′W in the middle of the Atlantic Ocean. These islets belong to Brazil. The largest islet covers an area about 130 m long and 50 m wide and it rises 18–20 m above sea level. There is no vegetation, and the only tree seen during our journey in 1997 was a small palm, probably imported by the Brazilians.

These islets are in the middle of the Equatorial Atlantic Ocean and they are among the few islands in the World that are not volcanic in origin. Most of the world's islands are volcanic and were formed by the repeated extrusion and piling up of volcanic rocks. The SPPR islets are of particular interest because they are the only emerged land mass in the Atlantic that exposes a tectonically emplaced, metamorphosed, mantle peridotite. The St. Peter and St. Paul's Rocks islets in the Atlantic and the Zabargad Island in the middle of the Red Sea are the only known islands in the World that consist entirely of ultramafic rocks. Geologists like to study these islets because they represent a displaced sample of our planet's upper mantle that is within their reach.

The first visit to the islets was during the scientific expedition of the *HMS BEAGLE* in 1843. Darwin in 1891 was the first to state that the islets were unlike any other island seen, and he reported that the rocks were olivine enriched, so he categorically denied their volcanic origin (Linklater 1972). Another scientific party landed on the St. Paul's Rock's during the HMS *CHALLENGER* expedition on August 19, 1873, and was reported by John Murray in 1882. Also in 1882, Renard wrote a detailed mineralogical description of the peridotites from the islets in the volume dedicated to the *CHALLENGER*'s expedition (Linklater 1972).

HMS (His Majesty's Ship) *CHALLENGER* was a 200 foot long, three mast, square-rigged, wooden ship, with a weight displacement of 2300 tons. It was the first ship to explore the sea floor. In the 1870s, the sea floor was basically unknown, so *Challenger* sounded and sampled all the oceans except the Arctic, thereby laying the foundation for modern oceanography. The captain of the ship was George Nares, born in Aberdeen. The director of the scientific team was Charles Wyville Thomson, a biologist and a professor of natural philosophy at Edinburgh. The ship left Portsmouth on December 21, 1872. John Murray, to whom we are indebted for publishing the first comprehensive results of the *CHALLENGER*'s ocean explorations, was born in Ontario of Scottish parents and pursued his studies in Scotland at the University of Edinburgh. He was 31 years old when the expedition started and he had an extensive background studying literature, chemistry and natural history. John Murray, using his family's funds, published the observations made during the HMS *CHALLENGER's* voyage in fifty volumes. When the ship arrived at the St Peter's and St. Paul's Rocks, all the crew landed on the islets and the report indicated that the only inhabitants of the islets, made-up of black looking rocks, were "noddies and boobies". Later on, in 1921, the *QUEST* Expedition also collected rocks from the SPPR Islets.

The HMS *OWEN* with captain Hall and chief scientist G. Evans landed in December 1960 on the St. Peter's and St. Paul's Rock's islets. J.D.H. Wiseman, a known specialist in land-based ophiolite complexes, accompanied the expedition. Wiseman (1966) found the rocks from the islets unusual when compared to other peridotites because of their mineralogical peculiarity. He noted the presence of abundant hornblende, a hydrated silicate mineral, and called the rocks "Challengerites", after the first discoveries made by the HMS *CHALLENGER* expedition.

The "volcanic versus metamorphic" origin of the rocks from the St. Peter's and St. Paul's islets has long been debated by geologists. Washington (1930) proposed that the origin of the rocks was from deep seated and deformed plutonic rocks. Tilley (1947) made further detailed studies of this expedition and suggested that the rocks forming the Islets were dynamically emplaced during the uplift of deep-seated material. Indeed, the presence of deformed and metamorphosed peridotite transformed into mylonite was observed, and this could very well indicate extreme pressure conditions.

More recently, Melson and Thompson (1973) published the first complete and comprehensive descriptions of the rocks forming the Islets and compared them to the samples collected from surrounding underwater structures during the *Atlantis II*'s twentieth cruise in 1966. Russian expeditions with the R.V. *AKADEMIC NIKOLAJ STRAKHOV* from the Geological Institute of the USSR Academy of Science also visited the area of the St. Paul's Rocks in 1988 and in 2000. Professor Leonid Dimitriev (1985) studied the material in detail and made the most accurate description of the mylonite samples from the St. Paul's Rocks submarine massif. Leonid stated that it is likely that the emplacement of mylonite, which is high temperature, probably around 800–900 °C, is the result of a plastic deformation of the upper mantle (Turner and Verhoogen 1960).

The St. Paul's Rocks Islets revisited in 1997–1998

One goal of our *St. Paul's Rocks cruise also called St. Paul's cruise* was to follow in the paths of the HMS *BEAGLE* and HMS *CHALLENGER* adventures and therefore to land on the islets in order to carry out geological reconnaissance with sampling stations and scientific observations. The main objectives were to collect oriented samples and drill cores for paleo-magnetic studies. Bertrand Sichler had brought a hand-carried coring device that could drill holes a few centimeters deep into the rocks. Thus, a landing expedition on the islets of the St. Peter's and St. Paul' Rocks was programmed. On January 3, 1998, we sent a zodiac (inflatable boat) with four people from the scientific team: Eulalia Gracia from the University of Barcelona (Spain), Bertrand Sichler (IFREMER), Thierry Juteau (University of Brest) and Ronan Apprioual (IFREMER) who was in charge of handling the zodiac and assisting the three geologists in their venture. A hand radio was assigned for communication. The support ship R.V. *Le NADIR* stayed nearby for safety reasons.

Since the shore party had left for the day, and in order to make good use of our time at sea, I decided to program a *Nautile* dive (dive # SP10) on the flank of the island while the other people were sampling its top. This would allow us to have additional observations along the northern flank of the structure on which the islets are formed near 0°50′S-24°10′W.

The party landed at 8:00 AM on the southwestern islet, called Belmonte according to a Brazilian sketched map. The landing was apparently risky and was facilitated thanks to a metallic ladder that the Brazilian Navy had put on board the support ship, LE NADIR, 2 weeks before. In order to carry out their objectives, the party had to split up. Thierry Juteau, Eulalia Gracia and Ronan Apprioual went to sample the rocks from the Belmonte landing area near the lighthouse where the site was most suitable to sample, while Bertrand went to collect drill cores with his hand-held coring machine. The best outcrops exposed are located at the northwestern corner near the lighthouse where 8 oriented samples and 15 oriented drill cores were taken. The area consisted of yellowish-green mylonitized peridotite. The rest of the area was extremely fractured by tectonic activity. Conglomerate breccias and hydrothermally altered veined material were sampled. Because it was risky to pass from one island to the other, the scientific party could only go from the Belmonte islet to the Challenger islet and they only had enough time to explore its southern portion. The southern part of this islet was difficult to reach therefore only three samples were taken. The scientific party was especially interested in this area because they found dark-colored, fine-grained amphibolite-enriched rocks, which were tentatively interpreted by Thierry Juteau as originally being dolerite (dykes) that had been extremely transformed during tectonic activity. The transformation (metamorphism) of these rocks into amphibolite suggests high temperatures (>400 °C) and high pressure. The main orientation of the vertical foliation for the mylonitized peridotite and amphibolite collected from the islets is roughly East–West, which is the same direction as the fracture zone itself.

Historical Background of the Cruise

The *St. Paul cruise* was made in conjunction with UBO (*Université de Bretagne Occidentale*) in Brest. We had been obliged to wait for 3 years before the IFREMER administration and the French National Scientific Community accepted our cruise. In addition, we encountered a great deal of harassment from the Brazilian authorities. Since the expedition was taking place in Brazilian territorial waters, we had to ask permission from the Brazilian Navy to carry out our study. This was not an easy task: the proposal had to be translated into Portuguese and some agreements had to be reached between the IFREMER and the Brazilian authorities.

The Brazilian authorities placed several hurdles before giving us permission to dive in their territorial waters. One of these was to have a Brazilian scientist onboard the submersible for each dive. This was quickly dismissed, since only one passenger is allowed in the submersible's sphere along with the pilot and co-pilot. However, we did agree to allow a Brazilian Navy officer on board the support ship, *Le Nadir*, at all times. The Brazilians also requested that part of the data obtained in their economic zone would be sent to the Brazilian Navy.

It took about a year to reach an agreement and to obtain all the proper authorization. I recall that 2 months before our departure, we still hadn't been granted our authorization to dive in Brazilian waters. I had to add a clause in our note of agreement to the authorities saying that if I did not receive the proper authorization prior to the departure for our cruise, scheduled for December 1997, I would change the program and we would pursue our dives outside the Brazilian economic zone. Also, I told them that if I did not obtain satisfaction, there would be no cooperation with any people from Brazilian Universities. This meant that they would miss the opportunity of having fresh data from the region. This clause was sent to the Brazilian Embassy in Paris via our IFREMER foreign relations department (Direction de la Communication International, DRCI). Happily, authorization for us to dive in Brazilian waters arrived soon afterwards.

It is becoming increasingly difficult for scientists to explore in the economic zones of any country because of the legal aspects of any such scientific exploration. Sometimes, the administrations of some countries want to be given all the data collected in their EZ (economic zone) and we often need to have hard discussions to convince them otherwise. It is frustrating when scientists no longer have the freedom to pursue their research because they have to deal with administrative constraints, which discourage us from working in any EZ. The day when all the oceans have been divided up among all the countries that have coastlines along the seas, it will be a nightmare for continued scientific exploration and human progress.

In addition to the French institutions mentioned above (IFREMER and the University of Brest), several other international institutes such as the University of Kiel (Germany, P. Stoffers), the CNR and the University of Bologna (Italy, E Bonatti), the Russian Academy of Science in Moscow in Russia (Gleb Udintsev) and Susanna Sichel from the University of FUMENCIA (Itiroi, in Brazil) were all involved in participating on the cruise and in studying the data obtained during our *St. Paul* mission.

However, we had a more serious problem to resolve if we were to carry out our objectives. We did not have any satisfactory map coverage for conducting a submersible program. Our IFREMER draftsmen, Daniel Carré and Serge Monti, had compiled the limited available bathymetric data at IFREMER in Brest by using the international General Bathymetry Chart of the Oceans (GEPCO) and some fragmented contour lines and sparse map data had been obtained from colleagues who had participated on American, Russian and German expeditions covering various portions of the St. Peter and Paul's Rocks fracture zone. For example the multichannel SeaBeam coverage of the western (near 1°N-29°W) and the eastern (0°50′N and 25°35′W) parts near the Ridge Transform Intersection Zones was carried out during both Russian (R.V. *STRAKOV*, Udintsev et al. 1990) and German (R.V. *SONNE* - Leg84, P. Stoffers) expeditions (1994) respectively. The American research vessel *ROBERT D. CONRAD* had obtained some scarce bathymetric data from the central portion of the transform fault between 1°N-27°40′W using conventional methods in 1987 (Courtesy of Jean Guy Shilling, from the University of Rhode Island, USA). It was in this context, with insufficient bathymetric coverage, that the diving expedition started. But this was a challenge and an opportunity for us to improvise in order to reach our scientific targets while at sea.

Diving in the Equatorial Atlantic

Finally after all the scientific and administrative procedures were overcome, the *Saint Paul cruise* in the Equatorial Atlantic was scheduled and I was very glad and eager to start the action. The *Saint Paul cruise* with the submersible *Nautile* and its support ship NO *NADIR* began on December 20, 1997 at 8:00 PM from the port of Recife in Brazil and ended in Fortaleza (Brazil). The total length of the cruise from port to port was 23 days and we were able to make 13 dives. The Saint Paul cruise ended on January 6, 1998 after the last dive. We left the site of the St. Peter and St. Paul Rock's islets for the next port of call, Fortaleza, about 60 h away.

In addition to the fact that this region of the Atlantic has the deepest and longest fracture zones going from coast to coast, a major reason for exploring this particular Equatorial region is because it is also among the areas of the World with the lowest volcanic activity and is therefore an area with a relatively thin crust. This enhances the possibility of observing deep-seated material from the lower-crust and upper-mantle. The tectonics of fracturing and shearing are still ongoing processes overriding that of magmatic upwelling during spreading. The localized volcanic structures, called intra-transform ridges (ITR), were also targets for our exploration because their location has an important implication in correlating the Earth's deep- seated mantle material exposed during fracturing with the upwelling of magma produced during lithosphere melting.

Thus, the St. Peter's and St. Paul's Rocks Fracture Zone, which has displaced *en echelon* segments due to several transform faults, is a site with small spreading ridges, called Intra-Transform Ridges (ITR). These small ridges are less than 20 km long and less than 10 km wide and are gashed by a central graben to depths of about 4100 m. We were confronted by the fact that we were able to explore two different geological activities (the emplacement of deep seated mantle and volcanism) within a relatively small area. Thus, it was going to be feasible to fulfill our double objectives during the limited time allotted to our project (only 15 days on site).

The geological setting and situation of the SPPR FZ is somewhat similar to what had been previously found in the Garrett transform fault where small sigmoidal ridges with recent volcanism were found inside a large fracture domain. The goal of the *St. Paul cruise* was to investigate two structurally different regions of the St. Paul's Rocks Fracture Zone: (1) The Intra-Transform Ridge (ITR) segments and, (2) The sigmoidal shaped St. Peter and St. Paul's Rocks (SPPR) massif on which the islets are constructed (Fig. 8.15a, b). We started with our exploration of the ITR, due to logistical reasons.

Diving in the Intra-Transform Ridges (ITR) of St. Peter and Paul's Rocks

The Intra-Transform Ridge (ITR) spreading centers represent short spreading ridge segments interrupted by East–West trending transform faults (<230 km) within the main St. Paul F.Z. On a regional scale, this reflects deep-seated colder convective systems than those existing to the north and south of the Equator, within a band of about 440 km (between 2°S and 2°N) in the Equatorial Atlantic. The three short Intra-Transform Ridge (ITR) segments at 25°27′W, 26°10′W and 27°42′W were seen on satellite altimetry data (Fig. 8.15a). The two intra transform ridges located at 25°27′W (ITR A) and 27°42′W (ITR C) were chosen for investigation because we had time to dive on them, while still remaining within our short schedule at sea (Hekinian et al. 2000). The short (<20 km in length) spreading segments show axial valleys about 4600–4700 m deep, are less than 2 km wide, and are bounded by walls that are 2000 m high. Like other major Mid-Atlantic Ridge segments, the Saint Peter's and St Paul's ITR have a general N 340°(NNW) trending orientation and sedimented nodal basins that are more than 5000 m deep (27°42′W and 25°27′W) at their spreading ridge segment ends (Fig. 8.15b).

AT 7 pm on the evening of December 23, 1997, we arrived on the first site of our dives, on the ITR system near 0°43′N and 25°27′W. Our first task was to try a new system of deep towed camera at a depth of 2750 m, in order to test the electric coaxial cable and the quality of the images. After 5 h on the bottom, the system had stopped working. Since we did not have any spare parts on board, the engineer (Cavarec) had to improvise the best he could in order to make the device work. This first lowering of the deep-towed camera enabled us to see the bottom landscape before the first dive.

The next morning, I was scheduled to make the first dive. I could see that my colleague scientists were pleased with my decision because they did not have to miss the Christmas Day's special meal. But, the steward (Michel, who was also called "Aldo") had given an order for the diving team to also be supplied with a better menu to take to the bottom.

The first task was to explore the depression, or Rift valley, and to find the type and the extent of any volcanic activity. As usual, on the morning of December 24, we were ready at 08h00. I went up the ladder on the side of the Nautile and down into the sphere where Olivier Cipriani, the co-pilot, was finishing his last check on the instruments. Then Max Dubois, the pilot, sat on the edge of the airlock, waiting for the Nautile to be pulled to its launching position on the back deck of the

Nadir's main deck. Max had been on other diving cruises with me; before he became a pilot he had also been working on the navigation team to deploy the transponders and the positioning systems for the submersible. Max was a good-looking young man with blond hair and a mustache. He was very much appreciated by some of the female scientists. He was a pleasant and experienced person, so I was pleased to do our Christmas Day dive with him.

At 11h54 we reached a sedimented bottom with partially buried collapsed pillow lava. The sediment contained traces of bio-perturbations and traces of holothurians. To the east, we saw an echo on our radar, probably a rocky outcrop. We decided to go to see what it was. I wanted to find out where the most recent volcanic or tectonic activity occurred, in order to know the state of evolution of the spreading axis.

At 12h12 we arrived at what seemed to be the foot of the East wall of the rift valley. We had to go along the strike of the scarp along several talus piles and make sure that we were on the first scarp of the wall. Then, after taking the first sample at 4670 m depth, we changed our course to a Northeast direction (310°) and went towards the center of the rift valley to try to find the area with the most recent eruptions. At 13h31, we were on a 2 m wide fissure cutting through a relatively fresh glassy lobated flow. I was happy to see that the sediment cover had decreased and that the fissure was recently formed. This suggested to me that we were on a tectonically active axis of the rift valley and that by taking a bearing along the same strike as the fissure, we might be able to follow the axis of the rift in order to get to where the spreading is currently taking place.

Hence, after taking a sample, we moved by zigzagging to N 300°. Indeed our approach turned out to pay off because we encountered several "en echelon" fissures oriented in a 340° direction. The current on the sea floor deviated our course more to the west. The fissures exposed lobated flows with glassy chilled margins. The presence of the sediment cover along our dive track was due to the fact this region of the Equatorial Atlantic has a large amount of plankton production. Indeed, through the porthole we could see white spots of dead foraminifera dropping down on the sea floor. The result of this dive was very conclusive because we found that this small ITR segment is still spreading, as was evidenced by the abundance of the 3–5 m fissures rather than by traces of recent volcanism.

Now the next step was to define how much volcanic activity had occurred in the past. In order to delineate the continuity and extent of volcanism over time, it is important to go to the sides of the rift valley to see if the same volcanic events took place in the past. Hence the goal of the next dive would be to climb the East wall of the ITR.

The next morning dive SP02 with Thierry Juteau, Patrick Ceilan as pilot and Patrice Lubin as co-pilot took place. They landed at the foot of ITR's East wall at 4700 m and followed an easterly course along the slope up to a depth of 3868 m, which they reached after 4 h and 58 min. At the beginning of the dives, Thierry noticed the presence of fresh flows at the intersection of the East wall and the Rift valley floor at 4700 m. At least two successive lobated flows were recognized, with the youngest flow being darker and shinier. He described a "glowing" of the glassy surface under the submersible's lights. The older flow was duller looking. The thin (<1 m thick) recent flows filling the interstices of the older pillows were limited in their extent (only a few meters wide and a few tens of meters long). Going further up the slope, which was made-up of staircase fault blocks between 4700 m and 3868 m depth, they crossed several other types of lithology.

This dive made a tremendous contribution to the project since they were able to find the contact zone between the intrusive mantle-material (serpentinized peridotite) and the volcanic lava flows. At 15h36, at depths between 3940 to 3868 m, the *Nautile* reached this boundary between the volcanic rocks and the exposed mantle peridotite. Before reaching the mantle peridotite outcrops, the divers had crossed between young volcanic rocks formed in the rift axis (at depths of 4630 and 4311 m), which had been moved aside during spreading. This series of lava flows showed a radial jointing exposed on inward-looking (toward the rift valley) faulted scarps and on talus piles at their feet. The most prominent (3) major fault scarps, with a relief of up to 226 m, occurred along the wall. The top of the scarps was covered with sediment and impressive, large talus piles, 15–74 m in height. At 12h44, the divers saw the first dyke, about 4 m thick and 8 m high, intruded into brecciated (fragmented and partially consolidated) basaltic debris. Between 12h44 and 13h59, the divers crossed a field of dolerite dykes between 4211 and 4019 m depth. These dykes were exposed during earthquakes related to the faulting and uplift of the rift valley wall and were oriented N010° (dipping 070° W). The dykes that had intruded into pillow lavas and volcanic breccia were 2–10 m in height and 1–4 m thick. Then, after a sudden break in the scarp, the submersible reached a flattish sedimented area at 4019–3940 m depth, lying on top of a thin volcanic pillow lava flow.

At 15h59, the submersible encountered the first serpentinized peridotite outcrop at 3940 m depth. It was exposed on a steep scarp. Thus, between the dyke complex and the mantle peridotite, there was only a 70 m-thick basaltic layer. The serpentinized peridotite continued to be observed during the continuation dive (SP06), which was made to complete the profile along the same slope. However, before continuing the profile to the top of the wall, we wanted to obtain more visual data along the east wall of the rift valley.

Therefore, on the same day as the SP02 dive, December 26, 1997, after recovering the *Nautile*, we planed to use our deep towed SCAMPI (#SC02) camera in order to extend our sea floor observations further north along the wall. The deep-towed camera was lowered to the bottom at a depth of 3890 m. Scampi's observations extended down to a depth of 4756 m at the foot of the rift valley. At 4150 m, the video camera showed an estimated bottom and then at around 4580 m, dark-colored massive fragmented blocks were observed which appeared to be slumped peridotite, which had rolled down from shallower depths. Deeper down slope (>4600 m), basaltic fragments forming large talus confirmed the previous dive's observations about the extent of volcanism as being all the way down to the Intra-Transform Ridge's rift valley floor.

I am not going to present the chronological order of the dives but rather I would like to use the structural setting of this area of the SPPR Fracture Zone in describing our dives, in order to provide some continuity for our geological discoveries.

Thus, the last dive (SP06) on the East wall of the rift valley was made in continuation of dive SP02, and was aimed at reaching the top of the wall. At 11h10, the pilot, P. Cheilan, P. Lubin, co-pilot, and I arrived on the sea floor at a depth of 4032 m on the top of a scarp made up of sediment that was partially burying the rocky outcrops. This sedimented plateau with sporadic outcrops of pillow lava represented a major break in the slope and was observed at 4020–3939 m. The first peridotite rocks associated with basalt and diabase (dyke) were seen at 3980–4000 m. After taking our sample of serpentinized peridotite (SP06-01), we moved on along the slope.

During our progression between 4032 m and 2870 m depths, we encountered several breaks marked by near vertical fault scarps, which exposed striated and massive serpentinized peridotite, suggestive of a forceful emplacement. Vertical outcrops of massive rectangular blocks of peridotite were observed and sampled at 3183-3321 m. These blocks were about 1 m thick (sample SP06-11), and showed sub-parallel prismatic fractures. They were oriented in a N 050° direction and contained white veins of gabbroic material (sample SP06-12) cross cutting the serpentinized peridotite. Other rodingitized (altered) gabbros (samples SP06-2, -7, -8, -9, and -12) intruding into the serpentinized peridotite were recovered in situ at various depths (3994, 3640, 3504, 3404, and 3219 m) along the slope.

To my surprise, at 15h23 and at a depth of 2873 m, we arrived at the foot of a scarp made up of a mixture of prismatic blocks looking like dykes and radial jointed pillow fragments. This marked the transition between the peridotite-gabbro association, which had been overlaid by a freshly slumped talus of prismatic and radial jointed pillow flows at about 2873 m deep (sample of basalt SP06-14). This zone is cut by a major fault oriented roughly north south, in the same direction as the rift valley floor. Thus, the transition between the peridotite-gabbro is short and the normal geological sequence was not observed. Indeed, the gabbroic units, which usually represented the solidified magma chamber (gabbro) prior to the eruption of basaltic rocks, were thin and discontinuous. In fact, most of the gabbros observed were columnar intrusions within the peridotite.

Another unexpected observation was the thin unit of a dolerite dyke intrusion that was exposed as large columnar blocks forming a talus pile. These were probably exposed during tectonic movements. Thus, between 2873 and 2710 m, which marks the top of the wall, only basaltic lava flows were seen along with occasional dyke intrusions.

From the reconstruction of this dive, I had estimated that the basalt-dolerite (dyke) complex did not exceed 150 m in thickness. This observation will be important for the geological reconstruction of the ITR spreading segment formations. Thus, a complete geological section from the base (4700 m) to the top (1850 m) of the entire East Wall of the ITR in the St. Paul's FZ area was completed during dives SP02 and SP06.

The other dives SP03 and SP04 essentially took place on the west wall of the rift valley, and they confirmed the large size of the serpentinized peridotite exposure and the scarcity of volcanic eruptions. Dive SP03 with Eulalie Gracia, went to the inside corner high of the western edge of the nodal basin, and showed the presence of two basaltic units sampled at 5100–4830 m, and 4630–4300 m. depths. These basaltic units, as was true for what we observed on the East wall (SP 06), are thin (<200 m thick) and are intercut by serpentinized peridotite at 4800–4730 m depth. Peridotite fragments intruded by massive rodingitized gabbro (4380 m) and partially buried by sediment are found at shallower depths (4330–4390 m).

Unfortunately, Dive SP04, with Susanna Sichel on board, did not last long. They were only able to remain for 1 h on the seafloor. At 4000–4286 m depths, on the lower part of the west wall, one sample of peridotite was recovered. As soon as the Nautile was pulled on board for repair, I decided to lower the deep-towed camera Scampi for another station (# scampi 4) in order to have a visual observation of the West wall. The deep-towed camera observations revealed the presence of peridotite between 4100 and 3500 m depths. This confirmed the extent of the peridotite as going all away to the top of the western wall and also revealed the scarcity of volcanism in the area.

Dive SP05, with Gleb Udintsev as observer, was of particular interest because it was made along the active transform fault bounding the northern part of the rift valley. Gleb is a well-known Russian geophysicist who has made valuable international contributions to marine geology. It was a good opportunity for us to have had him on board, so he could share his knowledge. At 75 years old, rather thin and tall (about 6 feet tall), he was an active sea-going scientist. After this cruise, he immediately joined a Russian ship in Antarctica for another 2 months at sea.

Dive SP05 was the first time that Gleb Udintsev had ever dived in a submersible. After all his years spent on surface ships imagining how the bottom of the ocean must be, he was finally going to be able fulfill his dream and take a real look at the sea floor. On December 29, 1997, at 11 h 27, Gleb was on the bottom at 4272 m depth with Max Dubois, and Olivier Cipriani. They landed on a sediment-covered apron with exposed, fragmented lava flows. The divers went north along the slope of the faulted terrain. Abundant staircase-like faulted blocks testified to the extent of tectonic activity along the East–West direction of the major strike-slip motion of the St. Peter and St. Paul's Transform fault's northern wall. The presence of compact sediment and conglomerates made up of broken and eroded volcanic rocks cemented by smaller sand-size debris indicated that there had been intense tectonic motion due to faulting. At 16 h 31, the dive ended after collecting 16 samples and spending 5 h on the sea floor. Gleb came on board, and when he descended the ladder from the submersible, he had a big smile. The co-pilot, Olivier, was waiting at the bottom of the staircase, with a tray carrying a glass of vodka, a T-shirt with a picture of the *Nautile* and a diploma to commemorate Gleb's first dive.

In summary, the dives in the 25°27'W ITR axial valley at 4700 m depth have revealed that this structure is less than 20 km long. The most recent volcanic activity of the 25°27'W ITR was found on the eastern side of the rift valley at 4720 m within a narrow band less than 200 m wide. The geological cross-section made on the rift valley floor going from the west to the east indicates a relatively thin volcanic unit (<150–200 m thick) made up of dolerite dykes and basalt overlying mantle peridotite. The adjacent East and West marginal walls consist of serpentinized mantle peridotite intruded by gabbro and dolerite. The extrusive lavas are found in the axial rift and on the top of the wall. The top of the East wall was an ancient spreading center less than 1 million years old, with a thin volcanic crust, which is the same as what was found in the recent axial valley; however, this ancient spreading center has been uplifted during the emplacement of a serpentinized diapir.

We left the first site on the evening of December 30, 1997, and stopped midway across the SPPR massif to make one more dive (SP07) on another ITR at 27°42′W. The next morning on December 31, Thierry Juteau was scheduled to make this dive on the ITR segment at 27°42′. The dive revealed the presence of fresh glassy pillow lava flows at about 4500 m depth along the eastern part of the rift valley. Unfortunately, because of technical problems encountered with the *Nautile,* this dive was short and lasted only 38 min on the sea floor. The problem was due to the main engine of the Nautile. It was not operational and they had to return to the surface for repairs. Nevertheless, we had acquired enough information on the ITR to satisfy our questions. Thus, we decided to leave this site in order to go on to the SPPR islets, another important priority. It was clear from our observations and sampling that these short ITR segments had erupted a limited amount of basaltic lava, which was mainly confined to their rift valleys. They are comparable to other magma-starved MAR segments found elsewhere.

Volcanism and Mode of Formation for the 25°27′W ITR

From the dive data and based on the samples collected, we were able to calculate the volume and extent of volcanism as well as make a solid hypothesis on the mode of emplacement of deep mantle material in the context of short-lived or magma starved spreading ridge segments. The volume of volcanic activity was estimated on the basis of the presence of dykes and lava flows outcropping on the East and West walls of the 25°27′W IRT. Our observations suggest intermittent, short-lived magmatic events.

The emplacement of the different lithology encountered on the ITR is inferred from the information gathered during both surface ship operations and dives. The difference between the rift valley depths at 4700 m and the shallow (<2500 m) topography of the rift valley walls at the 25°27′W ITR suggests an important uplift (>1000 m). A possible explanation for this uplift is attributed to the buoyancy effect of the loaded lithosphere and its rheology, which implies deformation and the flow of matter. Serpentinization weakens the oceanic lithosphere and enhances faulting during uplift. A colder lithosphere with heterogeneous rock types, such as serpentinized peridotite-gabbro with small amounts basaltic flow, will increase the amplitude of uplift. The emplacement of the residual peridotite is believed to have taken place during differential vertical motions of a partially serpentinized body forming the rift walls. The dyke and gabbroic intrusions that did not reach the sea floor represent the conduits that have produced the thin veneer of lava flows topping the residual peridotite. The sporadic occurrence of basalt associated with peridotite along the faulted walls of the rift valley suggests a successive stretching and uplift of the lithosphere in a magma-starved ridge segment (*see* chap. 4). This is comparable to the denudation hypothesis (Cannat 1993) suggested for spreading ridge segments with low magma supply.

Diving on St. Peter and Paul's Massif

The St. Peter's and St. Paul's Rocks (SPPR) massif on which the islets are built is about 90 km in length, 21 km wide at the 3000 m contour line, and has a sigmoidal ("S") shape according to the bathymetric studies of the area (Fig. 8.15a, b). This SPPR massif located between the northern MAR and the 27°42'W ITR, is associated with a seismically active 8–9 Ma old lithosphere (Wolfe et al. 1993). A compilation (Wolfe et al. 1993) of earthquake locations and their mechanisms of propagation along the SPPR massif indicates that the main seismic events have taken place within a band less than 2 km wide extending in an East–West direction. These seismic epicenters pass through a graben where basal-dolerite rocks (characterizing a spreading center) were observed during the submersible dives. Based on the distribution of the seismic epicenters and submersible observations, the SPPR massif was divided into two ridge systems: the North and South Ridges (Fig. 8.15a). Thus the SPPR massif is composed of two distinct structures: one trending N-E (North Ridge) and the other trending in a S–W direction (South Ridge) separated by an active fault coinciding with the concentration of the seismic epicenters (Fig. 8.15a, b). A near complete N–S geological section of the SPPR was conducted during six submersible dives.

Dives on the North Ridge of St. Peter and Paul's Massif

The North Ridge includes the islets of St. Peter and Paul's Rock's and, as mentioned before, they consist of extremely fractured and brecciated mylonites of mainly peridotite protoliths, tectonic breccias and mylonityzed conglomerates. The dives programmed on this ridge were aimed at determining the extension of such complex material at depth on the ocean floor and were intended to correlate the geology with that of the subaerial exposed formations on the Islets. How deep do the roots of the islets reach into the lithosphere?

Roger Hekinian (dive # SP08), Susanna Sichel (dive #SP09) and Gleb Udintsev (dive #SP10) did the dives on the North Ridge of the St. Peter's and St. Paul's Rocks Massif. Dive SP08 took place with the pilot Patrick Cheilan, co-pilot Patrice

Lubin, and arrived on the sea floor at 12:03 at 2795 m. Before we reached the bottom, I was nervous and at the same time curious about what I was going to see. We were going to the foot of the edifice on which the small islets are built, where only 1 % of the structure has emerged on the surface. The secret of its origin probably lies in the deep structural setting. After the *CHALLENGER*'s expedition in 1873, here we were, looking at the sea floor in the middle of the Atlantic Ocean where these islets were formed.

At about 11:50, as we approached the bottom, I started to turn around and lie down on my stomach with my eyes glued to the starboard porthole waiting to see the sea floor. The pilot turned on the floodlights and dropped 50 kg of weight from each side of the submersible in order to slow down our descent. At 50 m above the bottom, the water was a little blurry because of suspended particles in the water column. We continued to approach the sea floor: 40, 30, 25 m. Now I could almost see the bottom. There were no rocks but only sediment, no signs of current, and the sediment was abundantly disturbed by animal dwellers, forming strange designs, like circular thin ridges forming lines, small centimeter scale mounds and irregular lines that looked like "hieroglyph" writing. The slope of the ridge was to our left and our heading was N100° (that is, in a south-east direction). Patrick contacted the surface on the R.V. *Le Nadir* to tell them that we had landed and give them our depth.

After 4 min, the time needed to verify all instruments in the sphere, we started to move towards the slope and we saw the first rocky outcrop sticking out of the sediment. As soon as I saw this rocky outcrop, I asked Patrick to stop and take our first sample at 2792 m depth. This was a loose prismatic fragment of rock showing light and dark bands. Patrick turned the manipulating arm towards my side near the porthole in order for me to have a closer look at the sample. The rock was coated with a dark product but I could see alternating lighter bands. I knew then and there that our sample was a veined peridotite. I was happy, because this meant that further up we were going to find the outcrop from where the rock had slumped. Also, in light of the fact that the emerged St. Peter's and St. Paul's Rocks massif consisted of mantle peridotite, now we knew that the same material occurred all away down to 2792 m and maybe even deeper.

We progressed further up the slope, which varied in relief from gentle (10–15°) up to near vertical breaks (70–85°). At 13h09-13h12, we saw abundant avalanche debris. Soon after we were on freshly fractured outcrops showing white veins resembling gabbroic intrusions associated with talc-serpentinized peridotite and mylonite that we sampled at 2670 m. At 13h27 at 2601 m depth, we arrived on another near vertical scarp with a relief of about 100 m, made up of dark-colored mylonitized peridotite showing signs of striations and a fractured surface. At 15:25 at 1915 m, we encountered a pink colored octopus swimming on top of an avalanche of gravel slumped down slope that was detached from another near vertical outcrop that we were approaching. The color television was turned on to record a video sequence. At 15h38 at 1843 m, we climbed the vertical faulted scarp where the paved-like and foliated blocks of peridotite were at contact with prismatic and columnar rock. The sample taken from this scarp turned out to be a dolerite (intrusive basalt), which had intruded into the peridotite complex. This was an indication that we were in the presence of volcanic events that might have disrupted the mylonityzed (metamorphosed during tectonic events) peridotites. During the entire dive, we encountered serpentinized and mylonityzed peridotite with only a few intrusions of gabbro. When the dive ended at 16:39 (at 2551 m depth), we had sampled 16 rocks and we had traveled 4 km over a period of about 5 h on the sea floor.

The next morning dive SP 09, made by Susanna Sichel with Dubois as a pilot and Olivier Cipriani as a co-pilot, took place on January 2, 1998, between 3748 m up to 2841 m depths. This was Susanna's second dive and she was very enthusiastic about returning to the sea floor. During this dive, 19 samples were collected of which 14 were peridotites, 2 gabbros, 2 breccia of peridotite, and 1 indurate carbonate with recrystallized quartz veins.

Dive # 10 (SP10) made by Gleb Udintsev started late because the engineers had found a leak in the hydraulic system of the submersible and that had to be fixed. Gleb and the pilots reached the bottom at 11:49 on January 3, 1998 and stayed about 5 h covering a distance of about 5 km along the slope from 2563 to 1154 m depths. On his way up, Gleb observed a series of normal faults associated with talus rubble. The rubble, mostly peridotite at the foot of the scarps, was imbedded into compacted sediment and formed conglomerates. Gleb thought that this suggested that the area must have been a dynamically upraised mantle peridotite protruding into a sequence of sedimentary layers forming the ancient deep floor of the Equatorial Atlantic.

Three other dives were made on the south-facing flank of the St. Peter's and St. Paul's Rocks massif by Eulalia Gracia, Thierry Juteau and Bertrand Sichler, where peridotite and solidified carbonate sediment complexes were observed (Hekinian et al. 2000).

The common observations made during all the dives going from 1154 m down to 3627 m deep indicated that the slope forming the North Ridge, on which the St. Peter's and St. Paul's islets are built, is characterized by highly fractured and mylonitized terrain outcropping from small linear breaks deprived of abundant sediment cover. These mylonitized rocks represent vertically uplifted transformed mantle peridotite. The breaks that are observed on the slope of the North Ridge are normal faults with mild relief of less than 2 m and oriented in the same general East–west direction as the multiple equatorial fracture zones. These topographic steps generated fan-shaped avalanches made up of poorly sorted debris grading from sand to cobble size to large rectangular blocks (>50 cm in diameter).

The stratigraphic sequence consists of numerous serpentinized and mylonitized peridotite outcrops intruded by occasional gabbroic units, which were observed and sampled at depths of 3416 and 3514 m (SP09-8, -9) (Fig. 8.16). The ultramafic rocks are interrupted by basalts and dykes, which are shallower than 2700 m and occur at various depths (2365, 2000, 1700–1800 and 1593). Cobble-sized breccia of mylonite (SP10-14, SP9-16, and -17) and altered basalts (SP08-3, -12) are also found at various depths (1700-2965 m). The basaltic breccia (2727 m) consists of

Fig. 8.16 Photograph shows a mylonitized (tectonically metamorphosed) peridotite outcrop exposed on the northern flank of the St. Peter and Paul's (Island) massif at 3233 m depth (Fig. 8.15a, b). (Copyright IFREMER, *Saint Paul Cruise 1997, Nautile Dive SP10*)

fine-grained devitrified cement with zeolite (SP08-03). At depths shallower than 1000 m, only mylonitized peridotites were seen.

In addition, our deep-towed television camera instrument (Scampi station # 05) was lowered at night after my dive, at 22:00 on January 1,1998. This real time video imaging indicated that outcrops of peridotite on the slope of the St. Paul's Rocks North Ridge extend down to depths of at least 3927 m near 01°00 N-29°21 W (Fig. 8.15a, b). Thus, the peridotite outcrops continue deeper than the dives we made, which is to say, more than 4 km down the slope.

Origin of the St. Peter's and St. Paul's Rocks (SPPR) Massif

The SPPR has an "S"-like sigmoidal shape, due to the differential strike-slip movements of the two transform faults' walls, linking the two spreading ridges (MAR and the ITR at 27°41′W) (Fig. 8.15a, b). Such a strike-slip motion generated a tear in the lithosphere which accommodated volcanism and a serpentinized diapir on the South Ridge and as well as on the North Ridge. The strike-slip motion of the main transform fault walls was affected by sheering and thrust-faulting tectonic events accompanied by an uplift of the North Ridge and the emplacement of plastically deformed peridotite transformed into mylonite. This is similar to what happens when you squeeze toothpaste out of its tube. Based on geophysical data, bathymetric data and the field observations we obtained, the sequences of formation of the SPPR massif seems to include two processes or phases.

(1) The first phase was a major tectonic break due to the large scale shearing of two rigid plates forming the north and south walls, representing the uplifted transverse ridge of the St. Paul's Rocks fracture zone. This phase was followed by an extension or pull-apart within the two walls of the fracture zone giving rise to the small Intra- Transform spreading Ridges (ITR). During lithospheric extension, the serpentinized peridotites were emplaced, as has been observed elsewhere on slow and ultra-slow magma-starved spreading ridge segments (*see* chap. 4).

(2) The second phase giving rise to the SPPR massif was the result of the localized compression of a thrust-faulting and shearing event during plate readjustment that involved the northern transverse ridge bounding the fracture zone's northern wall. During this readjustment and the strike-slip tectonic motions, the transverse ridge was uplifted and mylonitized (tectonically metamorphosed).

Thus, the North Ridge, including the St. Peter's and St. Paul's Rocks islets, consists essentially of mylonites and was the site of both compression and thrust faulting events. The islets of St. Peter's and St. Paul's Rocks are part of the summit of a transverse ridge (Bonatti 1978, Bonatti et al. 1996), which runs parallel to the F.Z. Grinding and shearing during extreme strike-slip and thrust faulting motions produce the mylonites with their foliated structures found on the islets and the North Ridge. These thrust-faulting and shearing events were accompanied by a small rotational motion of the lithosphere, which gave rise to the "S" shaped (sigmoidal) structure of the SPPR.

The tectonics of the thrust-faulting phenomenon is an on-going process as has been suggested by the intense seismic activity of the area. Such seismicity on the SPPR massif is attributed mainly to strike-slip and thrust faulting motions (Wolfe et al. 1993). Earthquake analyses indicate that the depth of the focal mechanism of these tectonic faulting events is located at 7–14 km depth (Wolfe et al. 1993), which also corresponds to the brittle-ductile boundary (Solomon et al. 1988) of rock deformation (mylonitization) in the lithosphere. This boundary is believed to correspond to the region of the mantle where mylonitization takes place during episodic strike-slip shearing, thrust faulting and uplift, such as what has given rise to the SPPR massif. Mylonitization must have taken place at a high temperature (600°–800 °C) in the mantle when the original mantle peridotites were squeezed upward into the lithosphere and transformed into amphibole bearing rocks as suggested by their mineral deformation and transformation (Melson et al. 1967, Jaroslow et al. 1996).

In summary, the main results of the *Saint Paul cruise* have proven that the SPPR islets are not volcanic, unlike most other oceanic islands; on the contrary, these islets were formed during the uplift of deep-seated mantle peridotite which rose to the surface of the ocean from a depth of more than 7 km within the crust.

The St. Peter's and St. Paul's Rocks (SPPR) massif is made up of an "S" shaped sigmoidal structure resulting from the complex shearing and extension of the lithosphere plate with a small rotational component. The two structurally distinct ridges are attached and form a single unit: the St. Peter's and St. Paul's Rocks (SPPR) massif. The two structures are: (1) The North Ridge comprising the SPPR islets, which is composed essentially of mylonitized peridotite. The lithosphere uplift during strike-slip shearing tectonism is responsible for disrupting the volcanic stratigraphy and for producing a large amount of mass-wasted material, and (2) The South Ridge, which consists of a serpentinized diapir of peridotite (lherzolite) emplaced during plate extension. The few volcanics seen in this area occur in a graben-like structure, 1 km wide and less than 100 m deep, at 2200 m depth. This appears to be similar to the volcanism observed on other spreading ridge segments of the MAR.

Thus the SPPR massif is a fractured, composite structure with localized Intra-Transform Ridge (ITR) volcanism in its southern part (South Ridge). The ITR were all formed in the same way and are found along the St. Peter's and St. Paul's Rocks multiple fracture zones. The ITR were formed within the active transform zone where plate motion is responsible for enhancing an extensional component giving rise to intra-transform volcanism. Thus, volcanic rocks such as the basalt and the intrusions of dolerite dykes found within the SPPR massif have similar compositions to the northern branch of the present day MAR, and the eruption or intrusion of these mafic rocks took place during lithospheric extension. A similar situation where volcanism occurs within a transform fault (called a leaky transform) was seen in the Garrett transform fault in the South Pacific.

References

Arcyana (1975) Transform fault and Rift valley from bathyscaph and diving saucer. Science 190:108–116

Bonatti E (1978) Vertical tectonic in oceanic fracture zons. Earth Planet Sci Lett 37:369–379

Bonatti E, Ligi M, Borsetti AM, Gasperini L, Negri A, Sartori R (1996) Lower cretaceous deposits trapped near the Mid Atlantic Ridge and the opening of the equatorial Atlantic. Nature 380:518–520

Campsie J, Bailey JC, Rasmussen M (1973) Chemistry of tholeiites from the Galapagos islands and adjacent ridges. Nature 245(147):122–124

Cannat M (1993) Emplacement of mantle rocks in the sea floor at mid-ocean ridges. J Geophys Res Solid Earth 98:1978–2012

Casey JF, Bryan WB, Klitgord K, Dmitriev L, Silantyev S, Smith SE (1991) Long wavelength, inter-segment, and intra-segment scale geochemical variations in basalts between 30° and 34°N, Mid-Atlantic Ridge. EOS Trans Am Geophys Union 72:471

Constantin M, Hekinian R, Ackermand D, Stoffers P (1995) Mafic and ultramafic intrusions into upper mantle peridotites from fast spreading centers of the easter microplate (south east Pacific). In: Vissers RLM, Nicolas A (eds) Mantle and lower crust exposed in ocean ridges and in ophiolites, Kluwer Academic Publishers, The Netherlands, pp 71–120

DeMets C, Gordon RG, Argus DF, Stein S (1990) Current plate motions. Geophys J Int 101:425–478

Detrick RS, Needham HD, Renard V (1995) Gravity anomalies and crustal thickness variations along the Mid-Atlantic Ridge between 33EN and 40EN. J Geophys Res 10:3767–3787

Dmitriev L, Party Scientific (1985) Cruise report of the 2nd cruise of the R/V Akademik Boris Petrov. Vemadsky Institute Geochem, Moscow

Edwards SJ, Falloon TJ, Malpas J, Pedersen RB (1996) A review of the petrology of Harzburgites at Hess Deep and Garrett: implication for mantle processes beneath segments of the east Pacific rise. In: Macleod CJ, Tyler PA, Walker CL (eds.) Tectonic, magmatic, hydrothermal and biological egmentations of mid-ocean ridges. Geol Soc Special Publ 118:143–156

Fleming HS, Cherkis NZ, Heirtzler JR (1970) The Gibbbs fracture zone: double fracture at 52°330′ Nin the Atlantic Ocean. Marine Geophys Res 1:37–45

Fox J, Gallo DG (1989) Transform in the eastern central Pacific. In: Winterer NE, Hussong DM, Decker RW (eds) The geology of North America. Geol Soc America N:111–123

Francheteau J, Choukroune P, Hekinian R, Le Pichon X, Needham H.D (1976) Oceanic fracture zones do not provide deep sections in the crust. Can J Earth Sci Vol 13 N° 9:1223–1235

Francheteau J, Patriat P, Segoufin J, Armijo R. Doucoure M, Yelles-Chaouche A, Zukin J, Calmant S, Naar DF, Searle R (1988) Pito and Orongo fracture zones: the northern and southern boundaries of the easter microplate (southeast Pacific). Earth Planet Sci Lett 89:363–374

Francheteau J, Naar DF, Armijo R, Cogné JP, Constantin M, Girardeau J, Hekinian R, Hey R, Searle R (1994a) Black smoker discovered, Pito Seamount near Easter Microplate propagator tip. EOS Trans 75(16):322

Francheteau J, Armijo R, Cogné JP, Girardeau J, Constantin M, Hekinian R, Naar DF, Hey RN, Searle RC (1994b) Submersible observations of the Easter microplate and its boundary. EOS, Trans 75(44):582

Hébert R, Bideau D, Hekinian R (1983) Ultramafic and mafic rocks from the Garrett transform fault near 13°30′S on the east pacific rise: Igneous petrology. Earth Planet Sci Lett 65:107–125

Hekinian R, Aumento F (1973) Rocks from the Gibbs Fracture Zone and the Minia Seamount near 53°N in the Atlantic Ocean. Marine Geol 14:47–72

Hekinian R, Bideau D, Cannat M, Franchteau J, Hebert R (1992) Volcanic activity and crust-mantle exposure in ultrafast Garrett transform fault near 13°28′S in the Pacific. Earth, Planet Sci Let 108:259–276

Hekinian R, Bideau D, Francheteau J, Cheminée J-L, Armijo R, Lonsdale P, Blum N (1993) Petrology of the East Pacific-Rise Crust and Upper mantle exposed in the Hess Deep (Eastern equatorial pacific). J Geophys Res 98(B5):8069–8094

Hekinian R, Juteau T, Garcia E, Sichler B, Sichel B, Udintse G, Apprioual R, Ligi M (2000) Submersible observations of the equatorial Atlantic mantle: the st Paul fracture zone. Marine Geophys Res 21:529–560

Jaroslow GE, Hirth G, Dick HJB (1996) Abyssal peridotitemylonites: implications for grain-size sensitive flow and strain localization in oceanic lithosphere. Tectonophysics 256:17–37

Lagabrielle Y, Mamaloukas-Frangoulis V, Cannat M, Auzende J, Honnorez J, Mevel C, Bonatti E (1992a) Vema fracture zone (Central Atlantic): Tectonic and magmatic evolution of the median ridge and the eastern ridge¿transform intersection domain. J Geophys Res 97(B12):1978–2012

Lagabrielle Y, Mamaloukas-Frangoulis V, Cannat M,.Auzende J-M, Honnorez J, Mevel C, and Bonatti E, (1992b) Vema fracture zone (Central Atlantic): Tectonic and magmaticevolution of the median ridge and the eastern ridge-transform intersection domain. J Geophys Res 97:17331–17351

Linklater E (1972) The voyage of the challenger. Garden city. Doubleday and Company Inc, New York p 288

Lonsdale P (1978) Present day plate motion. J Geophys Res 83:5331–5354

Lonsdale P (1989) Segmentation of the Pacific-Nazca spreading center 1°N-20°S. J Geophys Res 94:12197–12225

Melson WG, Jorosewitch E, Bowen VT, Thompson G (1967) St. Peter and St. Paul's Rocks: a high temperature derived Intrusions. Science 155:1532

Melson WG, Thompson G (1973) Glassy abyssal basalts Atlantic sea floor near St. Paul's rocks: petrography and composition of secondary clay minerals. Bull Geological Soc America 84:703–716

Morgan JP, Chen Y (1993) Dependence of ridge-axis morphology on magma supply and spreading rate. Nature 364:706–708

Naar DF, Hey RN (1989) Recent Pacific-Easter-Nazca plate motions. In: Evolution of Mid-Ocean Ridges, IUGG, Symposium 8, AGU Geophys Monogr 57:9–30

Naar DF, Hey RN (1991) Tectonic evolution of the easter microplate. J Geophys Res 96:796–7993

Natland JH (1989) Partial melting of a lithologically heterogeneous mantle: inferences from crystallization histories of magnesian abyssal tholeiites from the Siqueiros Fracture Zone. In: Saunders AD, Norry MJ (eds) Magmatism in the ocean basins. Spec Publ Geol Soc London 42:41–70

Niu Y, Hekinian R (1997) Basaltic liquids and harzburgitic residues in the Garrett transform: a case study at fast-spreading ridges. Earth Planet Sci Let 146:243–258

Perfit M, Fornari DJ, Smith MC, Bender JP, Lagmuir CH, Haymon RM (1994) Fine scale petrographic variation along mid-oceanic ridge crest magmatism. Geology 22:375–379

Rusby RI (1993) Segmentation and reorganization of the equatorial Mid-Atlantic Ridge (5°-0°N) Bridge News 16:19

Searle RC (1983) Multiple, closely spaced transform faults in fast-slipping fracture zones. Geology 11:607–610

Solomon SC, Huang PY, Meinke L (1988) The seismic moment budget of slowly spreading ridges. Nature 334:58–61

Tilley CE (1947) The dunite mylonite of the St. Paul's rocks (Atlantic Ocean). Am J Sci 245:483–491

Turner FJ, Verhoogen J (1960) Igneous and metamorphic petrology. McGraw-Hill, New York p 694

Udintsev GB, Kurantsova NA, Pronina NV, Sminova SB, Ushakova MG (1990) Finds of continental rocks and sediment of anomalous age in the equatorial segment of the Mid-Atlantic Ridge. Trans USSR Acad Sci Dokl 312:111–114

Washington HS (1930) The petrology of St. Paul's rocks report on the geological collection made during the quest on the Shakelton-Rowet expȩdition. 1921-2. British Museum Natural History Capt XII :19 p

Wilson JT (1965) A new class of faults and their bearing on continental drift. Nature 207:343–347

Wiseman JDH (1966) St. Paul's rocks and the problem of upper mantle. Geophys J Royal Astron Soc 11:319–525

Wolfe CJ, Bergman EA, Solomon SC (1993) Oceanic transform earthquakes with unusual mechanism or locations: relation to fault geometry and state of stress in the adjacent lithosphere. J Geophys Res 98:16187–16211>

Chapter 9
Hotspots

Abstract Hotspots are regions of the lithosphere that are fed by the melting of a hot mantle plume, giving rise to volcanic activity on the Earth's surface. The source of hot mantle material (a "plume") could have originated in the lower mantle, at the boundary of the molten Earth's core at 2900 km depth. The mantle's thermal flow ceases when attaining the colder and more rigid lithosphere boundary, less than 100 km deep. The rising of hot mantle plumes contributes to partial melting and to a thinning of the rigid lithosphere's plates. Some volcanic edifices created during hotspot volcanism are formed on top of spreading ridges such as the islands of Iceland and Jan-Mayen in the Atlantic or Amsterdam Island in the Indian Ocean. Many other hotspot volcanoes occur in the intra-plate regions of the world's oceanic basins and form tall submarine or subaerial edifices rising more than 4000 m from the seafloor.

Introduction

The movement of the lithospheric plates over "fixed" thermal anomalies located in the upper mantle will form many linear volcanic seamounts and islands (Wilson 1963; Morgan 1972). Burke and Wilson (1976) identified up to 122 hotspots, which have been active during the past 10 Ma, as well as 52 hotspots located near the ridge axis. Crough (1983) detected about 140 hotspots located in continental and oceanic areas (Fig. 9.1). It is likely this number is no longer valid because, when looking at the World's gravity anomaly (Smith and Sandwell 1997), it appears that many unexplored intraplate structural highs may also be the result of hotspots.

The excess topography for volcanic constructions is proportional to the amount of melt delivered to the surface, and the amount of melt delivered is often considered to be greatest above hotspots. However, based on satellite altimetry data as well as on conventional bathymetry (depth sounding), it is difficult to distinguish between structures caused by hotspot volcanism from those inherited from crustal

R. Hekinian, *Sea Floor Exploration*, Springer Oceanography,
DOI: 10.1007/978-3-319-03203-0_9,

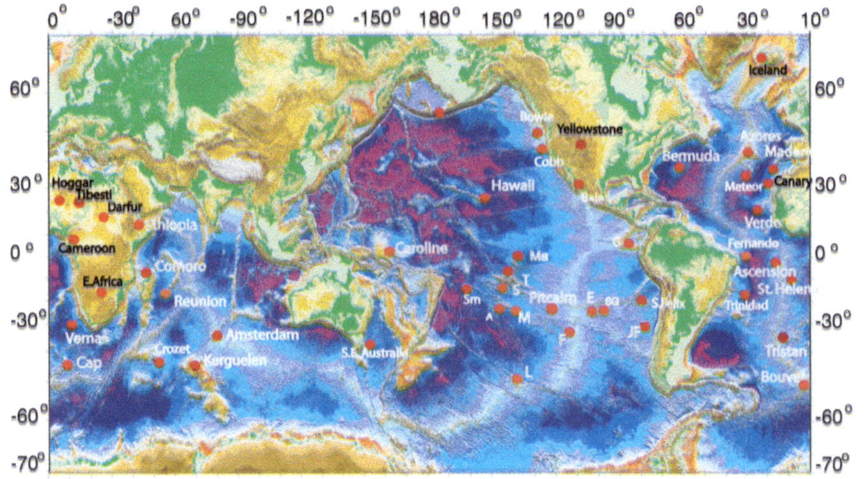

Fig. 9.1 World's map shows the submarine relief and the distribution of hotspots. *A* Arago smt., *E* Easter Isl., *M* Macdonald smt., *L* Louisville smt., *F* Foundation smt., *JF* Juan Fernandez Isl., *G* Galapagos, *Ma* Marquises Isl., *T* Tuamotu, *SG* Sal y Gomez Isl., *Sm* Samoa Isl., *S* Society Isl. Satellite altimetry map of Sandwell and Smith (1997) reproduced by D. Aslanian (IFREMER)

accretion at spreading ridges. Thanks to bottom imagery obtained from side-scan sonar coupled with deep towed bottom camera observations, we can also obtain an indication of "young" sea floor.

Even if we do not have a clear, general pattern for the distribution of volcanic edifices on the sea floor, nevertheless certain structural features such as the alignment of volcanoes and lithospheric swells can be used to define and predict the occurrence of newly created intraplate volcanoes having a "hotspot origin". Volcanic chains running perpendicularly or obliquely to oceanic ridge segments (such as the Foundation volcanic chain and the Easter Island volcanic line in the South Eastern Pacific) as well as undersea aseismic ridges (i.e. Walvis in the Atlantic, Ninetyeast in the Indian Ocean, and Cocos and Nazca Ridges in the central Pacific) are areas where upwelling mantle plumes have generated linear ridges topped by volcanic edifices. These volcanic chains are believed to be the result of the breaking up of the continental masses, which has thereby enhanced volcanism.

Thus, an important characteristic of hotspots is that they are usually linear features and/or they are well-developed isolated volcanic edifices that were built during magmatic upwelling and often form islands. In fact, most of the world's islands are the result of hotspot volcanism. However, hotspot volcanoes such as Iceland, the Azores, St. Helena, Ascension and Bouvet islands in the Atlantic and St. Paul, Crozet and Amsterdam islands in the Indian Ocean also occur in association with spreading ridges. Furthermore, hotspot volcanoes could occur as localized entities, often forming clusters of edifices, such as the Canaries and Cap Verde Island as well as the tall seamounts in the North Atlantic basin.

When mantle plumes forming hotspots are located in the vicinity of spreading ridges, this could interfere with the ridge segments and eventually change the geometry of a spreading ridge as well as the composition of the volcanics erupted at the ridge axis. Other chapters in this book (such as Chap. 7) will show how this type of interaction can influence the rock composition and geological structure when a spreading ridge meets a hotspot.

Why do we Study Hotspots?

Oceanic basins cover about 70 % of the World's oceans and seas. Along with spreading ridges, basins are also the sites of some of the major volcanic activity on Earth (Fig. 9.1). The intraplate regions have deep basins associated with hotspots where the exchange of matter between the Earth's mantle and the lithosphere is taking place. It is estimated that about 30 % of seafloor volcanism is due to hotspot activity.

It has only been since the 1970s that oceanic basins and their observed major linear structures were investigated. Before then, most of our efforts to understand oceanic basins were related to the study of their islands. Early geological studies had to concentrate on the most easily accessible, subaerial parts of the volcanic edifices. However, Brousse (Brousse and Maury 1980) and Menard (1986) accomplished pioneering work in Pacific Island studies.

Apart from being popular holiday destinations, oceanic-island volcanoes such as Hawaii, Tahiti, or the Canaries are built from magmas that yield valuable information about the interior of our planet. It is now well established that most of the Earth's volcanic activity occurs underneath sea level and often at great depths (>1000 m). Although the plate tectonic theory provides a framework for understanding the volcanic events on spreading ridge axes, it does not explain the presence of volcanism in ocean basins. Hence, a study of submarine volcanoes found at a distance from spreading ridges can provide critical information about the depth at which the mantle begins to melt and how mantle plumes rising towards the surface could generate "hotspots".

Hotspots have influenced the Earth's spreading ridge systems and learning about the processes controlling "hotspot versus spreading ridge interactions" could help us to better understand the global convection system of our planet's interior. If the formation of landmasses that went on to build today's continents started in the middle of ancient oceans, it would be worthwhile to understand how present-day mantle plumes form hotspots and oceanic islands (see Chap. 2). At the beginning of our planet's existence, emerged landmasses must have been the exposed summits of large, volcanic edifices, which appeared in the first "magma ocean". The importance of today's hotspots is that they produce volcanoes, which are the surface expression of upwelling convective cells of chemical fluxes having a correlation with deep mantle material (Morgan 1971, 1972). Jason Morgan's model (1971) was based on the idea that lower, primordial mantle material rises in the lithosphere as a plume or a "blob" having a lower density, due to its higher heat and its different composition when compared with the surrounding material. The plume rises from deep in the mantle, probably in the vicinity of (or exactly at) the core-mantle boundary region, about 2900 km deep (Turcotte and Oxbourgh 1978). The simplest hotspot model implies that the magmatic source is from the deep mantle and the hotspots are therefore relatively fixed with respect to each other, to a precision of about 1–2 cm/yr. An alternative view is that the plumes giving rise to hotspots are not "fixed", but rather they could be moved by convection currents in the mantle. The combined motion of the landmasses and the movement of the oceanic lithospheric plates could affect the mantle's convective system, which would then alter the trajectory of a rising mantle plume.

One important characteristic of hotspots is that they are relatively long-lived and their presence in a given place can be retraced up to 200 million years ago (Le Pichon and Huchon 1984). Thus, hotspots have existed prior to the formation of our present day ocean. The variation of "rock age versus distance" is an important notion for the definition of a hotspot. If we are dealing with a hotspot, the age of the volcanic rocks should increase at a distance from their point of eruption above the hotspot, since the lithosphere moves above and beyond the hotspot. There are notorious exceptions to this fact, but they were attributed to the existence of remainders of magmatic reservoirs for a period of 3–4 Ma such in Hawaii (Jackson 1976).

Within continental regions, upwelling magma underneath the lithosphere will contribute to the rupture of the plates and initiate oceanic opening. This happened more than 120 million years ago in the Atlantic Ocean, during the separation of America and Africa. Such an opening left a scar in the lithosphere thereby giving rise to volcanic chains such as the Walvis Ridge, the Rio Grande Rise, the Sierra Leone Rise in the Atlantic and to other similar structures in the Indian Ocean (i.e. the Ninetyeast Ridge) and in the Pacific (Nazca and Cocos ridges) at the beginning of the break-up of PANGEA. This phenomenon of rupture of the oceanic lithosphere is an on-going process.

French–German Cooperation: The *"Volcanisme Intra-Plaque"* Program

Franco-German collaboration for studying intraplate volcanism related to hotspots and their associated phenomena began in 1986. The project started after a telephone conversation between Professor Peter Stoffers (Fig. 9.2a) from the University of Kiel in Germany and Dr. Jean-Louis Cheminée (Fig. 9.2b) at Institut de Physique du Globe in Paris (IPGP). Peter asked Jean-Louis if he was interested in collaborating on the study of the Society hotspot, in the South Pacific. It was then that Jean-Louis told Peter to call me at IFREMER to ask my opinion and to see if I were interested. In fact, I was very interested in such an adventure because it could open a new perspective, since it entailed the exploration of a new area of the ocean floor that

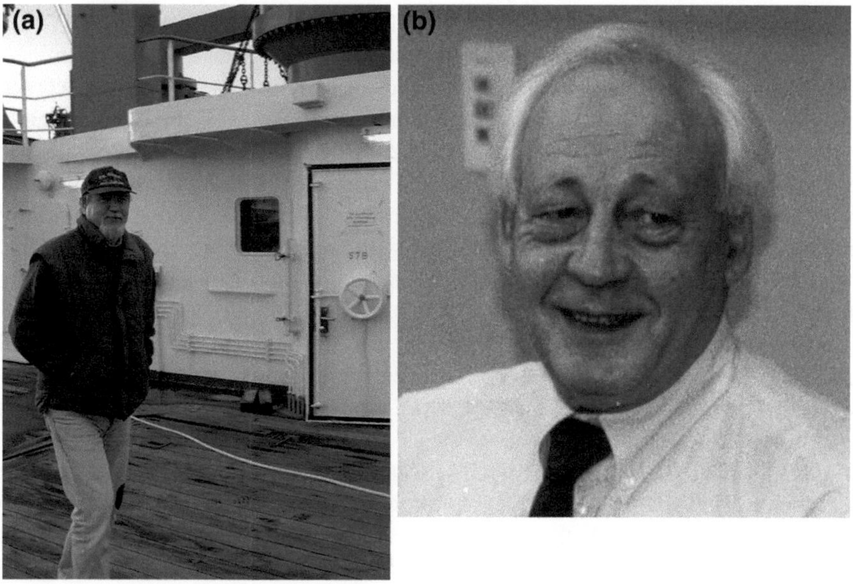

Fig. 9.2 **a** Professor Peter Stoffers (University of Kiel in Germany) on board the R.V. SONNE in the south Pacific Ocean, 1988, and **b** Dr. Jean-Louis Cheminée from the *CNRS* (Centre National de Recherche Scientifique, France)

I hadn't studied before. I was also pleased to be asked to join a joint Franco-German effort to share scientific and technical means. The VIP Program (**VIP** = ***Volcanisme Intra-Plaque*** = *Intraplate Volcanism*) was a welcome addition to the scientific projects that were currently happening in France. The funding agencies for the VIP Project included IFREMER (Institut Français pour l'Exploitation des Oceans), Centre National de Recherche Scientifique (CNRS), Deutshe Forschungsgemeinschaft (DFG), and Bundesministerium für Forschung and Technologie (BMFT). Several German institutions were involved in the program, such as the University of Bremen, Kiel, Hamburg and Regensburg.

However, at the time, this project was not a priority in the marine geosciences department of IFREMER, so it was difficult for me to mobilize my colleagues around such an endeavor. In order to have funding and ship time at sea, it was important that my institution accept the program on intraplate volcanism. Since I couldn't seem to raise much enthusiasm from the direction of IFREMER, I had to rely on other French institutions in order to be able to get this project launched. Indeed, the person responsible for the scientific programs at IFREMER was quite reluctant to approve the project. As has been true in many other cases, when a lack of interest in ocean exploration infected some people who had responsibilities in Paris and/or in my department, it was hard for me to pass my message. Therefore, because I had a lot of difficulty in promoting the project within my department, the only way for me to get involved in *VIP* was by means of help from outside institutions such the CNRS (IPG in Paris) and from some French universities.

Thus, the French-side of the international Intraplate Project (called "VIP": for the initials of ***Volcanism Intra-Plaque***) was headed by Jean-Louis Cheminée. Jean-Louis, whom I had known for years and with whom I had previously worked on the FAMOUS project, was very enthusiastic. He was an internationally renowned volcanologist and was responsible for the volcanological observatory at the *Institut de Physique du Globe* (IPG) in Paris. He was used to working in the field, and had started his career working on the volcanoes of the Afar region in Ethiopia, however he was also familiar with many oceanic islands. We had become friends during the FAMOUS diving project and had kept in touch ever since.

Most of the field exploration necessary for the hotspot project was carried out using German research vessels, however, three French sea-going expeditions were also involved. The N.O. *JEAN CHARCOT* worked in 1986 (when three days were assigned to the project because the ship was in the area for another reason). The N.O. *LE SUROIT* and the diving saucer *Cyana* were used in 1989 to explore the Society Hotspot. The French exploration of the Society hotspot continued ten years later in 1999, with the submersible *Nautile* and the N.O. *L'Atalante*. In addition, during the same cruise with the *Nautile* and N.O *L'Atalante* in 1999, some exploration of the Pitcairn hotspot also took place.

German institutions are well equipped to carry out long periods of oceanographic exploration. They have several outstanding vessels such as the *POLARSTERN*, an icebreaker, which is equipped to undertake expeditions in the Arctic and Antarctic regions. The F.S. *METEOR* is also a comfortable and well-equipped ocean-going vessel that does most of its work in the Atlantic Ocean, the Mediterranean Sea and the Red Sea. It was named after the first German research vessel, also called *METEOR*, which had an outstanding record of scientific data gathering during the beginning of the Nineteenth century when numerous physical oceanography and sedimentary studies were carried out and gave us important results that were later published by Georg Wüst and other German scientists. However, the F.S. *SONNE* (meaning "the Sun" in English) was the vessel that we used the most frequently during our project dedicated to Intraplate Volcanism (VIP) in the Pacific.

I have participated on more than 10 ocean going expeditions on the German Research Vessels F.S. *SONNE* and F.S. *METEOR*. Our VIP cooperation started with *SONNE*'s leg 47 in 1987, which was aimed at exploring the Society hotspot and the Austral volcanic chain, a line of volcanic islands and seamounts extending for a length of more than 1000 km and a width 500 km between latitude 18°S and Longitude 140°W. Our bi-national cooperation continued with eight other legs on the F.S. *SONNE*, extending exploration to the Pitcairn island area, Easter Island and the South Pacific-Antarctic spreading ridge segments. This fruitful Franco-German collaboration brought us to visit and explore uncharted areas of the Pacific Ocean and contributed to enhance the study of intraplate volcanism. The project helped make students more aware of the importance of hotspot volcanism, and increased our understanding of the volcanic processes taking place within individual oceanic plates, which had once been thought to be rigid and passive. The significance of the VIP project on a European scale, and even on an international level, was quickly recognized.

In 1987 and in 1992, the F.S. *SONNE* carried out geophysical and geological investigations of the Pitcairn and Easter Island hotspot chain. We continued our collaboration during a study on ridge-hotspot interactions in 1995, 1997 and 2001. This project was called the "Foundation Hot Line" and involved both the German ship *SONNE* in 1995 (*Leg SO 100*) and in 2001 (*Leg SO157*) both of which legs were organized by the University of Kiel in Germany, and also involved the French vessel *L'ATALANTE* in 1997. The cruises carried out by the research vessels *SONNE* and *L'ATALANTE* turned out to be very successful. During the *L'ATALANTE* mission, a multi-disciplinary approach was adopted and the VIP project assumed an even more international aspect than was originally expected. Several scientists from the USA, Canada, the United Kingdom and New Zealand were involved in addition to our core team of French and German scientists.

There are a few differences between being on board a German or a French oceanographic vessel, other than the languages spoken by the crew. For example, on the German vessels, once the instruments had been launched at sea by the ship's crew, then the dredging or the deep-towed instrument operations were all conducted by the scientists. Dredging can be very tedious and usually three people were assigned to work on each team. There was the person who was responsible for the dredging operation, once the dredge was on the sea floor, plus two assistants who were expected to take notes with any pertinent information during the operation, such as the amount of cable that had been let out, and the time that the dredge remained on the bottom. We needed to note the information given by a tension-meter with a pressure gauge, which was calibrated with the weight of the cable plus the dredge. The gauge gave a constant reading of the tension felt on the cable as it pulled the dredge along the ocean floor, or should it become snagged on something on the bottom.

The reason for choosing the Pacific ocean for our project on hotspot volcanism was because only a few scientific expeditions had been carried out on the seafloor of this large ocean, which covers one-third of the Earth's surface. The Pacific Ocean is also one of the most volcanically active regions and consequently has the largest amount of hotspot volcanoes. Furthermore, the South Pacific region was also selected because of the proximity of Tahiti for the logistics of moving scientists and material on and off the research vessels.

Polynesian Hotspots

The term Polynesia literally means "many islands" in Greek. James Cook and other explorers had discovered that despite the large variety of islands and their distance from one another, the inhabitants all seemed to speak a similar language. French authorities subsequently limited the term "Polynesia" to the islands where

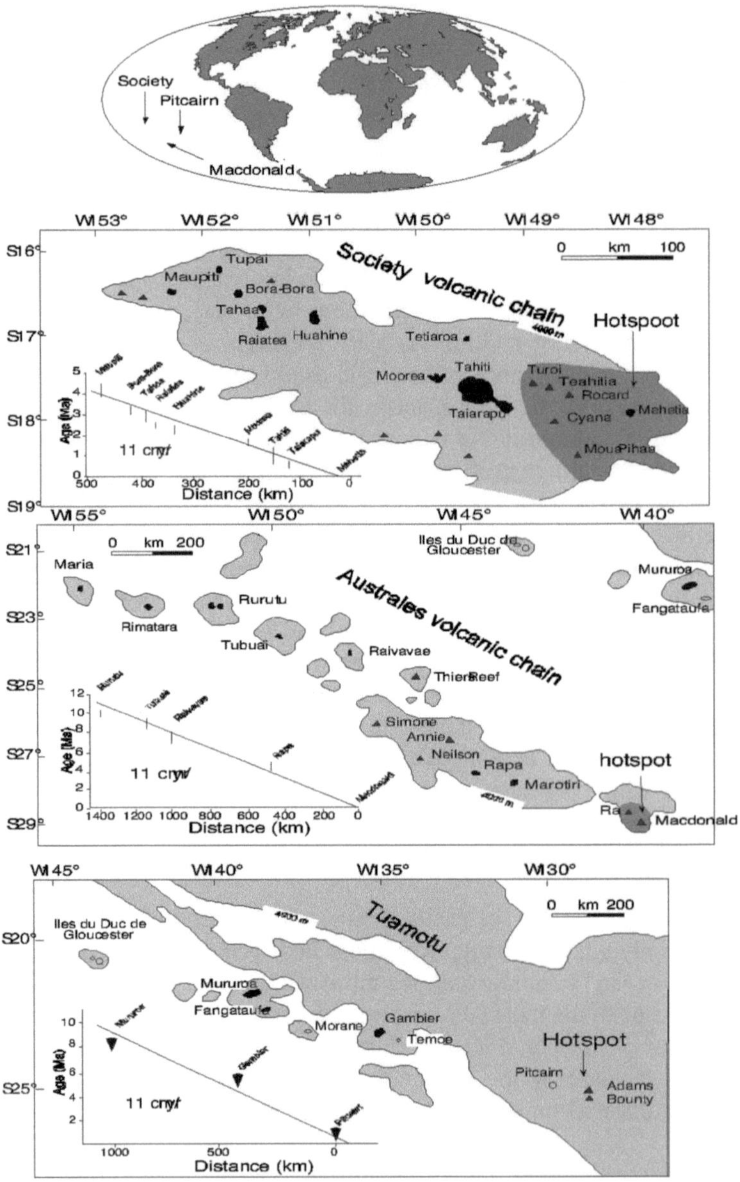

Fig. 9.3 Hotspot distribution and ages of the Society, Austral and Pitcairn volcanic chain in the South Pacific are reported. The age dating on the islands and seamounts helps to show the direction and rate of spreading of the Pacific Plate above the hotspots (Hekinian and Binard 2008)

the indigenous people all spoke the same language. The area of Polynesia comprises a triangle formed by the islands of Rapa Nui (Easter Island), Aotearoa (New Zealand) and Hawaii (Fig. 9.3).

In 1986, H.W. Menard wrote *The Europeans that had entered the Pacific for more than 300 years came from an unwashed, polluted, disease ridden culture that had passed from the Bronze Age to the Iron Age several millennia earlier. The people they found on the tiny isolated islands were clean, healthy, and generally friendly.*

Some of these islands were visited by the scientific team during our VIP expeditions, and several were used as a port of call for our research vessels. In 1987, during one of our stops in Tahiti (*SONNE*, Leg 100), we visited this island with Jacques Talandier who was, at the time, the director of the French seismic station in Papeete. Jacques took us on a tour to observe the volcanic landscape of the island. The last volcanic eruption in Tahiti was about 900000 years ago and despite the tropical vegetation, which grew abundantly, we were able to see remnants of large lava tubes forming tunnels. The tunnels were so large that two trucks could pass through them, side-by-side.

The submarine portion of the Society hotspot, east of Tahiti Island, was the target of several international missions. The deep-ocean field observations were carried out by the submersibles *Cyana* and *Nautile*, as well as by a deep-towed television camera system Ocean Floor Observation System (OFOS), plus side-looking sonar (GLORIA) and sea floor back-scatter imagery data collected by the F.S. *SONNE*, N.O. *L'ATALANTE*, N.O. *J. CHARCOT* and HMS *Charles DARWIN* of the United Kingdom, in 1988. The multichannel bathymetry coverage included SeaBeam (N.O. *J. CHARCOT*, 1986), Hydrosweep (F.S. *SONNE*, 1987), and the SIMRAD-EM12D systems (N.O. *L'ATALANTE*, 1999). However, it was during the *SONNE* Leg 47, in 1987, that an extended survey was first made using the German-constructed deep towed camera system (OFOS = Ocean floor observation system), plus dredging, plus the use of a large grab sampler equipped with a bottom camera system (GTV = Grab, TV camera). This initial exploratory mission permitted us to obtain our first knowledge about the volcanic and hydrothermal activity that was taking place on the Society hotspot.

The various expeditions of the VIP program had the additional advantage of obtaining information on the seismicity of the Society and Austral hotspots thanks to information provided by the Polynesian seismic stations, which centralized all the information on earthquakes and tsunami warnings with the Polynesian Seismic Network (*Réseau Sismique Polynésien RSP*) in Tahiti. Seismic stations had been installed on several islands such as Tahiti, Moorea, Rangiroa, Tubuai and Gambier. These stations were first built in the 1960s and in the late 1980s, they were skillfully being directed by Jacques Talandier from the Centre Enérgie Atomique (CEA, France).

In addition to Tahiti, during more than 20 years of the VIP Program's exploration of intraplate regions associated with hotspots, we made several stops at Easter Island (Rapa Nui). This Island was able to provide refueling facilities and had an international airport that facilitated traveling for our scientific teams and shipboard crew.

Easter Island (Rapa Nui)

Rapa Nui island is located at Latitude 27°09′S and Longitude 109°20W, in the South Pacific Ocean, about 3,700 km from the South American coast and 2,000 km from Pitcairn island (Fig. 9.3). The name Rapa Nui means "Big Bald Mountain" in Polynesian. (Rapa Nui means "Big Rapa" as opposed to Rapa-ita, meaning "small Rapa".) Probably, its first inhabitants gave the island its Polynesian name, because when they arrived they did not see any trees on it. Rapa Nui was colonized quite late, about 660–870 years ago, as shown by the carbon 14 dating on charcoal (Hunt and Lipo 2006). The charcoal strata were found at a depth of about 345 meters below the surface in a part of the island called Anakena. Jacob Roggeveen, on board the Dutch vessel *ARENA*, was the first western explorer to see the island, on Easter Day of the year 1722. The Chileans annexed the Island in 1888.

Easter Island is one of the places where Polynesians who were probably sailing from Papua, New Guinea, and from Fiji found a safe haven. It is an island of legends with mysterious statues of an unknown origin. The statues were most likely carved between the years 1000 to 1600 AD from the red and black tuff on the flank of the island's volcano, which had become a quarry. These strange, tall men with flat hats are called *mohai* in the Polynesian language. They can be up to 22 m tall. About 200 of the statues still lie unfinished in their quarry, while 500 others are scattered throughout the Island. The oldest statues are smaller, while the taller ones (180 tons), are the youngest. The population's memory of their ancient language and writing has vanished, so the local people can no longer read the hieroglyph tablets, copies of which are carved and sold to tourists (Decaud 1993).

Easter Island is a strategic place for marine geologists wanting to explore the South Pacific Ocean because of its position near the extension of the southern branches of the East Pacific Rise spreading centers, and also because of its relatively easy access by commercial planes. In addition, there is a refueling station for ships. The island is located 3,700 km from Valparaiso, Chile. The easiest way to get there when traveling from Europe is to travel by plane through Santiago de Chile.

I have been to Easter Island 8 times during the various cruises on which I participated during my career. Since my first visit in 1989 (*SONNE* cruise, Leg 65), there have been many changes on the island. The most drastic change was seen in 1997, after the departure of the film crew making the Hollywood production film called "RAPA NUI" with Kevin Costner. For the first time, I saw sidewalks and asphalt on the main road. I believe there must now be a few hundred automobiles on the island, whereas the first time I landed, there were only about 30 cars, and horses were the main means of transportation. Indeed, even now, there are more horses (5,000) on the island than human beings (4,000 local population).

When visiting Easter Island, we usually stayed with a family who lived in a house located on top of a hill, called the "Villa Tiki Hotel", in the village of Hangaroa. Maria Giorgina Paoa Hucki was our hostess, and she always received us with a warm welcome, which made all of us feel comfortable. Her husband was a teacher in the primary school while Maria remained at home to take care of her guest house and their two children.

Usually, Maria waited for us when we arrived at the small airport of Mataveri. The airport was built with U.S. funding and was adapted to receive the Space Shuttle in case of emergency. The landing strip was a strait highway and the largest size aircraft were able to land. The building for the travelers was relatively small, and was disproportionate with respect to the size of the landing airstrip. We arrived by plane on board a Chilean Airlines flight after departure from the town of Santiago De Chile. Three times a week, there was a plane that went to Easter Island and then continued on to Tahiti (Papeete). The Pasquean people (residents of Easter Island) have relatives in Tahiti, located about 4000 km to the west, which makes the airline very attractive for Polynesians wanting to visit family members as well as for tourists.

We would put our luggage in Maria's van and then be driven to her house located on a hill about five minutes from the airport. Breakfast and dinner were served during our stay. Usually they managed to serve a feast of lobster to us, as well as home-cooked local specialties, which consisted of fish-based meals and home-baked bread. The Easter Islanders speak Polynesian and Spanish but they cannot read the writing that is found on the wooden or stone tablets left by the early residents, copies of which are sold to the tourists. The Rapa Nui customs have been transmitted orally for several generations, however the mystery of the statues is still not yet solved.

The Island has a triangular shape. Each corner is formed by a large volcanic cone, which represents the coalescence of several smaller (<500 m in diameter) cones that have melded together to form the emerged landscape. The entire island is volcanic in origin and built on an old crust less than 10 million years old. The earliest and the latest eruptions on the island have been dated from 4 million to less than 1 million years ago, respectively. Recent age dating indicates that volcanism continued on Easter Island until at least 130 thousand years ago (O'Connor et al. 1995). The three volcanic cones that are visible on each corner of the Island are responsible for the shape of the landscape. The crust on which the Island is built is probably older, about 7 Ma if we extrapolate on the basis of the average rate of spreading (5 cm/yr).

Easter Island is located at about 350 km to the east of the East Rift spreading ridge of the Easter Microplate, which has been recently active. According to the reconstruction of the Easter microplate (Naar and Hey 1991), the Easter Island hotspot was formed about 3–4.8 million years ago near the southeastern boundary of the Easter microplate, an area which is now marked by a fracture zone called the Orongo transform fault (See Chap. 8). According to the hotspot hypothesis of Jason Morgan (1972 and 1971), the Sala y Gomez and the Easter Islands are part of a volcanic chain that was constructed when the Nazca Plate in the South Pacific moved above a mantle plume and gave rise to a volcanic chain.

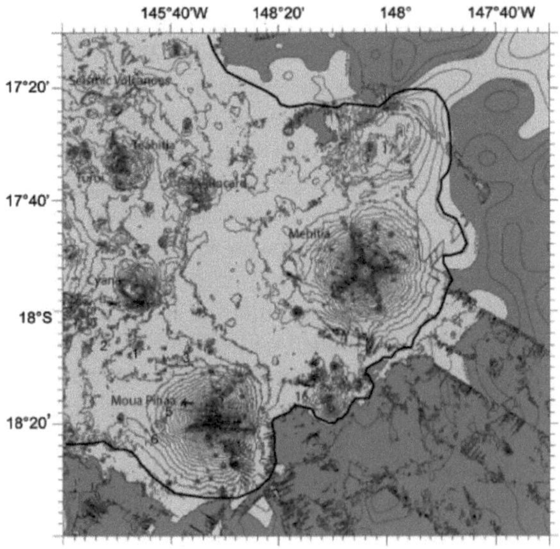

Fig. 9.4 Map of the Society volcanic chain in the South Pacific. The reconstruction of the bathymetry was done by A. Bonneville (personal communication) from a compilation of multichannel data. A series of volcanic edifices varying in size from 500 m to more than 1000 m have been constructed during hotspot activity, the limit of which is indicated by a thick, darker contour line (<4000 m)

The Society Hotspot

The Society Island chain and its hotspot region are located in the South Pacific Ocean and cover a total area of about 65,000 km^2. The volcanic edifices increase in age from the currently active Mehetia island (W148°04′-S17°52′), located at the border of the hotspot and surrounded by an abyssal hill region, to the Maupiti emerged volcano (W152°15′-S16°27′), located 440 km to the northwest, with an age of 4.34 Ma old (Duncan et al. 1974) (Fig. 9.4).

The area covered by the Society hotspot itself is about 15,000 km^2, and comprises about eighteen volcanic cones more than 500 meters tall formed on an ancient oceanic crust that is 65 million years old. From seismic monitoring (Talandier and Okal 1984; Okal et al. 1980) and based on field observations in the area, it was inferred that at least six volcanoes, located above the 3750 m contour line, are still active. Earthquake swarms were monitored by the Polynesian seismic station, which gave the location of the epicenters and was able to detect volcanic activities. Also, direct field observation and sampling of freshly erupted lava and hydrothermal venting helped us to localize recent volcanic episodes.

The individual cones are found to be grouped around seven major volcanic centers, which include: (1) the Teahitia volcano (2100 m high) at 17°32′S-148°50′W, (2) the newly discovered edifice called Cyana located to the south of Teahitia at 17°55′S-148°50′W, (3) The Rocard volcano east of Teahitia, (4) Turoi seamount west of Teahitia, (5) the Moua-Pihaa seamount rising to 3500 m above the sea floor, (6) Volcano # 17 north of Mehitia and (7) the island of Mehetia, which rises 450 m above sea level. In addition, other smaller edifices less than 500 m tall are scattered around the larger volcanoes. One group of small edifices, called the Seismic

Volcanoes, represents a relic of an ancient pre-hotspot edifice located on the northwestern corner of the Society hotspot (Figs. 9.3, 9.4). Their seismicity is due to crustal fracturing reactivated during hotspot activity.

All the large hotspot volcanoes show parasite cones and radial volcanic ridges or steep-sloped aprons. The aprons represent radial fissures of magma channels, often ending in small volcanic cones along the flanks of the edifices. The sea floor at a distance from the base of the major volcanic edifices is relatively flat and covered with fresh volcanic glass debris mixed with a thin sediment cover. A compilation of the multichannel bathymetric data (Bonneville personal communication; Jordahl et al. 2004; Cheminée et al. 1988; Stoffers et al. 1989) shows that the hotspot volcanoes are surrounded by topographic lows (3600–3950 m depth) bordered by deeper (>4000 m depth) abyssal hill regions representing an ancient, 65 million year-old oceanic crust.

When assuming that the plate moves at a speed of 11 cm/yr over a fixed hotspot (Heron 1972; Duncan et al. 1974; Minster and Jordan 1978), volcanic activities within the hotspot must have taken place during the last 900,000 years. Also, potassium-argon isotopic age dating carried out by Isabelle Le Roy in 1994 indicates that most submarine edifices are recent and go up in age to 550,000 years old. The youngest island of the chain is Mehetia, whose last eruption took place about 25000–73000 years ago (Leroy 1994).

Mehetia Island

The island formerly called “Meetia” or Mehetia lies 60 miles away from Tahiti and is part of the Society volcanic chain. Captain Samuel Wallis first discovered the Island on June 17, 1767. The island is now privately owned and is uninhabited except for the presence of a few goats. Mehetia (2 km^2) is a volcanic cone and rises 435 meters above the sea surface. It is about 4035 m high above the sea floor of the Society hotspot (Fig. 9.4). The last volcanic eruption on the island was dated as being 1 million years ago. During the last subaerial explosive eruption, a crater 150 m in diameter and 80 m deep was built. The only evidence of recent volcanism is found on the south flank of the volcano, at more than 1600 m depth, where seismic activity was detected (*Réseau Sismique Polynésien*, RSP) from December to April 1981. This was a target area for the submersible *Cyana*’s dive TH10 that took place in 1989. Dive TH10 found evidence of recent underwater explosive eruptions in the form of pyroclastic flows at 2301 m depth. The eruption took place from a small parasite cone located on the southern flank of the island at 2300–2400 m depths (Figs. 9.5, 9.6).

This uninhabited island has vertical, volcanic cliffs, which makes the island difficult to access. During the FS *SONNE* cruise of 1987, a group of German and French scientists landed on the SW side to sample the island. Because of the strong wave action on the rocky coast, the approach to the island was dangerous. A zodiac was first sent with the ship’s captain, Hartmut Andresen, to explore the

Fig. 9.5 Aerial photograph of Mehetia Island shows a small crater 200 m in diameter (courtesy of Nicolas Binard 1990)

Fig. 9.6 Lava tubes flowing down the slope of the Teahitia volcano at 2635 m depth (Hekinian and Binard 2008). (Copyright IFREMER, *Teahitia cruise 1989, Cyana dive Cy89-09*)

feasibility of landing. Captain Andresen was a tall man with a long face, grayish brown hair, deep blue eyes, and he always wore a red or white sweatband on his forehead (Fig. 9.7). He looked like a pirate coming from the time of the privateers of Saint Malo.

Fig. 9.7 Captain Hartmut Andresen wearing a white headband and Roger Hekinian in a zodiac, going to Mehetia island (Photo courtesy of N. Binard 1991)

The black rocks of basalt sticking up on the sea's surface near Mehetia were hazardous for the approach of the German research vessel. Nevertheless, the second officer on the bridge was trying to get closer to the zodiac, in order to pick up Captain Andresen and Nicolas Binard. Since their mobile telephone failed to work on board the zodiac, Captain Andresen was desperately waving his arms, trying to inform the officer on the bridge about the imminent danger of getting any closer to the shore. The FS *SONNE* was still heading towards the dark-black, rocky outcrops. In fact, the rocky coastline of the island at sea level was difficult to see from the bridge of the ship. Finally, the second officer understood the danger and gave an order to the machine room to stop the engines. The ship came to a halt only about 100 m from the shore and then reversed its course for a few tens of meters in order to bring the zodiac back on board.

When the captain returned on board, he agreed to allow permission for a few scientists and two crew members to land on a small, pebbled beach. The party went on shore for a couple hours and came back on board with samples of rocks taken from the west-facing cliff of the island.

Three years later, in 1990, my PhD student, Nicolas Binard, along with Professor René Maury from the University of Brest (UBO: *Université de Bretagne Occidentale*) went back to the island by helicopter in order to make a more detailed geological investigation (Binard et al. 1993). The goal of the expedition was to compare the surface lava morphology with that observed on the submerged flanks of the island. They found out that the upper part of the island was built during two major volcanic eruptions. The ancient edifices formed during explosive events had given rise to pyroclastic deposits 65,000 years ago. The last volcanic events took place about 3,000 years ago and formed basaltic flows that covered coral reef limestone. From the stratigraphic information, it was concluded that the island had been constructed during six successive volcanic building stages (Binard et al. 1993).

Diving on The Society Hotspot

The submarine exploration of the Society hotspot using the submersibles *Cyana* and *Nautile* started with the geological observation and sampling of the major and most recent active volcanic edifices in 1983 and continued in 1989 and 1999. During the 1983 and 1989 cruises, the use of *Cyana* permitted us to explore a portion of the tallest and most active volcanic edifices for the first time. The first submersible dives on the Society hotspot were done with *Cyana* in 1983 on the Teahitia volcano, using the support ship N.O. *Jean Charcot*. The name "Teahitia" was given by Jean-Louis Cheminée. In Polynesian "Teahitia" means "standing fire". The Teahitia submerged volcano (seamount) is located at 17°30′S-148°50′W, about 50 km northeast of Tahiti. It is approximately 3750 m high with a basal diameter of about 20 km. The submerged edifice culminates at a depth of 1400 m, and has three summit cones (Fig. 9.4).

In January 1989, a cruise called "Teahitia" took place with the submersible *Cyana* and the support ship N.O. *Le Suroit*. The cruise was headed by Jean Louis Cheminée and intended to explore the volcanic edifices that had shown recent signs of activity. Because of a conflicting schedule, I could not participate on this diving cruise, despite the fact that I had been very involved in its preparation. This is why my PhD student, Nicolas Binard, was able to go on the cruise, and he subsequently wrote his thesis on the morphology and style of eruption of the volcanoes that were explored.

The targets for exploration and the priority for each of the dives were determined according to the information gathered from previous exploration and from the earthquake monitoring in the area. Since *Cyana* was limited to diving to depths of less than 3,000 m, the submersible was only able to explore the tallest of the six major volcanic edifices of the Society hotspot: Teahitia, Turoi, Rocard, Cyana, Moua Pihaa and the flank of Mehetia Island. The deeper (>3750 m depth) portion of the seafloor was not accessible to *Cyana*. A later cruise, using the submersible *Nautile*, would be necessary to reach the deepest part of the hotspot. For the first time, the 1989 cruise with *Cyana* enabled us to conduct detailed observations of the sea floor formed during hotspot volcanism. The style of volcanism, as well as the types of rocks in correlation to their geological setting, could be defined.

Among the volcanic edifices, the Teahitia volcano is the most active on the Society hotspot. Eruptions have taken place at the summit as well as along lateral fissures or inside rift zones. Along the rift zones, there are a succession of small conical vents (<200 m in height) and "hornitos" of silica-enriched flow overlaying bulbous pillows and tubular flows as well as sheet flows (draped and flat flows) (Fig. 9.8a, b). The tubular flows with smooth surfaces are called "giant

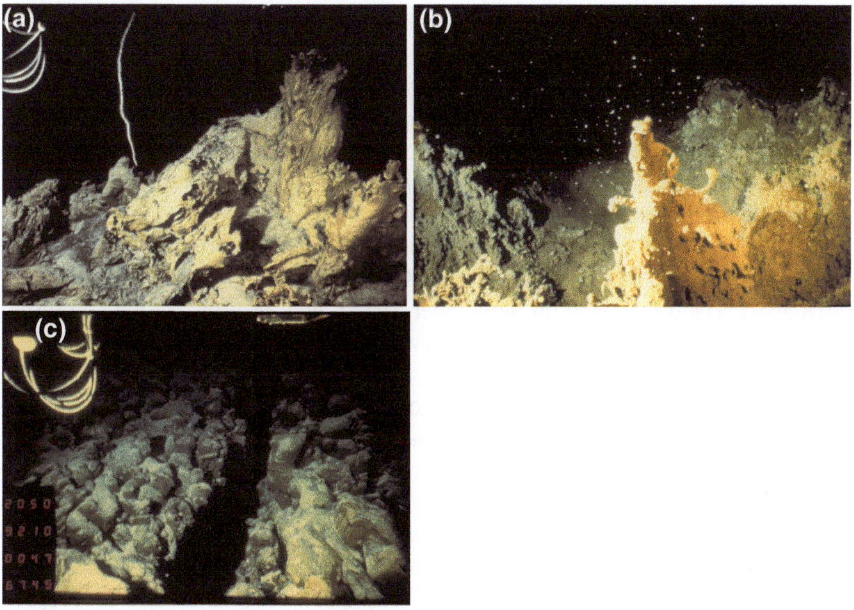

Fig. 9.8 *Bottom* photographs (Copyright IFREMER, *Teahitia cruise 1989, Cyana dives Cy 89-09, Cy89-04, Cy89-07*) from the Teahitia volcano in the Society hotspot (Hekinian and Binard 2008). **a** Photo from dive Cy89-09 at 2619 m shows a spiny silica-rich lava edifice (hornitos) on the flank of the Teahitia volcano. **b** Photo from dive Cy89-04 shows a hydrothermal chimney with clear colored exiting fluids. **c** Photo from dive Cy89-07 showing a fissure (2 m wide) on the flank of the volcano at 2906 m

flows" to distinguish them from those found on spreading ridges. They abound on the flank's eruptive sites (2600–3000 m depth) and are often associated with fissures near the base of the edifice (2900 m depth) (Fig. 9.8c). The rift zones showed the following characteristics of lava flow, according to observations during several dives (Binard et al. 1991):

(1) Bulbous and tubular pillows forming steep slopes (>50–60°);
(2) Brecciated flow fronts giving rise to vertical scarps, up to 50 m in height;
(3) Poorly-sorted talus material at the foot of these scarps with slopes of <30–40°.

The summits of the lateral vents (30–50 m in diameter) were flattened and consisted of large pillows, giant lava tubes, and haystacks (<3–5 m in height). Sheet flows occurred near the eastern base of the edifice (about 2800 m depth). The diversity in composition of the erupted lava includes alkali basalt (most abundant), trachybasalt, trachy-andesite, picritic basalt, basanite and ankaramite (Hekinian et al. 1991).

After the first cruise of 1989, another diving expedition, called the Polynaut cruise, took place ten years later, in 1999. This time, I had the opportunity to participate and dive in the Society hotspot. The primary objective of the Polynaut cruise was to study the Pitcairn hotspot, but we also concentrated a few of our dives in the Society region.

Two dives were made at the boundary marked by the 3200 m contour line limiting the hotspot. This area is also called the "bulge" (Fig. 9.4). The bulge is the ancient oceanic crust on which the hotspot volcanoes are formed. It is now covered with volcanic ash and fragmented debris (pyroclasts) that have been scattered from the eruptions of the edifices.

Three dives were made on Teahitia and two dives were done on the Rocard volcano of the Society hotspot. Dives PN17 (PN = Polynaut, N = Nautile), PN18 and PN19 were made on September 21, 22, and 23,1999, on the summit of the Teahitia Volcano to sample and observe the field of hydrothermal vents. Peter Stoffers (Kiel University in Germany), Steve Scott (University of Toronto, Canada) and Gindi French (Jet Propulsion Laboratory in Pasadena, California, USA) were the scientific observers on board the *Nautile*. Peter already knew the area because ten years before he had made his first dive here with the submersible *Cyana*. For Steve and Gindi, it was a new target and they were both curious to see if active hydrothermal activity still existed. The task was to observe the evolution of the hydrothermal field since the previous dives in 89. When comparing the data, it was found that hydrothermalism in the deposits located on the three summit cones was still active, however the presence of pyroclastic deposits was more prominent than before, and these deposits were now covering the flanks of the parasite volcanic cones.

Two other two dives, Dives PN20 and PN21, were dedicated to exploring another recent volcano called Rocard, which we believed was the site of abundant silica-enriched lava (Fig. 9.8a). We wanted to visualize the field of these types of lava flows with respect to the more commonly found basaltic flows. Very little is known about the mode and extent of silicic lava eruption on the seafloor.

Colin Devey, a geochemist from the University of Kiel, and I were scheduled to dive on Rocard on the 24th and the 25th of September 1999. Rocard is a "multiple volcano" made up of several (at least four) volcanic cones. Since we could not explore all of them, we decided to look at the two located on the hotspot extremities (Fig. 9.4): the most southern cone (dive PN20, with Colin Devey) and the most northern one (dive PN21, with Roger Hekinian). During our exploration of Rocard's volcanic cones, we found that the edifice consisted of silicic lava associated with hyaloclastites and iron manganese crust. The silica-enriched flows had given rise to multiple edifices less than 400 m in height that covered a surface of about 30 km on the sea floor.

Summary of Results

The important observations made during these expeditions on the Society hotspot volcanoes were that the "bulge" (or newly-formed sea floor on top of ancient, 60-million year old, oceanic crust) differs in its lava morphology, its composition and in the style of eruption. The volcanoes that were built on the ancient seafloor inherited from the EPR spreading plate boundary consist of conical edifices, which have erupted different types of lava than what is found on modern spreading ridges.

Petrological studies on samples collected from the Society hotspot volcanoes have revealed the presence of two major magma types: (1) A low-K tholeiitic basalt associated with ancient edifices, probably erupted during seafloor spreading at the ridge axis, and (2) an alkali-rich suite of basalt, trachybasalt and trachyte related to the more recent volcanic events of the Society hotspot (Hekinian et al. 1991). Other geochemical studies (Devey et al. 2003) have also shown the heterogeneous nature of the magma sources and their relationship to the alkali-basalt suites characterizing hotspot volcanoes. It was found that hydrothermal fluid circulation in hotspot volcanoes gives rise to precipitates on the sea floor that are deprived in metallic sulfides.

Austral Volcanic Chain

The Austral volcanic chain extends for about 1700 km in length and runs parallel to the Society Island Chain located 400 km to the south. As was true for the Society volcanic chain, the ages of the islands and atolls increase westward suggesting that they were originated by hotspot volcanism.

The Macdonald and Ra Seamount

In 1987, during the R.V. *SONNE* Leg 65 cruise, we had been able to entirely map the volcanically active Macdonald seamount for the first time. During the same cruise, when making a transect to the first island of the chain called Marotiri, a new volcanic edifice, about 50 km west of Macdonald and rising to 1000 m above the sea floor was discovered. This new edifice was named the "Ra" seamount. The name "Ra" means "Sun" in the Polynesian language as well as in ancient Egyptian.

Since our main goal was to explore the Macdonald seamount, we spent very little time on the Ra seamount. We only explored long enough to make a rock dredge and perform a deep-towed camera station on its summit. The deep-towed camera station revealed the presence of layered ash deposits covered by thin manganese crust and bulbous pillow lava suggesting that the seamount was an ancient, explosive volcano.

The Macdonald seamount located on the southeastern extension of the Austral volcanic chain at about 100 km away from the Ra seamount is one of the most active submarine volcanoes in the World (Fig. 9.9). The Macdonald seamount near 28°49°S-140°26°W was discovered in May 1967, following a strong seismic swarm

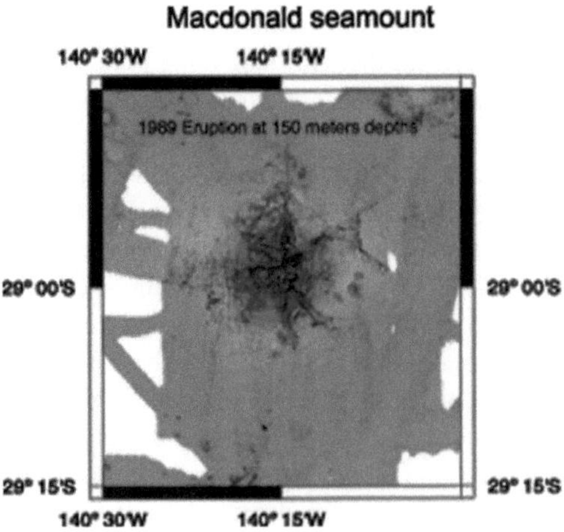

Fig. 9.9 Three-dimensional view of Macdonald Seamount in the Austral chain is shown. The rift zones representing lava channels show a star shaped form going down slope. *A. Bonneville compiled the bathymetry*

detected by the hydrophones of the Hawaiian Institute of Geophysics Network (Norris and Johnson 1969). The source of the seismic signal was localized near 28°48′S and 140°30′W in the southeasterly prolongation of the Austral islands. Geologist Rockne Johnson, while sailing with his wife and four children on board a yacht called the R.V. Havaiki, discovered the underwater edifice of the Macdonald Volcano in 1969. The volcano was named after a famous volcanologist, Gordon A. Macdonald, who was a professor at the University of Hawaii.

The Polynesian Seismic Network (*Réseau Sismique Polynésien* = RSP) of Tahiti closely monitors the activity of the Macdonald volcano. Because of its shallow depths of less than 50 m from the surface, it could become an emerged island at any time. Since its discovery, the Macdonald seamount has been frequently visited and mapped, and its eruptions were even directly observed on two occasions. During the visit of a French Navy-patrol boat "Enseigne de Vaisseau" (EV) *HENRY* on July 1983, a team of scuba divers explored the central part of the summit plateau at an average depth of 40 m. They identified a fissure with fresh walls and spatter cones made up of scoria-like lava. They observed a discoloration spot located approximately 2 km east of the summit, oriented NNW-SSE and about 700 m long. When the RV *MELVILLE* made further observations on October 11, 1987, the sea's surface was bubbling with rising steam and ash. The explosion of large bubbles with a rising gas column, and the formation of large green stains, was observed on the ocean's surface (Talandier et al. 1988).

Diving on Macdonald Volcano

On January 19,1989, after a brief pause of two weeks, the Macdonald seamount started to give off signs of activity that was monitored by the seismic network in Tahiti. Then, after a quiet period of four days, the volcano started to erupt again

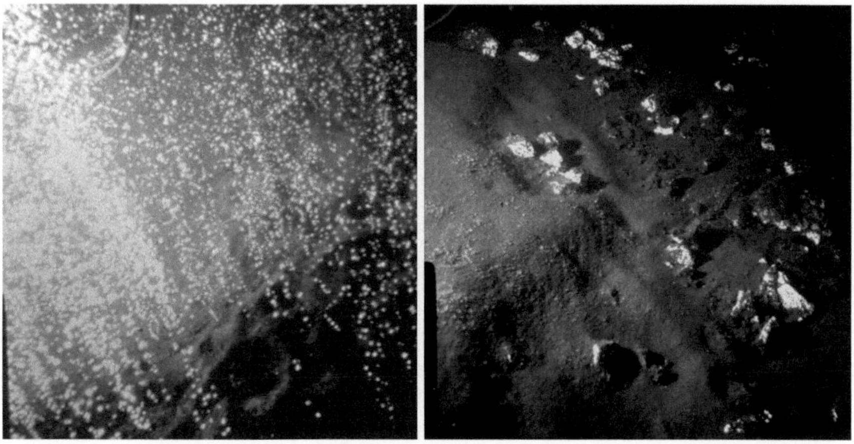

Fig. 9.10 Photographs of the Macdonald Seamount crater eruption of February 2, 1989, at 156 m depth. (Copyright IFREMER, *Teahitia cruise 1989, Cyana dive TH30*). **a** Bubbles of gas consisting of water vapor, carbon dioxide (CO_2) and sulfuric acid (H_2S) associated with pyroclasts (fragmented rocks), have exited from the crater. **b** Fragmented debris (pyroclasts) of intrusive rocks lying on volcanic ash at 59 m depth near the summit of the Macdonald seamount

just when the submersible and the support ship N.O. *Le SUROIT* arrived on the site, on the 24th of January 1989 (Figs. 9.9, 9.10a, 9.10b). The scientific expedition was directed by Jean-Louis Cheminée from IPGP (*Institut de Physique du Globe de Paris*) on board the *SUROIT*, commanded by captain G. Goasguen. Unfortunately I was not on board at the time, but my student Nicolas Binard was on the scientific team and he documented the eruption of Macdonald in detail in his Ph.D. thesis of 1992.

Despite the signs of volcanic activity, three dives were made on Macdonald from about 3011 m depth up to its summit, at less than 50 m below sea level. One dive (TH28) took place on the southern flank of the volcano from a depth of 1387 m up to 412 m. The second dive started deeper, at 3011 m, on the north-eastern side. These two dives noticed a large cover of dark ash and pyroclastic fragments that extended down to about 900 m depth.

It was during the third *Cyana* dive (TH30) on January 27,1989, at about 15:23 and at a depth of 160 m, near the summit of the Macdonald seamount, that the volcano started to erupt. While the *Cyana* was submerged at 140 m inside the crater of the volcano, Jean-Louis Cheminée and the two pilots in the sphere became unexpected spectators of the eruption.

The volcanic eruption was a potentially dramatic and unplanned event. Looking through their portholes, the divers suddenly observed magmatic gas, mainly a CO_2 release, associated with the eruption of volcanic ejecta (pyroclasts) bursting out from the crater's wall and landing all around the diving saucer. The submersible was caught in a turbulent flow of bubbles and sand-sized debris of pyroclastic material. In his report, Jean-Louis wrote that the visibility was reduced which

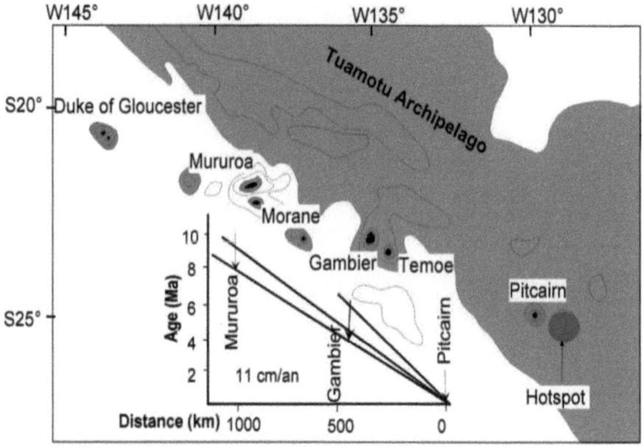

Fig. 9.11 General location map of the Pitcairn volcanic chain in the South Pacific Ocean showing the age distribution of the islands (Le Roy 1994). Age determination on the rock samples has enabled us to infer the spreading rate of the Pacific plate, which is 11 cm/year

made further observations impossible. The explosion forced the divers to leave the site and get to the sea's surface as quickly they could.

On the surface, the ship's hull was vibrating with a deep sound and volcanic ejecta (pyroclasts) were observed falling all around. The sea's surface was glowing with red gases, caused by the burning H_2 and colored by the high-temperature oxidation of FeO to Fe_2O_3, which was included in the volcanic ash. The smell of "rotten eggs" caused by hydrogen sulfide vapors was noted, especially during episodes of intense turbulence in the water. There were steam bursts and gas discharge. Also, there was a green discoloration on the water surface that spread over an area of at least one nautical mile in diameter (Cheminée et al. 1988). Water was sampled and for the first time, the presence of hyperthermophile archaebacteria, which could live at relatively high (80 °C) temperature within the crater of an erupting volcano, was revealed (Huber et al. 1990).

Several days later, the ship's hull vibrated with a deep sound that meant the Macdonald volcano was erupting once again underneath the vessel. Over a total of five days of eruption, gas and volcanic ashes were observed glowing on the sea's surface and dead fish were seen floating with their bellies up, or on their sides. The total length of each eruption lasted only few hours. If such an eruption had continued for a longer period, an island could have emerged on the water's surface, thereby endangering the vessel.

Pitcairn Hotspot

The Pitcairn hotspot is located at the southeastern tip of the ancient volcanic chain comprising the islands of Duke of Gloucester, Mururoa and Fangataufa and the Island of Pitcairn (Fig. 9.11). The Pitcairn hotspot region is built on a 30-million year old oceanic crust, between magnetic anomalies 7 and 9 (Heron 1972). The hotspot of Pitcairn is located 1500 km away from the East Pacific Rise, near 25°20′S-129°20′W, and extends up to about 100 km Southeast of Pitcairn Island.

Historical Background of the Expeditions

Prior to our 1999 diving expedition, several cruises had taken place in the framework of the Franco-German VIP collaboration in the area of the Pitcairn hotspot. The first comprehensive investigation comprising bathymetry and sampling of the Pitcairn hotspot was done in 1989 during the MIDPLATE cruise (Leg 65) of the FS *SONNE*. In addition, during the same cruise, it was planned to investigate a volcanic chain that extended further to the East prior to reaching Easter Island, which was our port of call. This cruise lasted about 6 weeks, with departure from Valpariso Chile on October 28, and arrival at Papeete Tahiti on December 6, 1989. The volcanoes that form the Pitcairn hotspot, which is located about 80 km east of the Island of Pitcairn in a previously uncharted region of the South Pacific ocean near 25°20′S and 129°19′W, were discovered during this 1989 cruise of the FS *SONNE* (Leg 65) (Stoffers et al. 1989).

The preliminary map of the Pitcairn hotspot that was made during this cruise was done with the German-made Hydrosweep multichannel bathymetry instrument. This gave us an overview on the distribution of the major volcanic edifices of the region. Further and more detailed investigation was conducted during the Polynaut cruise with the N.O. *L'ATALANTE* in 1999. The N.O. *L'ATALANTE* was equipped with a more sophisticated and more recent multichannel bathymetry system (Simrad EM 12), which enabled us to conduct a more detailed and extensive survey of the Pitcairn hotspot. Jacques Dubois, a geophysicist from IPGP (*Institut de Physique du Globe* in Paris), headed this cruise on the N.O. *L'ATALANTE*. Originally, the chief scientist was supposed to be Jean Louis Cheminée but he was unable to participate on the cruise due to family reasons. I had helped Jean-Louis prepare the original cruise proposal and was involved in the scientific planning.

A meeting prior to the departure of the cruise was held in Brest at IFREMER, presided by the port-captain, Armel Lestrat. He was worried about our cruise because it included diving on active submarine volcanoes. This was a potential problem since a few weeks earlier it had been reported (by the World's seismic network monitoring system) that a tremor was observed in the area of the Pitcairn hotspot. We all remembered the last experience in 1989 when an eruption had taken place while *Cyana* was diving on the Macdonald volcano in the Austral hotspot chain and we also remembered that the ship had had to quickly evacuate the site.

Despite this worry, the *Polynaut cruise* with the N.O. *L'ATALANTE* and the submersible *Nautile* started on August 22, and ended in Papeete (Tahiti) on September 26, 1999. The cruise was intended to investigate the geodynamic and volcanic setting of three hotspot regions: the Society, Pitcairn and Macdonald (Austral) hotspots. However this last target, had to be skipped because when we arrived on the site of the Macdonald seamount, on September 16, bad weather conditions prevented us from working. The bad weather lasted 72 h and did not seem to be getting better. The pounding of the waves under the keel was so great that it interfered with our bathymetry data acquisition and caused poor quality results. We decided to leave the site of Macdonald and go to work longer on the Society hotspot.

In my opinion, the *Polynaut cruise* seems to be an example of a very well prepared scientific expedition. Both the scientific and technical participants on the cruise were well adapted to life on board and to the experience of obtaining data at sea. They were respectful and cooperative with each other and the atmosphere on board remained very pleasant during the entire cruise. Captain Tredunic was a young officer who was eager to help and ready for any last minute changes in the program. This often happens when exploring an unknown domain. He took particular care and paid tremendous attention to his task of lowering and recovering the submersible safely on board. I also have to add that when a scientific vessel is in the vicinity of Pitcairn, and despite any plans to the contrary, it is extremely difficult to avoid making a stop on the island.

Pitcairn Island

Pitcairn Island is located at 25°05′S-130°05′W on the southeastern extension of the Duke of Gloucester-Mururoa-Gambier-Pitcairn volcanic alignment, with a N110° orientation (Duncan et al. 1974; Jarrard and Clague 1977).

The Island of Pitcairn was discovered July 2, 1767 by captain Carteret on board the British vessel HMS *Swallow*. The name Pitcairn was given after the midshipman who first sighted the island from the *Swallow*. No landing party was sent to shore. Because of an error in reporting the island's exact position on the navigation charts at the time, Pitcairn remained unknown and unvisited for the first 18 years after its discovery. Even today, the island is isolated from the routes of most normal sailing ships, which could be a blessing for the inhabitants. Only four scheduled supply ships per year will transit to bring mail and other supplies from the rest of the World.

When Pitcairn was first discovered in 1767, it was uninhabited, but the presence of stone axes, the remains of carved stone pillars resembling those of Easter Island, and skeletons wearing pearl mussel shells were found later on. The island is made up of steep slopes of basaltic lava and dykes standing 430 m high above the sea's surface (Fig. 9.12). It is 3.5 km long and about 2.5 km large, located about 2,000 km from Tahiti, 3,000 km from Easter Island and 5,500 km from New Zealand.

The island itself has a half moon shape that opens to the northeast with a volcanic crater of 3 km in diameter. The filled crater has a plateau with luxuriant tropical vegetation (Fig. 9.12). The apparent age of the surface lavas from Pitcairn Island is between 0.45 to about 1 million years (Duncan et al. 1974; Woodhead and McCulloch 1989; Devey et al. 2003). The last explosive eruption, which took place about 600,000 years ago, covered the island with ash and basaltic lavas that extended down to the ocean (Fig. 9.13). The base of the Island is at 3,800 m depth

Fig. 9.12 The Island of Pitcairn is seen from the northeastern side (photo courtesy of D. Ackermand) during our approach. The highest peak is at 341 m. The island was visited in 1999 during the Polynaut cruise with the N.O. *L'ATALANTE,* which gave us an opportunity to land and meet the inhabitants. In the sea, there is a Boston whaler sent by Pitcairn inhabitants to carry visiting scientists to their island

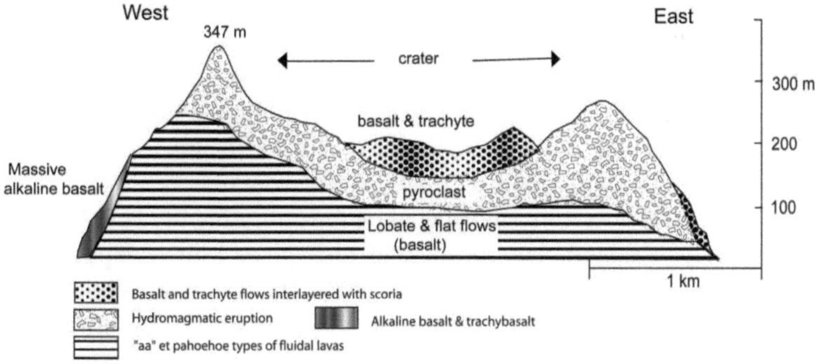

Fig. 9.13 A geological profile of Pitcairn was reconstructed after the map of Carter (1967). The profile shows that most of the island consists of volcanic flows of basalt surrounded by explosive material (pyroclasts)

under water. Because of its volcanic nature, the soil is fertile and the climate is moderate all year around.

Pitcairn Island became the place of refuge for the mutineers of the HMS *Bounty* on January 23, 1790, until it was found again in 1808, by captain Mathew Folger on board the US vessel *TOPAZ.*

Pitcairn Island entered into history when Fletcher Christian, the second in charge of the HMS *Bounty*, rebelled against Captain William Bligh in 1789, near the island of Tofoa in the vicinity of Fiji. Two thirds of the crew and officers rebelled and only 18 crewmen stayed with Bligh. However, there were only eight English sailors and officers along with Fletcher Christian, plus six Polynesian men and 12 Polynesian women, who went to settle on Pitcairn, because the others had perished on Tubuai Island during fighting with each other, or they had been found by the British and hung for their crime of mutiny. Before landing and settling the Island of Pitcairn, it took Fletcher and his little group about two months to find an appropriate place to hide.

During their journey, they stopped in Tahiti and persuaded a few Polynesian men and women to follow them on board the Bounty. In all, there were thirty-seven people on board. After finding the island, they burned the *Bounty* and settled down to communal life. The Island, because of its morphology with steep vertical scarps 200 m high and only one small rocky beach, called Bounty Bay, is very difficult to access by boat. It is constantly under large swells that are a few meters high. This makes landing quite hazardous for any navigator. Thus, Pitcairn was an ideal place to avoid having any unwelcome visitors.

By 1800, all the men were dead and only John Adams had managed to stay alive. He took care of the youths and the 10 women who were left. In 1887, the islanders had become members of the Seven Day Adventist religion. In 1823, sixty-six people were living on the Island. The peak of the population was in 1936 when more than 200 persons lived on the island.

Today, Pitcairn Island is the smallest democracy in the World and it has become dependent on modern technology, such as computers and Internet links, in order to remain connected to the outside World. However the islanders are very much self-sufficient. When I was first there in 1989, only about 41–43 islanders lived in their closed community having little contact with the exterior world. There were 13 adolescents and children who went to school with an Australian woman as their teacher. Contact with the rest of the World was mainly through the radio station run by several operators who stayed on the air during the day. Most contact was through New Zealand, but they could also place radio-telephone calls to Australia, the USA and Europe. It is thanks to the Pitcairn radio station that most direct contacts were made with passing ships traveling a few days away from the Island. That is how we were able to request a visit during our first trip to work in the area of Pitcairn Island.

All the transport to and from the Island is done on "longboats", which look like Boston whalers made of aluminum. The largest form of income for the islanders is in the worldwide sale of Pitcairn Island stamps for international collectors. Any personal income comes from the sale of hand-made products such as paintings, carvings, woodcraft, or woven baskets to visiting ships. The Pitcairn Islanders get their wood, called "miro wood", by traveling in their longboats to another inhabited island called Henderson Island that is located about 120 miles from Pitcairn. The trip usually takes more than a week. Pitcairn has electric power generators running about 7–8 h a day. Their diesel fuel comes from supply ships, and is carried to the island in drums by longboat. This permits the families to have

Fig. 9.14 Scientists and Pitcairn islanders in a "Boston Whaler" boat landing on the Island of Pitcairn. The woman waving her hand was the Mayor of the Island at the time of our visit. (Photo courtesy D. Ackermand)

home video, run tape players and radios. However the best entertainment is still the community's social get-togethers.

Visit to Pitcairn Island

My first visit near the island was made during the F.S. *SONNE* cruise of leg 65, and took place in November 1989. Captain Andresen was the master on board. The people from the Island wanted to come on board the *SONNE* to show us their products, mainly T-Shirts, postcards, carvings and stamps. They put all their material on the back deck of the ship for the crew to see, so we could buy things, or exchange them for other products on board such as paint or cans of food.

The physical appearance of the Pitcairn Islanders is quite variable. They could have Polynesian features and dark hair as well as typical Anglo-Saxon oval-shaped long faces, blue eyes and blond hair. It was at this time that six persons from the scientific party (Dietrich Ackerman, a mineralogist from Kiel University, Nicolas Binard, volcanologist, Silvain Pasqua, a camera man from France, Colin Devey, a geochemist from Kiel, Woodhead, a geochemist from Australia, and Jean-Paul Décriaud, a geophysicist at IFREMER) were invited to go and stay on the island for a couple days. In compensation, Captain Andresen invited three islanders to stay on board the SONNE. We planned to pick up the scientists after doing our survey prior to leaving the site. Meanwhile, the scientists were to spend their time getting acquainted with the inhabitants and looking at the geology of the island. They also collected rock samples to be compared to the data obtained from the submarine edifices (Figs. 9.14, 9.15).

Fig. 9.15 Steve Scott from the University of Toronto and Roger Hekinian looking at the geology of Pitcairn Island (1999). (Personal photo)

During our visit, the island's town mayor asked us if we could make a bathymetric chart of the area available to them. The reason was that he knew that tall volcanoes reaching shallow depths were good fishing grounds for his people. Indeed, the Pitcairn Islanders essentially survived by fishing. Several times a week they launched their whaling boats to the East, towards the recently formed volcanoes. Thus, we gave the islanders a recent map of the area and our gift was very much appreciated.

Our visit of the island was shortened because the weather conditions were worsening and the Captain wanted to be able to make a safe recovery of the people who had gone on shore. Thus, our scientific exploration of the Pitcairn hotspot was curtailed in order to pick up the 6 people we had left on land.

During another visit to Pitcairn, on September 11, 1999, with the N.O. *L'ATALANTE*, our goal was to make a gravimetric calibration that would serve as a reference station for the island in order to measure the gravity field near the volcanic edifices at depth. We hoped the measurements made in this remote area would increase the existing data we have concerning the Earth's gravity field. Since these measurements are more accurate on a solid, land-based area than at sea, we went into the village near the town hall where Jacques Dubois (geophysicist) and Jérome Amman, an engineer from IPG in Paris, made their measurements near the exposed anchor of the "Bounty". The town house, the post office and other one-story, wooden houses surrounded the square. The anchor of the "Bounty" had been previously recovered from Bounty Bay by divers from National Geographic Society, and then given to the islanders.

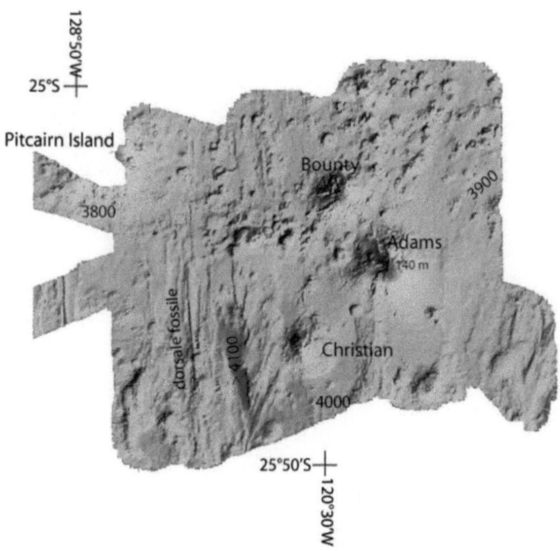

Fig. 9.16 A bathymetric survey covering a surface of 9500 km^2 obtained from multibeam data (SIMRAD EM12D system) enabled us to construct a 3D (three-dimensional) block diagram of the Pitcairn region in the South Pacific (Hekinian et al. 2003). The compilation of the bathymetric data was obtained from the N.O. *L'ATALANTE* during the 1999 cruise and the *R.V. SONNE* in 1989. The submersible *Nautile* with its support ship N.O. *L'ATALANTE* gathered geological and geophysical data from the sea floor of Pitcairn hotspot in 1999 during the POLYNAUT Cruise

Bathymetry and Structural Setting

The 1999 bathymetric studies covered an area of about 9,500 km^2. From the interpretation of bottom reflectivity (side scan sonar imagery) and the Simrad EM 12 multichannel bathymetry studies in the Pitcairn region, it was thought that submarine hotspot activity extends over an area of about 7022 km^2 (Hekinian et al. 2003). The most prominent feature of the Pitcairn hotspot is the large amount of volcanic cones. There is a total of about ninety volcanic cones and most of them are less than 500 m in height and less than 2 km in diameter. They have erupted a total of about 47 km^3 of volcanic rocks.

The sea floor forming the hotspot lies at 3750 m depth and covers a surface of 150 km in diameter. The floor surrounding the Pitcairn hotspot, which formed part of the ancient Farallon plate, is part of the older crust in the Pacific Ocean, estimated to be at about 4500 m depth (Parsons and Slater 1977; Crough 1978). Hence, all the recent volcanoes built from hotspot activity are above a depth of 3750 m. Thus, the recent volcanic activity within the Pitcairn hotspot region is limited to an area above the 3,700 m bathymetric contour line, where ancient crustal orientations are covered by more recent lava flows with large (>1000 m high) and small (<500 m high) volcanic edifices. Nevertheless, relics of ancient oceanic crust are still visible (Fig. 9.16).

Fig. 9.17 The *Nautile* dove on the Pitcairn hotspot's most eastern limit to measure the gravity anomalies at 3500–3700 m depth, at a distance from the base of the major volcanic edifices. Here we see the submersible's Pilot Justiniano (lying back) and Jacques Dubois (chief scientist) handling the gravimeter during dive. (Photo courtesy IFREMER, *POLYNAUT cruise 1999, Nautile dive PN07*)

The Pitcairn region includes two large edifices, which are located in the southeastern prolongation at about 90–110 km from Pitcairn Island. These large seamounts were named "Adams", after John Adams, a seaman who stayed loyal to Fletcher Christian, and "Bounty", after the ship HMS *Bounty*. They are approximately 3500 m in height, rising respectively to 55 m and to 450 m below sea level, and they have erupted a volume of at least 525 km^3 and 275 km^3 of lava respectively. It is not excluded that these edifices could have been found above sea level at one time. Age determination (K/Ar) on the volcanics from "Bounty" gave an age of about 344,000 $\pm$ 32,000 years (Guillou et al. 1997).

From earthquake indications and field observations, it is thought that the most eastern limit of the hot spot is now located near the Bounty and Adams volcanoes, at approximately 110 km east of Pitcairn Island, and that the spreading rate for the oceanic plate's motion is about 80–100 mm per year. Calculated on the basis of these two assumptions, the earliest volcanism forming the Pitcairn Island hotspot probably occurred 1.1 million years ago. The 5907 km^3 volume of volcanic construction over 1.1 million years gives a volcanic accumulation rate of about 0.0054 km^3/yr. This value is about 20–33 times less than that of the Kilauea volcano in Hawaii (0.1–0.18 km^3/yr) (Clague et al. 2000; Cayol et al 2000). The construction of large edifices such as the Adams and Bounty volcanoes (3,500 m high) with 310 km^3 of lava could have taken place in about 58,000 years.

The morphological interpretation from side-scan sonar images obtained with the N.O. *L'ATALANTE* in 1999 during the Polynaut cruise shows lava channels and dike propagation within the explored Pitcairn hotspot region. It was found that small volcanic cones (500 m high) are formed during episodic eruptions from a main magma reservoir, which is the same reservoir that fed the large volcanoes. In the Pitcairn hotspot, the only evidence of eventual feeder channels and eruptions are observed by the distribution of volcanic cones along preferential directions. Another indirect proof is suggested by the preferential occurrence of evolved (silica-enriched) lava on the small conical edifices surrounding the larger edifices.

Diving on the Pitcairn Hotspot

During the Polynaut cruise (1999), fourteen dives were performed on the Pitcairn hotspot and seven took place in the Society hotspot. In the Pitcairn area, 13 dredges and 14 hydrological stations (CTD = Current, Temperature and Density measurements) were also performed in addition to the dives. Our exploration was focused primarily on the geomorphology of the seafloor and on the composition of the erupted lava on the various, different-sized edifices.

From the surface ship observations, such as the bathymetry and deep-towed imaging systems, we were able to define the limit of the hotspots by the relative abundance of the sediment cover and the sampling carried out from previous cruises which gave us, by looking at the degree of rock alteration, the relative age of the sea floor. It was found that some seamounts were indeed ancient edifices probably having an age close to that of the sea floor on ancient spreading centers. Our estimate on the limits for the hotspot corresponded roughly to that of the 3500 m contour line observed on the bathymetry and the side-scan imagery.

One of the major objectives of the dives was to verify our surface observations and locate the boundary of the Pitcairn hotspot activity by direct observation and in situ sampling. This would be important in order for us to infer the volcanic budget associated with this hotspot. Also, due to the difficulty in acquiring continuous observations of oceanic outcrops, very little is known about the stratigraphy of submarine volcanoes. Thus, during the Polynaut cruise, a special effort was made to conduct detailed geological observations along several dive profiles from the bottom to the top of the volcanoes. Continuous sea floor observation of the morphology and direct sampling enabled us to determine the cyclic nature of the volcanic events that are responsible for building the tall edifices. The questions that we tried to answer were: What is the extent of the hotspot activity? Is there any magmatic and morpho-structural relationship between the various types of volcanic constructions?

Another objective was to obtain a measure of the gravity anomalies on the bottom. This was performed with a portable gravity instrument that Jacques Dubois (Fig. 9.17) had brought with him to be placed inside the submersible during each dive. The gravimetric data would serve to infer the flexure of the lithosphere and to determine the conditions of isostatic equilibrium that the lithospheric plate had undergone. This data would be contributed to the international geodetic database for an area of the World where, up to now, no in situ gravimetric data had been available. These measurements were also completed by others made along two submersible transects carried out at more than 3500 m depths in the vicinities of the two major volcanoes, Adams and Bounty (Dives 7 and 10).

The first gravity transect was made with Jacques Dubois as the observer on the northeastern limit of the hotspot and the abyssal plain, at more than 3500 m depth, starting at the base of the Bounty volcano. The second transect dive was done by Jérome Amman, an engineer at the IPGP (*Institut de Physique du Globe*, Paris), in the southern part of the hotspot at 3958 m depth. Continuous gravimetric measurements were carried out over a distance of 7.5 km.

There were three additional goals for our dives: (1) to determine the nature of any hydrothermal activity, (2) to carry out gas and fluid sampling of the hydrothermal vents and (3) to use a new sampling probe device in order to detect living organisms in a hydrothermal system.

Dives on Adams and Bounty Volcanoes

The Adams Volcano, which is the tallest edifice of the Pitcairn hotspot, consists of two summit cones at less than 100 m from sea level. The summit is covered with algae on top of a coral bed that is mixed with volcanic cinders and hydrothermal sediment. Three dives were made on this edifice with the submersible *Nautile* along the western flank of the volcano, from a depth of 3000 m up to the summit at less than 100 m form the surface. The dives have revealed that the volcano is still active.

I did not know that I was scheduled to make the first and the last dive of the Polynaut cruise. I was very pleased and honored to learn that my colleagues and Jacques Dubois, the chief scientist, had decided that I would make these dives. As usual, the briefing for the next morning's dive was made in the presence of the captain, the chief scientist, the head pilot (Jean-Paul Justiniano from La Seynes in Toulon), and all the scientific party. On the newly made map of the area, I pointed to my target area for the dive. This dive was the first of a series of dives that were planned to visit the flank of Adams in order to make a stratigraphic section. The first dive was to start at 2500 m depth, near the base of the volcano.

As soon as we arrived on the bottom on August 28 at 19:29 GMT (geographic Meridian Time), which corresponded to 11:29 local time, pilot Jean-Paul Justiniano stabilized the Nautile, or in other words, put the submersible into a state of neutral buoyancy, in order to take a gravity measurement. The seafloor at 2470 m depth was covered with sediment that was partially dusted by volcanic ash. The presence of animals, revealed by the worm burrows in the sediment, suggested that no recent volcanism had occurred. The gravimeter was placed between the legs of the co-pilot, Frank Rosazza, who was sitting in the back trying to stabilize the instrument by turning the screws with a screw-driver until the three eye bubbles would be horizontal. This usually took about 10 min and needed to be done without moving the submersible. This also required that nobody make move inside the sphere. Once the measurements were finished, I asked the pilot to take a sample of the rock that I had been looking at through the porthole while I was immobilized during the 10 min required for the gravity data acquisition. Jean-Paul was very fast in sampling, it took him only 5 min to grab a piece of the rock, and then off we went. During our progression on the bottom I was surprised to see so much volcanic debris in the form of ash and angular rocks suggesting explosive volcanism. This type of material at 2285–2700 m depth was associated with small volcanic cones less than a few tens of meters tall. The pillow lava flows that we encountered were enormous, reaching up to 4 m in diameter, and therefore they were much larger than those commonly found on spreading ridges. They also differed from typical spreading ridge pillow lava flows by their smoother surfaces and a lack of surface corrugation, which suggested that they were of a more fluidal nature. I called them "giant flows" because of their size and because they extended up to tens of meters in length down slope. As we climbed up the side of the volcano (<1900 m depth), there was an increase in the pyroclastic material, which was a mixture of ash and coarse gravel size debris scattered on top of the giant pillow lava flows. It was at a depth of less than 1700 m that I noticed hyaloclastite slabs, and patches of Fe–Mn oxyhydroxide, indicating the presence of hydrothermalism. When we reached a depth 1583 m, we heard an order on the radio telling us to come back up. Indeed, the time of our dive was over, so the pilot dropped some ballast and the submersible started to rise. It was now 00:35 GMT (Greenwich Meridian Time), that is to say 4:35 PM local time, and we had been on the sea floor for 5 ½ hours. We took our 10th rock sample and made another gravity station before leaving the sea floor.

Two other dives (PN 02 and -09) took place in continuity to the first one along the same flank of the Adams volcano. The scientists on board the *Nautile* were Colin Devey (from GEOMAR in Kiel) and Klaus Lachschewitz from the University of Bremen. They were both geochemists and it was the first dive for each of them. This means that they had to expect to have a ceremony of initiation after their return on board the ship. Indeed, during this cruise we had the most elaborate 'initiation ceremonies' for First Dives that I have ever seen. This was because we had several scientists and engineers that had never dived before. The initiation ceremonies were usually organized by the submarine group, assisted by other experienced divers. They generally consist in hosing the new diver with water when he comes down the stairs after getting out from the submarine. The first-time divers are prepared for what's going to happen. As soon as they come out of the submersible, they remove their fire-proof jump suits before coming down the steps. Sometimes the ceremony became more elaborate when a special mixture of water and other goodies such as leftover coffee and oil were thrown into a big garbage barrel and the diver was forced to step inside. In compensation for his discomfort and his patience, the novice-diver received a nice colorful diploma from the submersible group.

Colin Devey was ready on the deck on August 29, with his bag of warm clothes and his map with the track of the dive. His dive (PN02) was planed to start at a depth where my previous dive had stopped and then continue the ascension of the Adams Volcano to as shallow a height as possible. At 9 AM the *Nautile* was ready to dive. The pilot, Jean-Jacques Kaioun, and the co-pilot, Pierre Guyvarch, were finishing the last checklist when the bridge communicated over the loud speaker that they were on position above the target. The *Nautile* reached the bottom at 1637 m depth, and they saw sediment patches with volcanic gravel, volcanic ash and Fe–Mn crusts similar to that observed on the previous dive (PN01). After crossing several steps marking flow fronts, the dive ended at 942 m depth. They collected six samples, performed two gravity stations, and deployed a water bottle device to recover seawater. The dive lasted 5 h on the sea floor.

Thus, in order to complete the profile on the volcano by reaching the top it would be necessary to make still another dive. This was postponed because we first wanted to explore the other large volcano, Bounty, where we knew recent hydrothermal activity existed on the basis of previous surface observations.

Adams seamount is morphologically similar to the Bounty, and has a summit formed by two cones less than 100 m in height with a roughly north–south orientation. It is during the dive made by Klass Lachschewitz with Jean-Paul Justiniano and Frank Rosazza on September 5, that the top of the Adams Volcano was visited. After landing at 870 m depth, they climbed up to the summit. The summit was heavily sedimented with coral sand, volcanic ejecta, dark hyaloclastites, blocky lava flows and scoriaceous material emplaced during hydromagmatic eruptions. The sampling was successful and the divers brought up hydrothermal iron-manganese crust, and silicic lava fragments that turned out to be trachytes in composition. These massive looking flows are the last eruptive event. Also, because of the shallow depth at less than 100 m, red algae, red corals, sea urchin and white carbonate sand were observed. The fauna and the flora were more abundant than what had been seen deeper on the flank of the edifice.

The recognition of the volcanic stratigraphy was more successful along the Bounty Volcano during three successive dives (PN03, 04 and 14). Indeed, these dives were made so we could obtain continuous observations of the sea floor from 2500 m to the top of the volcano at 428 meters depth, with more extensive sampling than on the Adams volcano. Christèle Guivel, a structural geologist from the University of Nantes, was interested in doing a dive on this site. Thus, she went on the first (PN03) and the last (PN14) dives on the flank of the Bounty Volcano which took place with Jean-Paul Justiniano as pilot and Frank Rosazza as a co-pilot for the first dive, and with the pilot Jean Jaques Kaioun and Severine Beraud as co pilot. The first dive started on August 30, at a depth of 2566 m and ended at 1908 m. The last dive was added in order to continue the profile of dive PN03 and join the next profile of dive PN04, which was made at shallow depths (665 m), also on the western flank of the volcano.

Dive PN04 on August 31, was intended to identify the extent of the hydrothermal deposits on the Bounty volcano. Their existence was known but they had never been observed in their geological context. Steve Scott, a well-known ore geologist from the University of Toronto, was the first to dive on the active hydrothermal field of the Bounty volcano. Steve had come to Brest as a visiting professor in UBO in 1994. I had heard of Steve because of his scientific reputation, however over the years he has become a good friend since we have had many occasions to see each other socially and share our scientific thoughts and other ideas. Steve is a very tall man, almost 2 m in height, and he knew that he would not be very comfortable staying for several hours inside the sphere of *Nautile* (only about 2 m in diameter). Nevertheless, he was very exited to start his dive. With his yellow jump suit half hanging over his legs and wearing his Nautile tee shirt, at 9 AM he started to climb the ladder to get to the hatch where he would descend into the sphere. I said, "Good bye, I'll see you this evening for a beer!" Jean-Jacques Kaioun and Pierre Guyvarch (co-pilot) accompanied Steve during his journey.

They reached the sea floor at 10:49 at 665 m depth on top of a talus of fragmented rock debris. After climbing several vertical scarps representing a succession of lava flow fronts, at 11:50 (local time) the *Nautile* reached a cliff covered with a reddish-colored hydrothermal crust. Steve was pleased to see the first sign of hydrothermalism at 572 m depth. The first occurrence of hydrothermal material was noted at 665 m depth. Then they found the contact between the hydrothermal field and the volcanic rocks at 572–586 m depth, on the western flank of the volcano. They visited two hills about 30–50 m high, which represented two volcanic eruptive centers. The hill farthest to the west was entirely covered by ochreous colored hydrothermal deposits. Steve was particularly interested and wanted to verify the extent of the Fe–Mn–Si oxyhydroxide deposit. The thickness of the hydrothermal precipitates was estimated, from a previous surface coring, to be 120 m thick. Interestingly, since the discovery of high temperature sulfide precipitates on spreading ridge segments, the search for similar deposits on hotspot volcanoes had been un-productive. However, we had evidence that sulfide precipitates existed inside the volcanic edifices since they had been found in veins of volcanic rocks collected from various hotspot seamounts.

Steve Scott was the first to point out that the abundance of such deposits on hotspot volcanoes was the result of mixing between reduced ascending sulfide-rich fluids and descending seawater. The divers left the summit of the volcano at 18:16 and around 19:30 they were back on deck. When the pilot opened the hatch, Steve was the first to emerge. He had a big smile and told us how fantastic it was to be down there. In fact, he was so pleased with his dive, he had forgotten his discomfort at being in such a small space. I handed him a can of beer while a cameramen was snapping pictures and taking video footage.

Other dives on top of the same volcano, Bounty, were programmed to make detailed observations and to sample the active hydrothermal vents. We also tried to make a special effort for getting water samples from the active vents. Thus dives PN05, PN06 and PN12 were done with the participation of water chemists such as Carl-Dieter Schonberg and Olaf Thiessen from the University of Kiel, and Gary McMurtry, a geochemist from the University of Hawaii.

Two other dives (PN 11 and PN 12) on the Bounty were devoted to testing a laser deep-sea probe device for the first time at sea, under the responsibility of Dr. Arthur Lane, an engineer from the Jet Propulsion Laboratory (JPL) of Pasadena California. The instrument designed at JPL was supposed to be able to detect and record spectral or fluorescent signatures of bio-luminescent life and minerals. This preliminary deep-sea experiment was intended to gather data under the extreme conditions of a high-pressure environment. This was a stepping-stone to the further development of a prototype that could eventually penetrate the ice caps of Mars or explore other extraterrestrial oceans in search of life, such as on the surface of Europa, one of Jupiter's moons. Dr. Lane was a short man with white and gray hair, glasses and a small beard. He often wore a worried expression since he was probably thinking about the outcome of his experiment. Indeed, a failure of the experiment could jeopardize his entire project. He was also a practicing Jew, so he honored the "Sabbath" meaning he refused to dive or work on Saturday. He used to spend most of his time in his cabin. Our chief scientist, Jacques Dubois, in respect for Dr Lane's beliefs, had to re-schedule his dive for another day. Dr. Lane made two consecutive dives on September 9 and 10. Luckily, his experiments went well, since he was able to penetrate his probe into the hydrothermal field and make appropriate measurements.

The last dive on Bounty, PN14, took place on September 13, 1999 and was aimed at continuing the profile of dive PN03 which had also been performed by Chrystèle Guivel. The dive reached the bottom at 11:19 (local time) at 1890 m depth near where her previous dive had stopped. Since Chrystèle had the experience of dive PN03 and was therefore familiar with the landscape, she was able to pick up where she had left off. As on the previous dives, they observed a succession of lava flow and volcanic debris in the form of ash and fragments of pyroclasts. This was the last dive on the Pitcairn hotspot so the ship moved to the next target, which was dedicated to the Teahitia hotspot in the Society region.

Dives on Bounty Volcano: Volcanic Stratigraphy

The main goal of the dives conducted on the Bounty Volcano was to be able to show that continuous sequential volcanism is responsible for building the tall hotspot edifices on the sea floor. Although the identification of sequential volcanic activities has been observed on subaerial volcanoes, it was never before done for volcanoes on the deep seafloor. Geological observations were made during five dives (PN03, 06, -04, -12, and -14) along the slope of a large volcanic edifice, the Bounty Volcano (Fig. 9.18a, b).

Along the dive profiles that were reconstructed from observations on the western flank of the Bounty volcano, at least nine volcanic sequences (units) have been recognized. Each of these units varies from about 50 to about 500 m thick and averages about 300 m in thickness. Each volcanic unit is constructed with a succession of high temperature (>1100 °C) events giving rise to basalt and

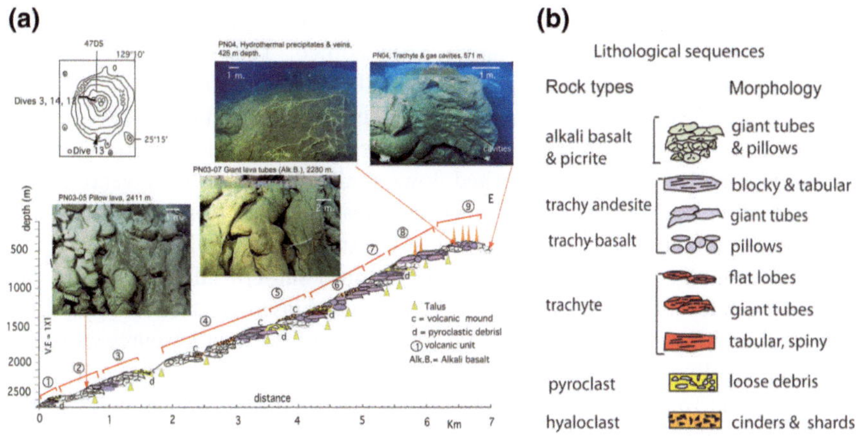

Fig. 9.18 **a** Geological profile constructed from observations made during four dives (PN03, -06, -12 and -14, see small inset map in upper left corner) along the western slope of the Bounty Volcano in the Pitcairn hotspot. The geological sequences of the nine different volcanic units (*circled* numbers) encountered along the slope were drawn from visual observation and were based on the composition of the lava sampled (photos with arrows). **b** shows the sequences of the different lava flows on a schematic profile that was determined by combining dive observations and data on the composition of the rocks collected

related rocks followed by lower temperature (<1000 °C), more viscous flows of silica-enriched lava such as trachyandesite and trachyte. These latter flows have ended their eruption with explosive activity giving rise to pyroclastic debris.

As differentiation takes place in a magma chamber, the volatile-enriched and low-temperature melt will be concentrated in a higher level of the reservoir. This will lower the melt's density and facilitate the extrusion of low temperature lava. During each volcanic cycle, after their extrusion the individual flows have extended for a relatively short distance, less than 300 m in length, when compared to subaerial eruptions, which could reach distances covering one or more kilometers. This is due to the limited extent of magma delivery and to the sea floor environment where there is seawater and a higher pressure that could cause a more rapid cooling of the flows. The observed sequential volcanism forming the volcanic edifices is due to individualized volcanic pulses taking place on the flanks and the summits of the volcanoes. This observation has reinforced the idea of the importance of sequential volcanism as being responsible for the construction of the tall edifices.

My dive (PN13) was made along the southern slope of a small (<500 meters tall) adventive cone located near Bounty's base, at 3000 m depth, and on the lower flank of the Bounty. The reason for this dive was to find out if the recent volcanic activity known to occur on the summit of the edifice was also taking place on the volcano's flank. The other reason for Dive PN13 was to sample more rocks and compare their composition to those found on the main edifice. To my surprise, the adventive cone was essentially made of silica-rich lava, that is to say, trachyte and trachy-andesite. Indeed, there was no evidence of any high temperature basaltic lava, such as was seen on the larger main edifice. This observation became important, and pushed us to sample other small volcanic edifices.

During our survey, we found that all the other small volcanic edifices less than 500 m tall, less than 2 km in diameter, and located at a distance from the larger cones, were made up of silica-enriched flows similar to that of the adventive cone of the Bounty Volcano. Thus an important question arose from our findings: Is such silica-rich and viscous lava formed as a result of crystal-liquid fractionation taking place during the channeling of magma in a sub-crustal environment at a distance from its source? Or, are these silica-enriched lavas produced directly from the partial melting of a low temperature component of the mantle source material?

The channeling of subcrustal magma is similar to a plumbing system. It could supply magma from either a main reservoir underneath the larger edifices and/or directly from the mantle plume piercing the lithosphere. It is believed that the shallow magma reservoir underneath the large volcanic edifices is supplied from the partial melting of the upper mantle located underneath the hotspot. Then, a more fractionated magma, the silica enriched melt, could flow laterally through crustal weaknesses, fractures and fissures, to supply small eruptive cones. This also agrees with the Sr and Pb isotopic data published and interpreted by Colin Devey and his co-workers (2003), which suggested that the volcanics on the smaller edifices have their origin directly from the partial melting of a mantle plume source. It is likely that repeated melt injection within the main magma reservoir underneath the large edifices and then subsequent crystal-liquid fractionation will facilitate the lateral flow of the magma supplying the small edifices.

Volcanic Cyclicity

On the ocean floor it is difficult to be able to observe the sequential nature of volcanic events. The most reliable way is to conduct in situ detailed submersible and/or ROV observations and to sample the different types of material exposed (Fig. 9.18a, b). The dives made on the Bounty and Adams volcanoes helped us to understand the succession of volcanic events constructing the edifices. The most obvious sequential variability is observed when there is a drastic change in the mode of extrusion and/or related to a major tectonic disruption of the volcanic terrain. For example, explosive events generating collapsed structures and other types of forceful injections (i.e. dykes) exposed during faulting will give rise to contrasting volcanic landscapes, such as intrusive rocks overlying extrusive flows. Examples of fracture zones and uplifted blocks during tectonic events are given in Chap. 8 which deals with the Hess Deep and the Garrett transform fault.

Other evidence of sequential variability is due to the compositional changes of the extruded material. In fact, the compositional changes observed among the different rock types helped us to recognize the existence of volcanic stratigraphy and determine the sequential eruptive events. Following the logic of crystal-liquid fractionation, pyroclasts and silica-enriched lava should erupt prior to basalt. As differentiation takes place, the volatiles forming the lighter components of the molten material will tend to aggregate and concentrate in the higher levels of the reservoir or magma column. When the pressure exercised by the volatiles exceeds the lithostatic pressure, violent eruptions giving rise to pyroclasts will occur. Thus, during subaerial volcanism, it is often reported that an eruptive cycle starts with the most evolved and fractionated material and terminates with a quieter event extruding basaltic lava. In the presence of sequential units of composite lavas, the explosive events giving rise to the pyroclasts and hyaloclasts are believed to precede the quieter eruptions of basaltic flow. Indeed, in a magma conduit, the less viscous mafic melts such as basalt ascend more rapidly. If they are mixed with the more viscous silica-rich melt, this will lower the bulk density of melt and facilitate its extrusion (Gibson and Walker 1963; Yoder 1973). This is true when both the silica-rich and the mafic magma coexist together.

Generally, the melting of evolved basalt (i.e. basaltic andesite) inside a magma chamber produces silica–enriched minerals during the transit of a hotter mafic melt (see Chap. 5). Hence, during an eruption, it is possible that pyroclasts and hyaloclasts could mark the end of volcanic events (Batiza 1989; Hekinian et al. 2000) rather than their commencement. This is mainly observed for edifices whose summits are built by only pyroclasts and hyaloclasts (Fig. 9.18a, b). Whether the volcanism terminates with either explosive pyroclastic material or a quieter basaltic flow, knowing what is, in fact, the final stage is very useful for defining volcanic cyclicity.

The changes in lava morphology and rock composition are often difficult to interpret without a continuous observation conducted on the sea floor. Sequential differences in morphology could be seen only when deep-towed cameras and/or submersible explorations have been able to conduct detailed and continuous geological observations. Unfortunately, there are still very few visual observations on the deep-sea environment that have been able to map the stratigraphy of sequential eruptions. Thus, the direct field observations on Bounty and Adams have enabled us to recognize several volcanic events that are responsible for building tall (>3000 m) and small (<500 m high) volcanoes. It was inferred that these edifices were built on a geological time scale of less than one million years, and more likely in just a few thousand years.

The two volcanic edifices on which we were able to identify the cyclicity of sequential volcanism are the Bounty and the Adams volcanoes constructed on the Pitcairn hotspot. As shown from the geological profiles made during the submersible dive observations on these edifices, we can say that the exposed lava units have different morphologies and composition. The least evolved lava of basaltic composition occurs essentially as pillows and giant tubes. Some dull-looking, older flows are partially covered by more recent glassy flows (PN3-03). Occasionally, basaltic flows form small (<20 m high) rootless mounds (e.g. haystacks and hornitos) supplied by short tubular lava-channels along the slopes of the edifice, disturbing the sequential variability. This is an indication of a replenished magma chamber giving rise to a new eruptive event.

More viscous evolved lava (trachy-andesite and trachyte) forms flattened tubes and tabular blocks with a thick (up to 10–20 cm thick) glassy crust. These flows commonly have a scoria-like surface due to the cracking of the outer crust followed by melt spill and quenching during drainage of the lava tubes. Also they do not show radial joints but, rather, reveal a layered internal structure with elongated cavities (Fig. 9.18a).

The explosive events marked by the extrusion of pyroclasts in the form of dark colored ash mixed with rock debris (pyroclast) and hyaloclasts are associated with the evolved flows of trachybasalt and trachy-andesite at various depths (i.e. 500–2691 m depth). Sometimes, these flows are overlaid by hyaloclastites, which were seen up to about 3000 m depth during dives PN14 and PN13 (Fig. 9.18a, b).

In summary, the observed sequential diversities of the exposed lava are related to individualized volcanic pulses taking place during the construction of the volcanic edifices. Each individual flow extruded during each cycle has flowed over a relatively short distance (<300 m) when compared to subaerial volcanism. This is due to the smaller magma delivery as well as to the sea floor environment (seawater pressure and rapidly quenched lava cooling).

Hydrothermal Activity of Intraplate Hotspot Volcanoes

Since hydrothermal activity was abundantly found in most spreading ridge segments in the World's oceans, it was important to look for such phenomena in intraplate regions associated with hotspots. *Cyana* dive Cy83-62 in 1983 during the Cyacite cruise of the N.O. *JEAN CHARCOT* found the first hydrothermal field discovered on the south Pacific intraplate region of the Society hotspot. Jean Louis Cheminée was the first scientist to see an active hydrothermal field on top of the Teahitia volcano at 1454 meters depth.

The recently active hotspot seamounts contain hydrothermal material made up of red and yellow Fe-oxyhydroxide, plus chimneys and slabs forming layered sequences of semi-consolidated deposits on top of lava flows and/or volcanic ash (lapilli) formations. Hydrothermal chimneys varying in size from less than 20 cm up to 15 m in height were seen, mainly on the Teahitia and Turoi volcanoes. On the Turoi seamount, which consists of extinct fields and older looking lava flows, the chimneys are coated by Fe–Mn crust. The active fields on Teahitia consist of smaller (<20 m) chimneys, which show no signs of Fe–Mn surface coatings.

Deep towed camera and submersible surveys showed that the summit (1400–1700 m depth) of the Teahitia volcano is made up of reddish-orange hydrothermal chimneys associated with flat-lying, powdery, ochreous, discontinuous Fe-rich deposits covering an area of more than 1 km^2. The hydrothermal fluids are charged with Fe, Si and Mn and minor amounts of other transitional metals (Hoffert et al. 1987; Puteanus et al. 1991; Michard et al. 1989). They have a similar morphology and composition to those encountered on the off-axial seamounts near the EPR at 12°43′N and 11°30′N. Low temperature (<30–60 °C) diffuse fluids are spewed out from the central orifices (<3–4 cm in diameter) and through the surrounding basal material made up of thin slabs of Fe-oxyhydroxides. In addition to the chimney fields, other flat lying and layered, partially indurate hydrothermal material have completely covered the lava flow and the cinders on a rift zone at 2800 m depths on the Teahitia volcano. These deposits are composed of a few centimeter-thick, layered or loosely packed volcaniclastic debris and/or volcanic ash, which alternate with thin layers of purple-red Fe-oxyhydroxides. The hyaloclastites are encountered in the vicinity of the loosely packed, flat lying volcaniclastic-bearing deposits at about 1600 m depths near the summit of the Teahitia volcano.

More scattered Fe-oxyhydroxide powdery material coating the top of pillow lava at a depth of about 200 m occurs on the Moua-Pihaa volcano. The diffusion of water through rocks and between pillow interstices is associated with living colonies of small coral. Ochreous Fe-rich semi-indurate crusts were also recovered at two dredge stations (stations DR28 and DR29) (one on the summit of Moua-Pihaa at less than 200 m depth, and the other one on the northern flank at 2000 m depth). These Fe-rich crusts have the same composition as the flat-lying, layered deposits described on the summit of the Teahitia volcano.

The lack of sulfide (metallic) deposits on all the intraplate hotspot volcanic edifices that were explored is puzzling. The fluids from the chimneys that were collected and analyzed indicate that that they are deprived in transitional metals, except for iron and manganese, and they are less acid with a lower pH = 5–6 when compared to spreading ridge sulfide deposits. Nevertheless, the hydrothermal fluids from hotspot edifices are enriched in iron, manganese, silica, lithium, barium and Rare Earth elements when compared to seawater. The discharge of such a type of fluid gives rise to iron, silica and manganese hydroxide. These products are formed in a more oxygenated environment with respect to the sulfide deposits found on spreading ridges. Assuming that the process of rock alteration is the same on hotspot volcanoes and in the spreading ridge system, then the hydrothermal fluids formed in intraplate regions have undergone chemical changes prior to their exit on the seafloor. These fluids have been mixed with seawater, within the volcanic edifice, during their ascent. This will lower their temperature and oxygenate their systems. The metallic sulfides will either precipitate within the interior of the edifice, or form oxides.

The fact that high temperature (>250 °C) polymetallic sulfides precipitate within the conduits of the edifices is evidenced by the presence of sulfide veins and veinlets forming pyrite, chalcopyrite, and sphalerite, all of which are associated with hydrated low temperature (<60 °C) products. These veins and veinlets are exposed from underneath the hydrothermal deposits during tectonic fracturing, or are found within the fragmented precipitates of magmato-phreatic explosive eruptions. Thus, polymetallic sulfides are indeed present in intraplate hotspot volcanoes, but they are just not visible on the surface. Instead, they seem to occur in the interior of the volcanic edifices, in the form of veins and veinlets, also called "stockwork formations" in the mining industry.

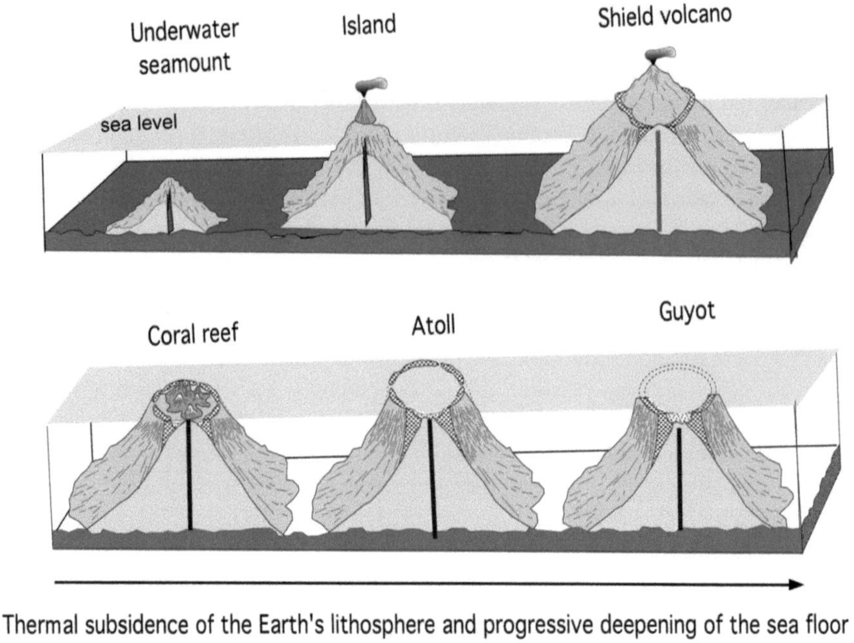

Fig. 9.19 The formation and evolution of submarine volcanic edifices suggest that the birth of an island is a long process of submarine volcanic construction. (1) It starts on the sea floor with a small volcanic mound a few tens of meters in height and reaches the surface to form an island in less than one million years. (2) The first emerged cone is constructed during hydromagmatic explosions (when seawater mixes with magma) associated with occasional lava flows. (3) Subaerial volcanism forms a shield volcano, whose height could reach several thousand meters above the sea surface. (4) If volcanism ceases, then the volcanic edifice will start to erode. At warm latitudes a coral reef develops. (5) An atoll is formed. (6) If the sea floor subsides due to age, this will cause the entire volcanic edifice to sink, so it becomes a guyot

Formation and Evolution of a Volcanic Island

The creation of an oceanic island takes place within a relatively short time when compared to its evolution once it has been formed. The activity of a hotspot volcano is related to its position with respect to the mantle plume supplying the magma. By knowing the rate of spreading, we could estimate the speed of the movement of a plate over a hotspot. In the central Pacific, by calculating the average speed of a plate's motion, it is estimated that the time for creating a volcanic island is about 1 million years.

As soon as an island is formed, it will begin to be eroded by the waves and the alteration of its rocks. If the magma supply is sustained and continues to provide lava to the surface, the island will grow (Fig. 9.19). The lava flowing on the flanks will form a domed shaped edifice with a gentle slope of 6°–10°. A sudden halt of volcanic activity indicates that the volcano has moved too far from its hotspot mantle source and the magma has started to cool. The eroding processes will then take over so the volcanic edifices forming the island will start to crumble and break apart. However, with time the eroding processes will diminish, and the island will become more stable. This is true when coral grow on the outside edge of the islands since coral reefs can protect an island from intense wave erosion. During plate drifting, as the lithosphere ages and cools, thermal subsidence of the lithosphere will affect the structure of the island, which will sink. The island and the coral reef will disappear except for the coral growing on top of each other. An atoll is a sunken, flat-topped island whose center is surrounded by coral. Atolls are found in the equatorial zone between about 28°N and 28°S, which corresponds to the latitude of coral growth. When atolls sink with an increase in the age of the lithosphere, then the volcanic edifices become guyots, which are mainly found on ancient crust (>80 million year old areas of the Pacific Ocean basin).

References

Batiza R (1989) Petrology and geochemistry of Eastern Pacific spreading center. In: Winterer EL, Hussong, DM, Decker RW (Eds) The Eastern Pacific Ocean and Hawaii: geology of North America. Geological Society of America, Boulder, pp 145–160

Batiza R, Niu Y (1992). Petrology of magma chamber processes at the East Pacific Rise 9°30′ N. J Geophys 97:6779–6797

Binard N, Hekinian R, Cheminée J-L, Searle RC (1991) Morphological and structural studies of the society and austral hot spot regions in the South Pacific. Tectonophysics 186:293–312

Binard N, Hekinian R, Cheminée J-L, Stoffers P (1992) Styles of eruptive activity on intraplate volcanoes. J Geophys Res 97:13999–14015

Binard N, Maury RC, Guille G, Talandier J, Gillot PY, Cotten J (1993) Mehetia Island, South Pacific: geology and petrology of the emerged part of the society hot spot. J Volcanol Geothermal Res 55:239–260

Brousse R, Maury RC (1980) Volcanisme et pétrologie de lîle de Tubuaï dans les îles Australes (Pacifique Sud): In: Cahiers de L'Indopacfique, Published by Fondation Singer-Polignac, 43 Avenue George-Mandel, 75016, Paris :131–193

Burke KC, Wilson JT (1976) Hot spots on earth's surface. Sci Am 235(2):46–57

Cayol V, Dietrick JH, Davis AS (2000) High magma storage rates before the 1983 eruption of the Kilauea, Hawaii. Science 288:2343–2346

Clague DA, Reynolds JR, Davis AS (2000) Near ridge seamount chains in the northeastern Pacific Ocean. J Geophys Res 105:16541–16561

Cheminée J-L, Hekinian R, Talandier J et al (1988) Geology of an active hot spot: Teahitia-Meahetia region in the south central Pacific. Marine Geophys Res 11:27–50

Crough ST (1978) Thermal origin of hotspot swells. Geophys J Royal Astr Soc 55:451–469

Crough ST (1983) Hotspot swells. Ann Rev Earth planet Sci 11:165–193

Decaud A (1993) Histoires extraordinaires. Perrin, Paris, p 296

Devey CW, Lackschewitz KS, Mertz DF, Bourdon B, Cheminée J-L, Dubois J, Guivel C, Hekinian R, Stoffers P (2003) Giving birth to hotspot volcanoes: distribution and composition of young seamounts from the seafloor near Tahiti and Pitcairn Islands. Geology 31(5):395–398

Duncan R, McDougall AI, Carter RM, Coombs DS (1974) Pitcairn Island- another Pitcairn hotspot? Nature 251:679–682

Gibson IL, Walker GPL (1963) Some composite rhyolite/basalt lavas and related composite dykes in Iceland. Proc Geol Assoc 74:301–318
Guillou H, Garcia MO, Turpin I (1997) Unspiked K/Ar datingof young volcanic rocks from Loihi and Pitcairn hotspot seamounts. J Volcanol Geochem Res 78:239–249
Heron EM (1972) Sea floor spreading and the Cenozoic history of the East-central Pacific. Bull Geol Soc Am 83:1671–1692
Hekinian R, Bideau D, Stoffers P, Chiminee J-L, Muhe M, Binard N (1991) Submarine intraplate volcanism in the South Pacific: geological setting and petrology. J Geophys Res 96:2109–2138
Hekinian R, Pineau F, Shilobreeva S, Bideau D, Gracia E, Javoy M (2000) Deep sea explosive activity on the Mid-Atlantic ridge near 34°50′N: magma composition, vesicularity and volatile content. J Volc Geotherm Res 98:49–77
Hekinian R, Cheminée J-L, Dubois J, Stoffers P, Scott S, Guivel C, Garbe-Schönberg D, Devey C, Bourdon B, Lackschewitz K, McMurtry G, Le Drezen E (2003) The Pitcairn hotspot in the South Pacific: distribution and composition of submarine volcanic sequences. J Volcanol Geotherm Res 121:219–245
Hoffert M, Cheminée J-L, Person A, Larque P (1987) Dépot hydrothermal associé au volcanisme sous marin intraplaque : Prélévement effectué sur le volcan actif de Teahitia (Polynésie française). Comptes Rendu de l'Académie des Sciences Paris, !II, 304(14):829–832
Huber R, Stoffers P, Cheminée J-L, Richow HH, Stetter NO (1990) Hyperthermophhilic archaebacteria within the crater and open-sea plume of erupting Macdonald Seamount. Nature 345(6271):178–182
Hunt TL, Lipo CP (2006) Late colonization of Easter Island. Science 311:1603–1606
Jarrard RD, Clague DA (1977) Implications of Pacific island and seamount ages for the origin of volcanic chains. Rev Geophys Space Phys 15:57–76
Jackson PD (1976) Comments on new results in resistivity well logging. Geophys Prospect 24:407–408
Jordahl KA, McNutt MK, Caress DW (2004) Multiple episodes of volcanism in the Southern Austral Islands: flexural constraints from bathymetry, seismic reflection and gravity data. J Geophys Res 109:9–28
Leroy I (1994) Evolution des volcans en systeme de point chaud: ile de Tahiti, archipel de la Societé (Polynésie Française). Thesis Université Paris Sud Centre d'Orsay N° d'ordre 3368:271
LePichon X, Huchon P (1984) Geod, Pangea and convection. Earth Planet Sci Lett 67:123–135
Menard HW (1986) Islands. Scientific American Books Inc. New York, p 230
Michard HW, Michard G, Javoy M, Cheminée J-L, Binard N (1989) Chemistry of submarine springs from the Teahitia seamount. EOS, Trans Am Geophys Union 70:495
Minster JB, Jordan TG (1978) Present day plate motion. J Geophy Res 83:5331–5354
Morgan WJ (1971) Convection plumes. Nature 230:42–43
Morgan WJ (1972) Plate motion and deep mantle convection. Geol Soc Am Memoir 132:7–22
Naar DF, Hey RN (1991) Tectonic evolution of the Easter Microplate. J Geophys Res 96(B5):7961–7993
Norris A, Jonhson RH (1969) Submarine volcanic eruptions recently located in the pacific by sofar hydrophones. J Geophys Res 74:650–664
O'Connor JM, Stoffers P, McWilliams MO (1995) Time-space mapping of Easter chain volcanism. Earth Planet Sci Lett 136:197–212
Okal EA, Taladier J, Sverdrup KA, Jordan TH (1980) Seismicity and tectonic stress in the South Central Pacific. J Geophys Res 85:6479–6495
Parsons B, Slater JG (1977) An analysis of the variation of ocean floor bathymetry with age. J Geophys Res 82:803–827
Puteanus D, Glasby GP, Stoffers P, Kunzendorf H (1991) Hydrothermal iron rich deposit from the Teahitia, Mehitia and Macdonald hot spot areas, South Pacific. Mar Geol 98:389–409
Smith WHF, Sandwell DTl (1997) Global sea floor topography from satellite altimetry and ship depth sounding. Science 277:1956–1962

Stoffers P, Botz R. Cheminée J-L. Devey CW, Froger V, Glasby GP, Hartmann M, Hekinian R (1989) Geology of the MacDonald Seamount region, Austral Islands: recent hot spot volcanism in the South Pacific. Marine Geophys Res 11:101–112
Talandier J, Okal EA (1984) The volcanoseismic swarm of 1981–1983 in Tahiti-Mehetia area, French Polynesia. J Geophys Res 89:11216–11234
Talandier J, Okal EA, Craig H (1988) Seismic and in situ observation of Macdonald seamount eruption, 11 October 1987. EOS, Trans Am geophys Union 69:258–259
Turcotte DL, Oxbourgh ER (1978) Intraplate volcanism. Phil Trans Royal Soc London (A) 288:561–579
Wilson JT (1963) A possible origin of the Hawaiian islands. Can J Phys 41:863–870
Woodhead JD, McCulloch MT (1989) Ancient sea-floor signal in Pitcairn Island and evidence of large amplitude, small-scale mantle heterogeneity. Earth Planet Sci Lett 94:257–273
Yoder HS (1973) Contemporaneous basaltic and rhyolitic magma. Am Mineral 58:153–173

Chapter 10
Subduction Zones

Abstract After its creation at spreading ridge axes, the oceanic lithosphere will move with the drift of the plates until it encounters the border of another plate. These "converging plate" areas are the locus of lithosphere shrinking and large amounts of tectonic activity. During convergence, one plate may slide under another plate; this is called "subduction". Subduction plates sink into the Earth's mantle as convergence takes place. Deep trenches, island arcs and/or high mountain ranges mark the areas of plate convergence. Depending on the type of rocks that have built the surface of the plates, density differences mean that some plates will be pushed up, while others are subducted. Converging plate regions are highly affected by earthquakes and active volcanism.

Trenches mark the boundary where tectonic plates converge and interact with each other until one of the plates, the heaviest one, is subducted underneath the other and plunges into the mantle (Fig. 10.1). This is called the subduction zone. During plate convergence, the penetration of cold lithosphere into the asthenosphere would cause heat to be conducted into the solid slab (McKenzie and Sclater 1971). The frictional heat that is induced is sufficient to generate volcanism. Anderson et al. (1976) have shown that heat flow profiles in the Japanese and Java trenches reveal a thermal increase on the plate towards the arc and back arc basins. The bending, deformation and migration of the trenches will depend on the nature of the lithosphere (ancient vs. young) and on the velocity of plate motion. For example, the forces acting at the trenches will have a different effect on old (stiff) subducting plates, which will facilitate trench advance towards the back arc regions since the stiff plates will resist bending (Lallemand et al. 2008). On the other hand, the forces acting on the trenches, when associated with younger, more flexible subducting plates, will move the trenches towards the open ocean.

R. Hekinian, *Sea Floor Exploration*, Springer Oceanography,
DOI: 10.1007/978-3-319-03203-0_10,

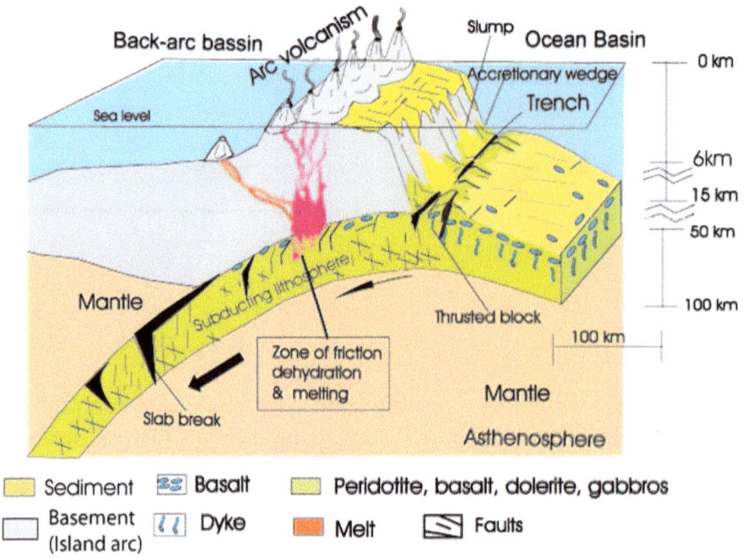

Fig. 10.1 Subduction zones are formed when two moving lithospheric plates meet. The denser plate (i.e. the heavier, basalt-peridotite formation) tends to plunge or to be subducted underneath the lighter or less dense (silica-enriched) formation. The convergence of two plates with different densities creates a zone of weakness in the lithosphere forming a depression or a trench

Nomenclature of Subduction Zones

Convergent plate boundary regions are the place where the oceanic plates created on spreading ridge axes have drifted away from their point of origin and are subducted beneath intra-oceanic and/continental volcanic chains. The trenches, and the volcanoes that are associated with subduction zones, are all part of what is referred to as an "active margin". The geological setting of the convergent plate boundary regions is comprised of several provinces. These provinces include the "fore-arc" area, separating the trench from the island arc, the "island arc" and/or the "volcanic chain" and the "back-arc basin". Because basalt, which is the prominent type of rock of the oceanic lithosphere and crust, is denser than the continental and/or island arc rocks, basalt will sink underneath the lighter types of rock formations and thereby generate volcanic activity and earthquakes.

During subduction, new crust can be created in the fore-arc, or within the arc or in back-arc area. In the fore-arc, a "mixture" of lighter oceanic sediment plus the debris of broken and altered silica-enriched rocks, which result from the degradation of older continental shore lines or island arcs, can be deformed, folded and faulted during subduction (Fig. 10.1). In subduction zones, this lighter silicic rock material will be pushed upward until it rides above oceanic crust and forms "accretionary prisms", which are wedge-shapes of deformed sediment that has been scraped off the oceanic lithosphere and added to island arcs and/or continental margins. Even if these subduction systems constitute <15 % of the crust created (Lallemand et al. 2008), they are important areas since this is where any eventual new crust will be accreted onto the continents.

Subduction zones are located above the inclined lithospheric slabs that plunge under a crust of lighter density (silica-enriched and metamorphosed rocks). The descent of the lithospheric slabs, which are about 60–90 km thick, is defined by the traces of a seismic zone that is called the "Benioff zone", named after a US. geologist named Hugo Benioff. The Benioff zone refers to a seismic area where slabs of a plate are subducted under the continents and/or island arcs producing deep-seated earthquakes. A bending of the oceanic plates is accompanied by their breaking up when slabs under thrust the overriding landmasses. During subduction, earthquakes are caused by strain accumulation taking place during the plate motions and will depend on several factors such as: plate velocity, plate age, the length of the down-going slab, the dip angle of the Benioff-zone, and the structure and nature of the plate. Also, the presence of earthquakes will be influenced by fault-breaks in the subducting plate, by the degree of rock alteration, and by the presence of large seamounts and/or sediment. All these factors have an impact on slab-subduction. For example, the fractured oceanic crust forming a horst and graben type of structure can develop a heterogeneous contact plane that will decrease the strength of the subducted slab. Also, a plate's older age, therefore making the plate denser, also means that the plate will have a stronger tendency to sink with less strain accumulation, thus generating smaller earthquakes. A structurally uniform subducting plate will form a more resistant slab and thus be likely to accumulate more strength in resisting subduction. More details on the subject of subduction are available in the work done by Hamilton (1974), Kanamori (1986) and in a publication edited by Larter and Leat (2003).

Volcanism and Metamorphism in Subduction Zone

During subduction, the plunging slab will drag indurate sediment and altered rocks, thereby generating frictional heating of the rocks, so they will melt and form local pockets of magma. The magma pockets will rise, due to the effects of buoyancy, and arrive in a shallower level in the crust to form magma pools. It is also known that the plunging slab contains water due to its release from compacted sediment and the alteration of ancient oceanic lithosphere. This water enters into the lattice of hydrated minerals during metamorphism. The water content of altered oceanic rocks (metamorphosed rocks = metabasalt, metagabbros, serpentinized peridotite) can be as much 2–6 % of a rock's weight (H_2O = 2–6 wt%). This water content will lower the melting temperature of the oceanic rock to about 600° at shallow depths or will create a lower pressure in an anhydrous condition. The depth of seawater penetration and the subsequent hydrothermal circulation favoring rock alteration could be as deep as several kilometers (10–15 km), on the basis of inferences of earthquake analyses defining the limit between the brittle/

ductile transition zone in the lithosphere. Water's contribution to metamorphism means that progressively less water will remain for penetration in the deep mantle. Heating and dehydration during the metamorphism of the subducting oceanic slabs will give rise to submarine as well as island arc explosive volcanism. Some examples of the most devastating subaerial volcanic eruptions such as the Krakatau volcano (Indonesia) in 1883, Tambora (Indonesia) in 1815, El Chichon (Mexico) in 1982, Ruiz (Columbia) in 1985, and Mont Pelée (Martinique) in 1902 were the result of subduction processes at plate convergence areas (See Chap. 5).

The formation of volcanoes and the production of earthquakes along subduction plate margins have a common origin. The loci of the major earthquakes are clustered in the Benioff zone during the breaking up of a slab penetrating into the Earth's mantle below 50 km depth at a variable angle of about 20–50°. These deep earthquakes often occur in the marginal basins at a distance (300–1000 km) from the trenches. In the Indonesian region, the epicenters of deep earthquakes are often located 500–600 km north of the Java trench. The strength is most intense in the rigid lithosphere and vanishes slowly at depth within the mantle. In the immediate vicinity of the Sumatra-Java (Indonesian) trench system, the greatest earthquakes are shallow due to a low angle thrust during the breaking of the slabs (Kanamori 1971). The World's most active earthquake regions are found in the Pacific convergence margins, which are rimed by deep trenches and arc volcanism.

During the Cenozoic period (<50 million years ago), the Pacific plate converged in the west at contact with the Asian Plate, and on the south at contact with the Indian-Australia-New-Guinean plates that were moving northward. The areas of Eastern Australia as well as New Guinea and Indonesia have all been affected by such plate rearrangements. Earlier, during the late Paleozoic (>250 million ago) and Triassic (200–250 Ma) periods, the granitic and volcanic terrain of the Eastern Australia-New Guinea-Indonesian regions (including the Sumatra-Java area) were separated from the Indian Ocean by a deep trench, with a depth of about 6000 m, bordering the islands of Java and Sumatra in Indonesia. Convergence of the Pacific plate has created one of the World's longest and deepest subducting trenches. An example of this type of system of Pacific plate convergence and subduction is seen on the Tonga-Kermadec volcanic arc and trench.

Tonga-Kermadec Volcanic Arc

The Tonga-Kermadec volcanic arc system is the locus of volcanic activity related to the convergence of two plates: the Pacific and the Indo-Australian plates. The Tonga-Kermadec system is 2,530 km long, making it one of the longest, continuous intra-oceanic arcs in the World. It exists in an area going from New Zealand to the island of Samoa near latitude 15°S (de Ronde et al. 2005; Garry Massoth 2005 personal communication). This region is marked by an extensive felsic type (rhyolite) of volcanism associated with considerable amounts of silica-enriched products (Bloomer et al. 1994). The subducting plate is nearly perpendicular to the

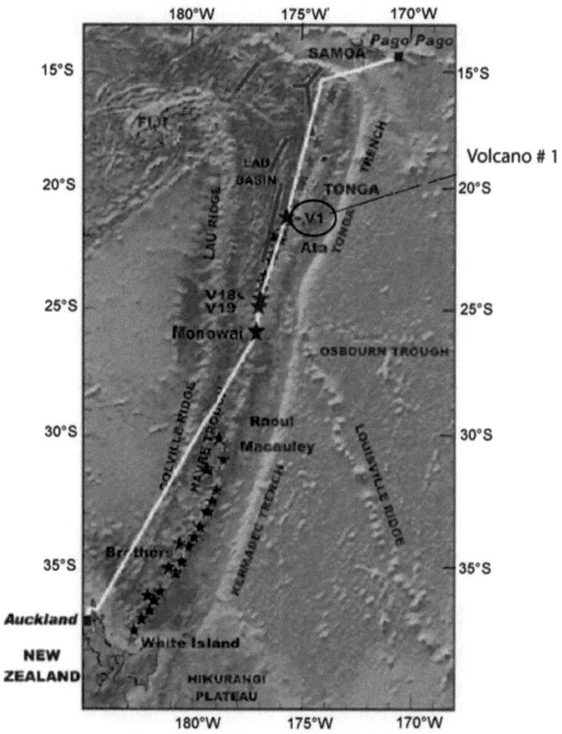

Fig. 10.2 Regional map of the Tonga-Kermadec Volcanic Arc in the southeastern Pacific Ocean has been retraced from the General Bathymetric Chart of the Oceans (GEBCO). The Subduction Zone is *light-grey*, immediately to the *left* of the Kermadec and Tonga trenches. The ship track (KOK) is shown with a *bright white line*. Volcano #1 and # 19 are shown on the map and were explored by the Submersibles Pisces IV and V (Hawaiian Undersea Research Laboratory, Univ. of Hawaii) in collaboration with the geology department of the University of Kiel, Germany. Other subaerial and submarine volcanoes are marked by *small* and *large stars*; some are listed by name

trench and is sinking at a rate of about 160 mm/year. The overall region is considered as being an area of intense volcanic and seismic activity. However, because the Tonga-Kermadec area is mostly uninhabited, it has received very little attention. Although several US and French expeditions have worked on the back-arc basins of the Tonga-Kermadec Island arc in the past, very little is known about the arc itself. Scientists have investigated the southern portion of the Tonga-Kermadec Ridge only after the 1990s (Wright 1994; Wright et al.1996; de Ronde et al. 2005; Stoffers et al. 2006). The Tonga-Kermadec Trench borders the Tonga-Kermadec volcanic arc to the east, and is bordered by the Havre trough and the Lau basin to the west (Fig. 10.2). The northern part of the system (the Tonga Trench) is the location of one of the World's deepest troughs, since it is about 11,000 m deep. This is about the same depth as the Marianna trench, found further north in the western Pacific.

The Tonga Trench

The Tonga Trench is a nearly straight north-south linear structure east of the Tonga-Kermadec Volcanic Ridge. The Trench's margin is a typical example of a convergent zone on a young crust, since there is only a thin-crusted intra-oceanic setting where the Pacific plate lithosphere is subducted underneath the volcanic island arc chain, inside the Tonga-Kermadec Ridge. The subducted lithosphere's slab descends at a 50° angle to about 660–700 km depth in the Tonga Trench (Barazanki and Isacks 1971). In the deepest central area, the trench narrows to <10 km wide. It is likely that the slab thins out with depth, until it is only about 2–10 km thick.

It was in the middle of 1950s, during a mission of the RV. HORIZON from Scripps Institution, when peridotites and dunites were first dredged from the Tonga Trench at 9000–9400 m depths (Fisher and Engel 1969). This is another indication that the crust in the subducting area of the trench is relatively thin, since deep-seated rock formations were exposed on the trench's floor surface.

Tonga Islands

The northern area of the Tonga-Kermadec Ridge, also simply called the Tonga Ridge, forms a large volcanic plateau or platform. The Tonga Ridge consists essentially of coralline-capped Eocene-Miocene volcanic islands and shallow (<100–1000 m depth) seamounts. The basement of the Tonga platform exposed on Eua Island is composed of depleted tholeiite, andesite and dacite with an age of about 40–46 Ma (Ewart and Bryan 1972).

There are a total of 169 emerged islands but only 36 are inhabited. Thus, these islands are mostly unpopulated and those with human beings living on them are often used as mere vacation grounds or are cultivated for crops due to their rich, volcanic soil. The population of Tongans is about 100,000 people, most of whom live on the Island of Tonga itself. Tonga Island is part of Melanesia (the so called "dark islands" whose name was coined by the French explorer, Dumont d'Urville).

The original Tongans came from the Fiji islands and settled in Tonga and Samoa about 800 B.C. It is also near one of these islands, the volcano of Tofoa, that Captain Bligh and Fletcher Christian had their altercation on board the HMS Bounty, which later led to a rebellion (mutiny) on board the ship. Charles Wilkes, commander of the 1838–1842 US Exploration Expedition around the World, landed on Fiji and on the Island of Tonga to conduct geological observations. Today most Tongans are Catholics, Protestants, Pentecostal and/or Seven Days Adventists, as are the inhabitants of Pitcairn Island.

Volcanoes of the Tonga-Kermadec Ridge

As part of the Tonga-Kermadec Arc, it has been determined that the Tonga-Kermadec Ridge hosts at least 94 volcanoes (de Ronde et al. 2005). A more conservative estimate of 12 subaerial and 71 submarine edifices was given by

Garry Massoth (personal communication). About 87 % of the volcanoes are located in the Tonga-Kermadec Ridge (de Ronde et al. 2003). Several times a year, it is common for the population of the volcanic islands to feel large seismic quakes. Some historical eruptions along the Tonga-Kermadec Ridge were reported from Raoul Island in 1800, and from the Monawai volcano in 1998.

The first submarine eruption observed by scientists was on May 5, 2009 in the northern area of the Tonga-Kermadec ridge near the island of Tonga, to the west of the Mata volcano. Scientists on aboard the R.V. *Thompson* using an ROV witnessed two active volcanic eruptions during two of Jason's ROV dives at depths of 1174 and 1208 m. A report written by David A. Clague, senior scientist from the Monterey Aquarium Research Institute in California, described the events and was published on the Web (see Chap. 5).

Cooperative Programs and Sea-Going Expeditions

It was after the discovery of hydrothermalism and the recovery of the first sulfide samples dredged from the calderas of the Brothers and Rumble II volcanoes from the southern Kermadec arc (de Ronde et al. 2005), that an international cooperative program organized by New-Zealand and Germany took place. This program was headed by professor Peter Stoffers from the University of Kiel and by Ian Wright from the National Institute of Water and Atmospheric Research in Wellington, New Zealand. The program was designed to study several volcanic edifices along the arc.

The first F.S. *SONNE* cruise (Leg 135) took place in September 1998 and discovered black-smoker venting as well as low-temperature, diffuse venting within the Brothers seamount caldera, on the Monowai, Rumble III and Clark volcanoes (Stoffers et al. 1999a, b) located along the Tonga-Kermadec ridge (Fig. 10.2). The Brothers seamount was observed in detail with a deep-towed bottom video camera system (OFOS = Ocean floor Observation System) and a geological map was compiled. This seamount revealed the presence of an extensive hydrothermal sulphide field covering of more than 200 m in diameter that was concentrated on the southern rim of the caldera. The success of this cruise gave an incentive to the international community, mostly the US, Japan and New Zealand, to take the lead in pursuing research for mineral deposits in the area. Between1999 until 2005, sea-going expeditions surveyed more than 6 new volcanoes along the Tonga-Kermadec Ridge system (De Ronde et al. 2005).

The second F.S. *SONNE* cruise (Leg 167) took place between October 12 and December 2, 2002, to carry out additional bathymetric surveys and to sample rocks from the Tonga-Kermadec volcanic chain. As mentioned before, it is believed that buoyancy in the deep-seated mantle of the Earth creates an ideal situation where two colliding plates could form a deep trench and then subsequent lithospheric uplift could form a volcanic ridge, where volcanic islands would eventually be formed. The Tonga-Kermadec Island arc would seem to be the result of such phenomena, and in fact, there are hundreds of islands.

On October 18, 2002 we stopped the *SONNE* near the island of Ota where a team of scientists landed to pick up some samples from the island's recently active volcano. Before we arrived, it was also on this island that about 50 young Tongans had been isolated from the rest of the World during a period of a month for bad behavior in the capital city of the Tongan kingdom.

During the FS *SONNE* cruise, about 20 submarine volcanoes (seamounts) were sampled. Based on the samples recovered from the twenty edifices, six undersea volcanoes were selected for further detailed investigations with deep-towed vehicles to take video films and bottom photos. It was found that most of the seamounts surveyed along the Tonga-Kermadec volcanic chain showed multiple cones and variable (1–3 km in diameter) sized caldera that were usually a few hundred meters deep. Only a small amount of these calderas were deeper, going up 800 m depth.

The composition of the volcanic rocks forming the Tonga-Kermadec volcanoes ranges from basalt and dolerite to silica-enriched lava (diorite, tonalite, rhyodacite). However, most of the seamounts erupted felsic lava consisting of pumice and andesite. Most of the caldera walls revealed an abundance of volcaniclastic deposits composed of black and light colored scoria debris of pumice that had been mixed with occasional sub-rounded sand to boulder-sized rock fragments. The blocks are accidental, fragmented material that had been released during explosive eruptions.

Another cooperative program was planned for June 2005, between the Hawaiian Undersea Research Laboratory (HURL) of the University of Hawaii and the University of Kiel (Germany) to carry out a diving expedition on several active volcanoes in the northern part of the Tonga-Kermadec volcanic chain, between 25°S–177°W and 20°S–175°W. Thirteen submersible dives using Pisces IV and Pisces V were scheduled to take place using the support vessel R.V. *KA' IMIKAI-O-KAMALOA* (called KOK for short).

The ship *Ka'imikai-O-Kanaloa* (meaning in Polynesian "searcher of the sea of Kanaloaé and called K-O-K for short) was launched in July 1992 for SOEST (School of Ocean Earth Science and Technology) in Honolulu and was operated by the Hawaii Undersea Research Laboratory (HURL). KOK is the mother ship for the submersibles *Pisces* 4 and 5. It is 224 feet long and has an "A" frame with a 30-ton hydraulic crane for launching the submersible. The ship carries a crew of about 13 and has facilities for 15–20 visiting scientists and technicians. In addition, the ship carried an ROV with a depth capability of 900 m.

We left Auckland, New Zealand on Sunday June 6, 2005 at 16 h and headed north towards the Tonga-Kermadec volcanic arc. The wintery weather was calm but cold during the first 2 days then it started to change so it became rough and windy. The sea state varied between 4 and went up to 7, with winds of 25–70 km/h. This type of weather lasted until June 16. At this point, the weather started to ameliorate so normal diving could take place every day from June 16–22.

From the previous surface ships' exploration, it was found that several volcanic and hydrothermal activities had occurred in shallow water (<1000 m depths). The volcanic edifices chosen for diving consisted of large, collapsed calderas with volcanic cones, which had erupted considerable amounts of pyroclasts and both basalt and silica-enriched lava flows.

Most of the dives were made on two volcanoes called Volcano #19 and Volcano #1. These volcanoes were chosen because they were considered to be in a dormant stage. I will describe our endeavor on volcano # 1 and suggest further reading on Volcano #19 in a paper published by Stoffers et al. (2006).

Submersible Pisces Dives: Dive on Volcano # 1

Volcano # 1 is located at about 40 km north of Volcano # 19 and 50 km south of the island of Tonga on the northern part of Tonga-Kermadec Ridge. Because of the extensive fishing activities around Tonga by the local fishermen, some precautions had to be taken during our dives. In order to prevent any accidents, the evening before each dive, two ROV (Remote-controlled Ocean Vehicle) immersions were done in order to have a visual observation of the sea floor along the caldera wall and its summit (Figs. 10.2 and 10.3a, b).

A total of four dives (PIV 139, −140, −141 and −143) took place within Volcano # 1. The objectives of the first three dives were to explore hydrothermal venting and do a detailed study of the biological communities associated with the thick beds of sulfur cemented ash at water depths of 160–200 m depths located around the centrally located explosive craters on the caldera floor. A densely populated field of mussels (200 × 400 m) covered the sea floor in a warm hydrothermal water environment with temperatures of 30–70 °C (Stoffers et al. 2006). The highest recorded temperatures were around 130 °C. The pH of the clear venting fluid was between 3.2 and 5.8, and there was a concentration of H_2S and CO_2.

After exploring the centrally located, hydrothermally active mound of the caldera, we wanted to explore the caldera wall in order to make a stratigraphic geological section of the eruptive sequences contributing to the construction of volcano #1.

Dive PIV 143 (Pisces 4, dive N°143) was also the last dive of the cruise and I had the privilege of making this dive at about 4 km to the northwest of the previous dive sites (Fig. 10.3a, b). On June 24, 2005, at 9:30 AM I was ready and waiting on the deck. I was looking forward to hearing the pilot give us the go-sign. After half an hour of waiting for the dive to take place, the pilot informed me that we would have to postpone the dive because of a hydraulic problem. After repairing the hydraulic leak, 2 days later, on Sunday, June 26, 2005, Terry Kerby, Max Cremer and I were again scheduled to dive on the caldera wall. Dive PIV 143 was launched during moderately good weather conditions and lasted about 7 h during which time geological observations and sampling were successfully carried

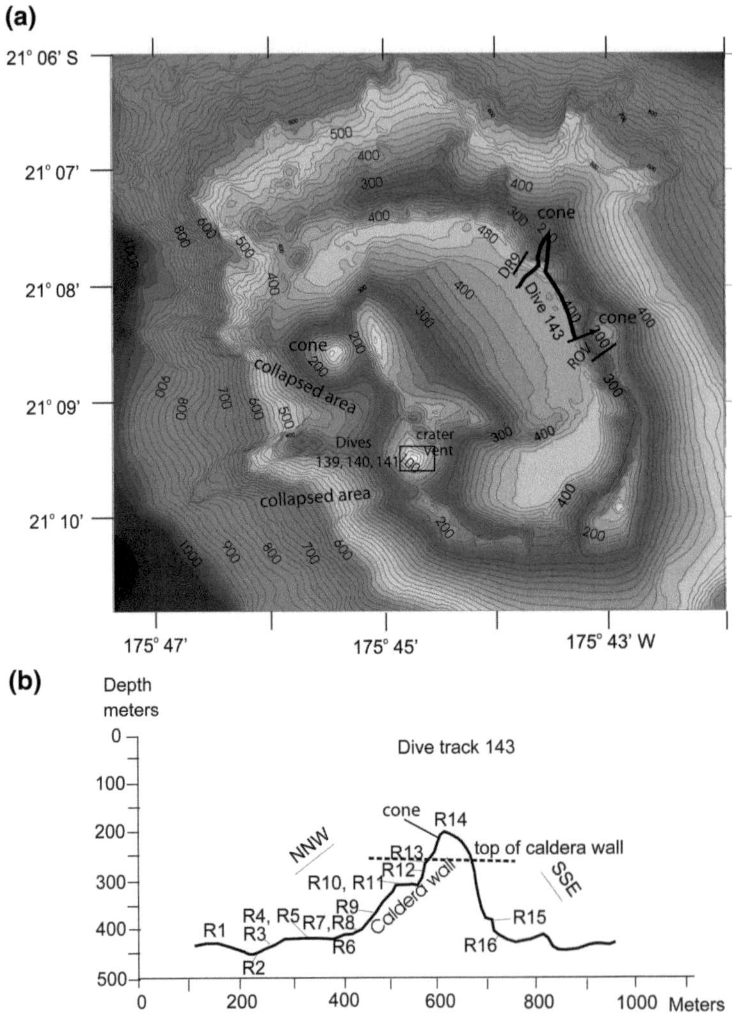

Fig. 10.3 **a** Bathymetry of Volcano #1 showing the caldera's small volcanic cone and wall. The Dive tracks (PIV 143) of the submersible Pisces in 2005 (Research Vessel KOK) and the location of the ROV tracks are shown as well as the previous R.V. SONNE Leg 167 dredge sites (Stoffers et al. 2003, 2006). **b** The dive track shows depth profiles of the two submersible crossings, one going up-slope and the other going down-slope on the caldera wall. The dive track converges at the summit of the volcanic hill. The sample sites (R1, −16) and the various lithologies are illustrated in Figs. 10.4 and 10.5

out. The submersible landed on the caldera floor near a previous FS *SONNE* (Leg 167) dredge site (DR9) at 460 m depth, about 300 m from the northeastern corner of the caldera wall (Fig. 10.3a). A complete geological stratigraphic section was conducted along the northern wall between 450–210 m depth near 21°07.86′S and

175°43.58′W. This stratigraphic section could be completed because we made two crossings:

(1) **The first section**, done on the way up the scarp of the caldera wall, was made from the foot of the caldera near 460 m depth up to 260 m high at the summit of the caldera. However, when reaching the ledge of the caldera wall, the dive continued further up to explore the summit of a small volcanic cone about 400 m in basal diameter, 40 m. high, culminating at a depth of 217 m. When reaching the summit and after sampling, we descended a gentle slope in a south-south east direction to the edge of the caldera wall with its vertical scarp about 259 m deep.
(2) **The second section** descended the steep scarp of the caldera, and took place about 60 m away from the first one (Fig. 10.3b). Moving downward on a scarp with a submersible is a difficult task because the pilot has to keep the submersible facing the scarp in order to have a visual contact at all time with the sea floor. It is like backing down a mountain. This is critical and important for the geologist who is supposed to describe the rock formations. So, the submersible descended slowly, and this was stressful for the pilot because he had to use the vertical engine and constantly keep the submersible in neutral buoyancy. The dexterity of the pilot was a great help for me, since it meant I could have continuous visibility of the near vertical scarp. I was able to observe the bedded-ash layers and pyroclastic flows that were alternated by lava flow and intruded by dykes, just as I had seen in the previous crossing during the upward trip on the wall. However, this second section also revealed more intrusive and extrusive flows than what I had seen on the first crossing (Fig. 10.3b). In addition, we observed an extensive dislocation and slumping of the poorly consolidated pyroclastic deposits, which exposed a dyke complex.

Before leaving the bottom, Max asked me if I wanted to take one last souvenir. Quickly he grabbed a last sample of a loose rock fragment lying on the sedimented sea floor. After reaching the bottom of the scarp we traveled for about 1 km along the base of the wall until we were called from the ship and told to resurface because the weather had started to change.

Volcanic Stratigraphy

The geological section shown in Fig. 10.4a–f was constructed according to the in situ observations made during dive PIV 143. The section consisted of several volcanic sequences formed of four types of volcaniclastic deposits associated with several types of intrusive and extrusive flows (Fig. 10.4a).

Several stratigraphic sequences were identified along the wall of the caldera. They are listed below:

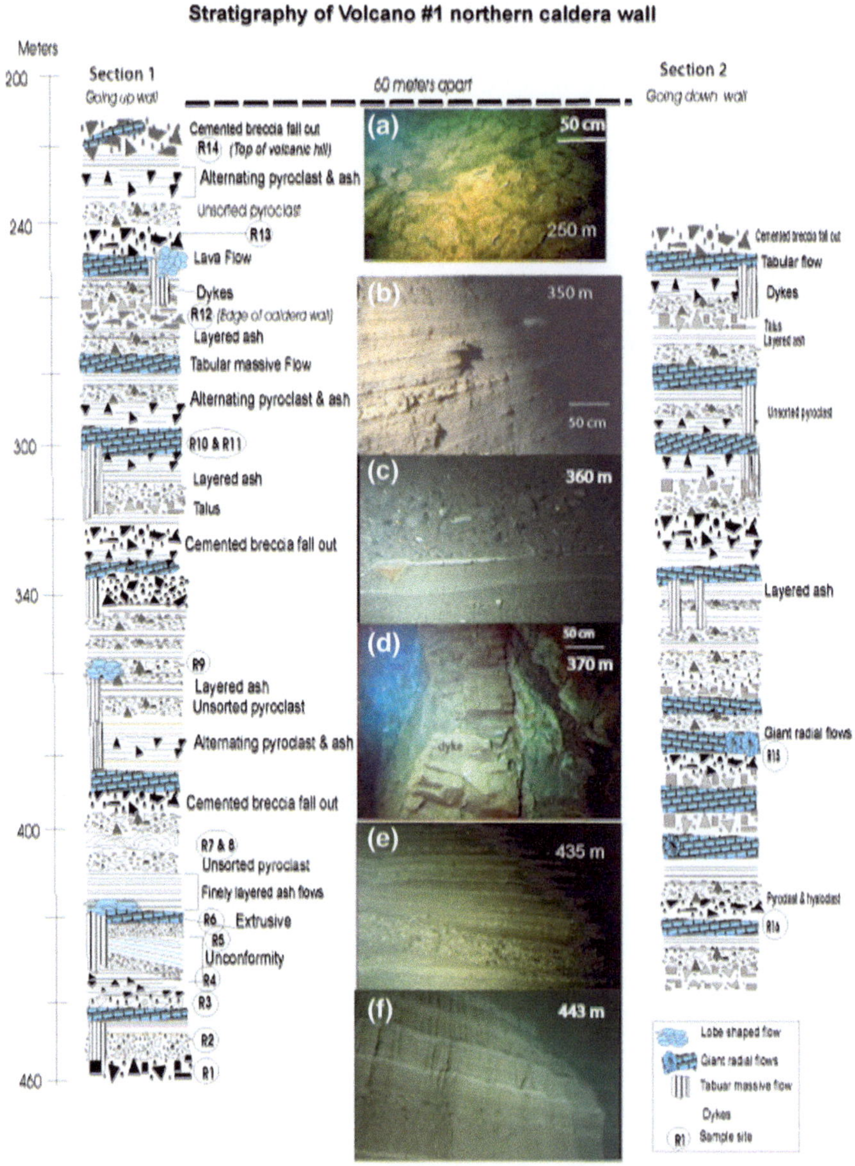

Fig. 10.4 Stratigraphic section of Volcano # 1 caldera's wall in the Tonga-Kermadec volcanic arc shows several volcanic sequences reconstructed after dive PIV 143 of the submersible Pisces

◀ IV (Fig. 10.3). The two tracks, one upward and the other one downward, made at a distance of 60 m apart, converge at the summit of the volcanic hill (Fig. 10.3a, b). The *bottom* photographs shown in the middle of the two stratigraphic sections illustrates the sample sites and the lithology. Assuming that each pyroclastic sequence going from coarse-grained and/or unsorted layers to layered ash deposits represents a single event, then at least twenty eruptions have occurred. The deposits are mainly oriented horizontally, unless otherwise indicated. In-situ photographs. **a** Rusty colored cemented pyroclastic debris exposed on a faulted-step near the *top* (260 m depth) of the caldera wall. **b** Alternating sequence of graded sand-, cobble- and silt- sized pyroclastic deposits at 350 m depth. **c** Unsorted and poorly-sorted pyroclastic and ash layers at 360 m depth overlaying finer grained ash-pyroclast beds. **d** Vertical dyke near an andesitic lava flow in the vicinity of sample R9. **e** Layered ash and silt- to sand-sized pyroclastic deposits with a tilted unconformity at 435 m depth. **f** Finely layered ash and sorted fine-grained (silt-sand size) pyroclasts overlaying coarser ash-pyroclast units at 440–443 m depth near the base of the caldera wall (21°07.8′S–175°43.6′W)

(a) Thin ash layers consisting of clay to silt-sand-size (<5 mm in diameter, white colored layers) alternating with darker and thicker coarse-grained ash beds (centimeter scale) (Fig. 10.4b, f).
(b) Pyroclastic deposits from only a few cm up 10 cm in thickness, consisting of angular lapilli (fragments of 5–70 mm in diameters) (Fig. 10.4b). The larger blocks are gabbroic tonalite and andesite embedded in silt- and sand-sized material.
(c) Closely alternated sequences of pyroclast and ash layers of a few meters thick containing cobble-sized accidental debris. These alternating units represent a more violent eruption than that of the fine sized ash beds (Fig. 10.4c).
(d) Rusted and cemented breccia formed of sand-, gravel- and lapilli-sized debris embedded in iron-oxyhydroxide hydrothermal precipitates that occur on the top of the caldera wall and floor (Fig. 10.4a).

We could see that shallow water (<600 m depth) and subaeriel island-arc volcanism differs from deep-sea volcanism by the production of pumice and due to the abundance of finely layered dark and light colored deposits. The magma that generated the light grey and dark ash layers is the result of mixed basaltic (dark colored) and more silica-enriched (white colored) products. The silica-enriched material is produced from a more evolved magma residing in the magma chamber prior to when the eruption started.

The processes giving rise to the mixed layered light and dark colored volcanic units has been attributed to crystal grinding and to the absorption of a solidified lava flow on the wall of the magma conduit (Polacci et al. 2005). The change in viscosity during an upwelling of fluidal basaltic magma plays a role in lubricating the crystallized and evolved wall rock formations. If the temperature is high enough, remelted the evolved crystallized products will mix with the basaltic melt. Thus, a sequential eruption of dark and light layered units will take place. Similar phenomena could also happen on deep-sea (>1000 m depths) volcanic edifices, such in those found on the MAR axis at 34°504 N and in intraplate regions of the Pitcairn, Society and Austral hotspots volcanoes (See Chap. 9).

Fig. 10.5 Columnar jointed massive flows are observed on the caldera wall of Volcano # 1 in the Tonga-Kermadec arc between 377 and 415 m depths. Similar subaerial structures are seen forming volcanic necks and columnar flows in the Columbia River basalts (Long and Wood 1986) and in the Coastal Range of Taiwan (Juang and Chen 2004). These massive columnar jointed flows with straight tabular shapes and parallel sides show near horizontal cooling fractures at 380 m depth near giant radial jointed pillow flows. The individual blocks forming the flow are mostly rectangular, about 10 cm in diameter, forming a staircase formation extending from a few meters up to tens of meters thick

The intrusive and extrusive volcanic flows associated with the volcaniclastic deposits show several types of morphology as indicated below.

(1) **The flattened, lobated and vesicular flows**. They often have a pancake-like pitted surface and a semi-circular scoria-like appearance with chilled margins and they contain large gas cavities, which are about 1 cm in length. These flows are associated with the rusty-colored breccia cementing material (i.e. R10, R14) at 324 m and 255 m depth respectively (Fig. 10.4a).
(2) **Vertical dykes** form separate units about 1 m thick that often terminate at the base of the pillow lava and lobated flows (Fig. 10.4d).). They are holocrystalline (completely crystallized) and porphyritic plagioclase-rich felsic rocks. They differ from other massive tabular flows (see below) by their shorter and smaller (centimeter scale) parallel cooling joints characterizing each individual unit, as well as by their lack of large cavities and abundant vesicles, and their more crystalline texture.
(3) **Massive tabular flows developing radial and columnar jointing.** The columns are either linear or curved with variable orientations such as radial and fan-shaped units reaching heights up to 10–15 m high (Fig. 10.5) These joints are cooling cracks that are perpendicular to the surface of the flow and are caused during the rapid cooling of a hot lava flow in contact with seawater in a

submarine environment. The individual columns have a diameter varying from 10 to 20 cm and show hexagonal and cubic cross sections. From field observations, these flows show different morpho-structural patterns. They consist of tabular layered flows forming a horizontal and slightly inclined "staircase" with sill-like complexes or with a blocky appearance. They are dark-grey fine-grained intermediate ($SiO_2 = 55–60\ \%$) silicic lava. A sample (R15) of a massive columnar flow collected at 380 m depth consists of a fine-grained andesite containing an assemblage of plagioclase-clinopyroxene-titanomagnetite. The rock's parallel oriented cracks are filled by a dark chilled goundmass formed during its cooling prior to the complete solidification of the rock. More details on the geology and formation of these lava flows are found in Hekinian et al. (2008).

In summary, Volcano # 1 consists of an oval-shaped collapsed caldera that is 30 km^2 and whose floor is located at 450 m depth. A portion of the caldera wall was visited and the stratigraphy was studied during one dive (PIV 143). The results of the dive indicate that the caldera wall's stratigraphy is the same as an andesitic strato-cone volcano growing as a result of numerous repeated small eruptions of pyroclast-ash deposits, alternating with an occasional outpouring of basaltic tubes and pillow lava (Fig. 10.4b, c, e). In other words, Volcano # 1 was constructed during successive and short-lived volcanic explosive eruptions of pyroclastic deposits, which alternated with a quieter outpouring of massive lava flows. The repeated, violent, explosive volcanic eruptions formed thin (centimeter thick) layers of ash and pyroclastic debris. The floor and the top of the caldera wall showed occasional small volcanic constructions covered by consolidated ash-lapilli (volcaniclastic) deposits. These volcanic events gave rise to the successive eruption of silica-rich intrusives and viscous extrusive flows followed by, or preceded by, volcaniclastic deposits, which represent the sequential explosive events forming alternate layers of fine-ash beds (5–10 cm thick) and coarser pyroclastic deposits varying between 50 and 100 cm up to 20 m thick. The massive flow units forming columnar, tabular and giant radial-jointed (GRJ) flows represent viscous magma conduits or pipes. The magma conduits are several tens of meters in diameter and form horizontal columnar flows with "staircase-like" and fan-shaped appearances giving rise to the GRJ flows. The giant radial-jointed (GRJ) flows are created by magma issued from larger conduits and protrude outward up to several tens of meters away from the intrusive pipes. The columnar flows are comparable to volcanic-neck magma intrusions like those encountered in the subaerial regions of the Coastal Range area of the Sierra Nevada and in Eastern Taiwan (Juang and Chen 2004; Huber and Rinehart 1967). However, since not all the massive columnar flows encountered on the caldera wall of Volcano #1 are seen terminating with a GRJ, it is not excluded that some of the tabular flows were originally horizontally injected sills.

Final Thoughts

It seems fitting that my last sea going experience terminated with this cruise on the KOK, and that I participated on the last dive of the mission on Volcano #1. I find it quite ironic that I began documenting my studies of the ocean floor with a paper about rocks from the Indian Ocean's spreading ridge where the crust is being created, and my last sea going experience took place in a subduction zone near Tonga. During my profession as an oceanographer, I have been privileged to visit all the world's oceans and I have tried to answer some of our questions about how our planet was formed.

I find it comforting to think that once the tectonic plates have been subducted, at the end of their journey drifting on the asthenosphere from their place of creation at a spreading ridge, their rocks will be fused once again. It is likely that some day, in the distant future, a subducted plate's rocks may once again return as fresh lava at another spreading ridge somewhere on our planet. It appears that living organisms and even the rocks of our planet have a period of birth (creation), existence and death, but in the case of seafloor rocks, we know that their "death" (by melting) in a subduction zone is not at all the end of their existence.

References

Anderson RN, Uyeda S, Miyashiro A (1976) Geophysical and geochemical constraints at convegent plate boundaries-part I: dehydration of down going slab. Geophys J Roy Soc London 44:333–357

Barazangi M, Isacks B (1971) Lateral variations of seismic-wave attenuation in the upper mantle above the inclined earthquake zone of the Tonga Island arc: deep anomaly in the upper mantle. J Geophys Res 76:8493–8516

Bloomer SH, Ewart A, Hergt JM, Bryan WB, (1994) Geochemistry and origin of igneous rocks from the outer Tonga fore arc (Site 841). In: Hawkins J, Parson L, Allan J et al. (eds) Proceedings of the ocean drilling program, scientific results, 135:625–646

de Ronde CEJ, Massoth GJ, Baker ET, Lupton JE (2003) Submarine hydrothermal venting related to volcanic arcs. In: Simons SF, Graham IJ (eds) Gingebach memorial volume: volcanic geothermal and ore forming fluids: rulers and processes within the Earth. Soc Econ Geol Spec Publ 10:91–109

de Ronde CEJ, Hannington MD, Stoffers P, Wright IC, Ditchburn RG, Reyes AG, Baker ET, Massoth GJ, Lupton JL, Walker SL, Greene RR, Soong J, Ishibashi CWR, Lebon GT, Bray CJ (2005) Evolution of a submarine magmatic-hydrothermal system: brothers volcano, Southern Kermadec arc, New Zealand. Econ Geol Bull Soc Econ Geol 100:1097–1134

Ewart A, Bryan WB (1972) Petrography and geochemistry of the igneous rocks from Eua, Tonga Island. Geol Soc Am 83:3281–3296

Fisher RL, Engel CG (1969) Ultramafic and basaltic rocks dredged from the nearshore flank of the Tonga trench. Geol Soc Am Bull 80:1373–1378

Hamilton W, (1974) Earthquake map of the Indonesian region.US Geol Surv Map I-875-C

Hekinian R, Mühe R, Worthington T, Stoffers P (2008) Geology of a submarine volcanic caldera in the Tonga arc: dive results. Jour Volcanol Geoth Res 176:571–582

Huber NK, and Rinehart CD (1967) Cenozoic volcanic rocks of the Devils Postpile Quadrangle, Eastern Sierra Nevada, California. U.S. Geol. Survey Prof Paper 554-D:D1–D19

Juang W-S, Chen J-C (2004) The topographic landscape of volcanic necks in coastal range and Lutao. East Taiwan Coll Res 17:83–92

Kanamori H (1986) Rupture process in subduction-zone earthquakes. Ann Rev Earth Planet Sci 14:293–322

Kanamori H (1971) Great earthquakes at island arcs and the lithosphere. Tectonophysics 12:187–198

Lallemand S, Heuret A, Faccenna C, Funiciello F (2008) Subduction dynamics as revealed by trench migration. Tectonics 27(3) (TC 3014). doi:10.1029/2007TC00212)

Larter RD, Leat PT (2003) Intra-Oceanic subduction system: tectonic and magmatic processes. Geol Soc London Spec Publ 219:1–17

Long PE, Wood BJ (1986) Structures, textures, and cooling histories of Columbia River basalt flows. Geol Soc America Bull 97:1144–1155

McKenzie D, Sclater JG (1971) The evolution of the Indian Ocean since the late cretaceous. Geophys J Roy Astron Soc 24:437–528

Polacci M, Di Muro A, Papale P (2005) Novel interpretation for shift between eruptive styles in some volcanoes. EOS 86(37):333–336

Stoffers P, Wright IC, Scientific party (1999a) Longitudinal transect of the Kermadec-Havre arc-back-arc system: initial result of the Sonne cruise (SO 135). Interidge News 8(1):45–50

Stoffers P, Hannington M, Herzig P, Wright IC, de Ronde CEJ, Shipboard Scientific Party (1999b) Elemental mercury at submarine hydrothermal vents in the Bay of Plenty, Taupo volcanic zone, New Zealand. Geology 27:931–934

Stoffers P, Worthington T, Wrigth IC, Ship Board Scientific Party (2003) Cruise Report SONNE 167, Louisville Ridge:Dynamic and magmatismof mantle plume and its influence on the Tonga-Kermadec subduction system. Report of the Institute für Geowhissenschaften, Universiyat Kiel n20:276

Stoffers P, Worthington T, Scharz-Shampera RU, Hannington M, Massoth J, Hekinian R, Schmidt ML, Lundsten J (2006) Submarine volcanoes and high temperature venting on the Tonga arc. Geology 34(6):453–456

Wright IC (1994) Nature and tectonic setting of the southern Kermadec submarine arc volcanoes. Mar Geol 118:179–236

Wright IC, Parson LM, Gamble JA (1996) Evolution and interaction of migrating cross-arc volcanism and back arc rifting: an example from the Southern Havre Trough (35°20′–347°S). J Geophys Res 101:22071–22086

Glossary

Accretionary prism When deformed sediment is scraped off the oceanic lithosphere and deposited in wedge-shapes at the edge of descending, subducting slabs under island arcs and/or continents

Active margin Area marked by faults and earthquakes where two tectonic plates converge

Adiabatically If mantle rises adiabatically, there is compression or expansion of volatiles with no heat exchange. Compression causes a rise in temperature, expansion causes a decrease in temperature

Adventive cone Also called a *parasite cone*, is a smaller volcanic edifice formed on the flank of a main volcano

Amphibole Fibrous mineral that is a hydrated silicate of iron, calcium, magnesium

Anatexis Recrystallization of previously hardened rocks

Ankaramite Olivine-enriched basalt containing mainly pyroxene and olivine

Anorthite Mineral that is a silicate of aluminum and calcium

Antigorite Laminar brown serpentine mineral

Aphyric Rock without visible crystals

Aseismic ridge Ridge formed during continental opening and that does not have seismic activity. It represents an ancient volcanic ridge and/or a fragment of a continent

Augite Mineral of pyroxene (silicate of aluminum, magnesium and calcium)

Authigenic Refers to minerals that have not been transported, but have been crystallized locally, in the same place where they are found

Basanite Volcanic rock enriched in alkali compounds

Benioff zone Seismic area under continents and/or island arcs at which place a subducted slab produces deep-seated earthquakes

R. Hekinian, *Sea Floor Exploration*, Springer Oceanography,
DOI: 10.1007/978-3-319-03203-0,

Birnessite Manganese-bearing mineral containing calcium, potassium and sodium

Chrysotile Hydrated silicate of aluminum, iron and magnesium associated with serpentinite

Curie Isotherm Temperature of 550 ± 30 °C that it is assumed to correspond to the depth at which magma chambers exist

Conchoidal fracture Fracture of solidified flow with a smooth surface

D" Layer of the Earth ~200 km thick located in the lower mantle directly above the Mantle-Core boundary

Diabase Intrusive igneous rock having the composition of basalt

Diagenesis Transformation (alteration) of sediment or fragmented debris after its deposition on the sea floor

Diapir Large intrusion within a pre-existing formation

Ductile Soft-material, not solidified, capable of being deformed

Eon refers to a time scale of half a million years or more

Eclogite Rock composed of garnet and sodium-enriched pyroxene

Fault Is the result of crustal fracturing along which there has been displacement of the sides of the fault with respect to each other. This displacement is parallel to the direction of the fracture

Felsic Light-colored silicate minerals such as feldspar, plagioclase

Ferrosilite Component ($FeSiO_3$) of an iron-rich pyroxene mineral group

Fore-arc Region between the subduction zone (trench) and the volcanic arc

Gabbro Coarse-grained intrusive rock made up of olivine, pyroxene and feldspar

Gondwana Single, large continent existing in the Southern Hemisphere during the Paleozoic period (280 million years ago)

Gossan Iron-bearing hydroxide deposit formed by the oxidation of sulfide minerals

Hackly Term used to define lava flows having jagged surfaces

Harzburgite Altered and serpentinized peridotite made up of 40–90 % olivine and pyroxene

Haystacks Conical shaped structures of lava extruded from tubes that are piling on top of each other. Haystacks are formed from a central orifice similar to a miniature volcano

Hypersthene A mineral of pyroxene. Silicate of aluminum iron and magnesium

Incompatible elements elements that are too large sized to enter into the lattice of a mineral

Intrusion Emplacement of magma within the lithosphere, by means of dykes or sills, as opposed to extrusion, which is a volcanic outpouring of melt onto the surface of the lithosphere

Ions Electrically charged atoms formed by the loss or gain of electrons

Jean Charcot Sea-going research vessel used by CNEXO and IFREMER for scientific exploration. She was named after the Captain Jean-Baptiste Charcot master of the Ship *Pourquoi Pas?* Who explored Antarctica's shore in 1908. He spent the winter on Petermann Island, located on the west side of the Antarctica Peninsula

Lapilli fragmented debris of volcanic rock during volcanic eruption

Large Ion Lithophile Elements (LILE) Elements such as potassium (K), strontium (Sr), sodium (Na) and zirconium (Zr) that have large-size radii and therefore take up more space than other elements in the lattice of minerals. The range of variability for the LILE in basalts is as it follows: $Na_2O + K_2O$ (<3–4 %), K/Ti < 0.15–0.50, Zr/Y (<3–16), Zr (50–200 ppm), Y (50–200 ppm), Sr (50–200 ppm) and Nb (50–200 ppm)

Lithology Refers to the nature of the rocks within a geological formation

Mafic Dark-colored rocks composed of olivine, pyroxene and other dark colored minerals. Term used for basalt, dolerite and gabbros

Magma Liquid silicate-melt at high temperature (>1,000 °C). "Lava" is a word to describe magma that is extruded during volcanism

Magmato-phreatic Refers to a volcanic eruption that is triggered by the contact of hot magma and cold seawater

Manganese nodules Rounded, ball-shaped concretions of iron-manganese formed on the sea floor after seawater has transported and precipitated metallic elements

Metamorphism Transformation of a rock that was already in a solid state, under the influence of temperature and/or pressure, giving rise to newly formed minerals

Metasomatism An alteration process (metamorphism) by which an original rock is transformed by mineralized solutions (liquid or vapor) to produce new material

Microplate Small rigid plate of about 300–600 km in diameter that was formed during the motion of major plate reorganization, and that is now trapped between larger plates of more than a thousand square kilometers

Mylonite Crushed and recrystallized rock formed during tectonic activity

Neoblast Crystal grain in a metamorphic rock

Normal fault Also called a *slump fault*, is a fault in which the hanging wall appears to have moved downward in relation to the footwall

Pangaea Large landmass, considered as being the original continent that has existed for 200 million years and included all the continents we see today

Peridotite alteration Chemical reaction during serpentinization (alteration) of peridotite in the presence of water: Peridotite $+ 2H_2O = Mg^{2+} + SO_2 + H_2O$ $2Mg_2SiO_4$(forsterite) $+ 3H_2O = Mg_3Si_2O_5(OH)_4$(Antigorite) $+$ MgO (OH) (Brucite)

Petrology Study of rocks by *petrologists*, who are scientists able to determine the origin of rocks through a study of their mineralogy, texture, structure and chemical composition

Phenocryst Large (>0.1 mm in diameter) crystal in a rock

Picritic basalt Type of basalt that is enriched in olivine (silicate of iron and magnesium) minerals

Pillow mound Piling up of pillow lava flows found in all underwater volcanic terrains

Plagioclase Mineral that is a silicate of aluminum, calcium and sodium

Platinoid The platinum group metals abbreviated as the PGM including: platinum (Pt), osmium (Os), iridium (Ir), rhuthenium (Ru), rhodium (Rh) and palladium (Pd). Alternatively called platinoid, platinum group, platinum metals, platinum family or platinum group elements (PGEs)

Porphyritic Texture of an igneous coarse-grained rock, containing large sized crystal grains or phenocrysts

Pyroclast Volcanic debris having a grain size more than 64 mm in diameter extruded during explosive eruption. The name pyroclasts comes from the Greek *pyros* meaning fire and *klastos* meaning fragment

Pyroxene Mineral made up of a silicate of aluminum, calcium, magnesium, and iron

Pyroxenite Intrusive rock associated with peridotite that consists of clinopyroxene and orthopyroxene, with or without plagioclase. Pyroxentite is found forming veins and lenses (0.3 up to 20 cm thick) that are variably serpentinized. Also contains the minerals amphibole, spinel and ilmenite. They can be classified as being orthopyroxenite and websterite

Rheology The study of the deformation and flow of matter

Rhyodacite Silicate-enriched rock close in composition to granite

R.V. Letters preceding the name of a ship, abbreviation for the term *Research Vessel*. If the ship is French, I have used the letters N.O. meaning *Navire Oceanographic*, and for German oceanographic vessels, the letters FS (*Forschungsschiff*) are used

Scoria Fragmented lava with a spiny surface formed during explosive eruptions

Seismograph From the Greek words "seismos" meaning earthquake and "graphos" meaning writing. A device that serves to measure time and the ground motion of the Earth's interior. The geologist reads and interprets a "seismogram", where he can see the type of waves, their path, and their change. The difference in travel time of these types of waves, which will also vary with depth, has revealed a picture of the Earth's interior layers

Serpentine Foliated and hydrated silicate of calcium, magnesium, iron mineral

Serpentinization Transformation of olivine minerals into serpentine minerals during hydration. The density of the rock could be obtained from the modal analyses of the serpentine fraction in peridotite through the formula: density (gr/cm^3) = 3.300–0785* S (serpentine)

Sepiolite Clay mineral of hydrated silicate magnesium $(Mg_4Si_2O_5)3\ (OH)_2\ 6H_2O$

Silicic lavas Alkali- and silica-enriched rocks having high silica content ($SiO_2 > 53$ %). They include andesite, trachyte, trachy-andesite, and trachybasalt

Sodalite Sodium hydrated-silicate mineral

Stockwork Term used by mining companies and referring to a mineral deposit formed by a closely spaced network of veins and veinlets

Tectonite Deformed and metamorphosed peridotite during stress-induced motion of the lithosphere

Thin section A thin (1 mm thick) sliced section of a rock or mineral glued onto a glass support in order to be viewed under a microscope. Used in mineralogy, the study of mineral crystals

Tonalite Igneous rock of felsic composition. Contains feldspar, quartz and amphibole

Thrust-fault Fault with a dip of 45° or less on which the hanging wall seems to have moved upward with respect to the footwall

Trachybasalt Alkali-enriched basalt (silicic basalt) with moderately rich silica content

Trachy-andesite Alkali-enriched basalt with moderately rich silica content and also showing fluidal texture

Tumuli (tumulus, singular) Also called pressure ridges, tumuli are characterized by domed shape structures due to the channeling of large lava flows along a gentle slope. A tumulus represents a small-sized volcanic dome where the doming and bursting of the chilled crust occurs during the flowing of hot lava in a conduit. A single and homogeneous type of flow will form a tumulus. The bursting and cracking of the convex shaped portion is the result of pressure differences between the gaseous phases in the hot conduit and the exterior

Ultramafic Dark-green colored intrusive rocks composed of ferromagnesian minerals. As opposed to mafic extrusives and intrusives. No ultramafic rock are extruded, but they are only intruded into the lithosphere where they may later be uplifted due to tectonic activities

Vesicles Holes or pockmarks in solid rocks formed when gas (volatiles) escaped from gas-filled cavities before the melt solidified

Volcaniclastics Fragmented deposits of explosive origin, including bombs, shards and lapilli

Wilson Cycle Cycle of Continent-Ocean plate motions so-called in honor of Tuzo Wilson who first hypothesized the existence of several stages of opening and closing of the oceanic basins due to "continental drift"

Xenolith and xenocryst Foreign rocks (xenolith) and mineral crystals (xenocryst) included in another rock type such as basalt, dolerite, gabbro or peridotite. Their importance resides in the fact that they represent solids that were torn-off from an outcrop and remobilized during magma circulation in the lithosphere therefore they can provide information on the composition of the substratum. They are recognized by their chilled contour lines due to reaction with the host rock, giving rise to corroded edges and chemical disequilibrium with their host rock

Zeolite Group of minerals including natrolite (hydrated sodium aluminum silicate) and heulandite (hydrated calcium aluminum silicate). They are a colorless or white water-enriched (hydrous) aluminum silicate that contains calcium, sodium and potassium. They are known to form crystals in the cavities of rocks such as geodes

MIX
Papier aus verantwortungsvollen Quellen
Paper from responsible sources
FSC® C105338

If you have any concerns about our products,
you can contact us on
ProductSafety@springernature.com

In case Publisher is established outside the EU,
the EU authorized representative is:
Springer Nature Customer Service Center GmbH
Europaplatz 3, 69115 Heidelberg, Germany

Printed by Libri Plureos GmbH
in Hamburg, Germany